Advances in Intelligent Systems and Computing

Volume 415

Series editor

Janusz Kacprzyk, Polish Academy of Sciences, Warsaw, Poland
e-mail: kacprzyk@ibspan.waw.pl

About this Series

The series "Advances in Intelligent Systems and Computing" contains publications on theory, applications, and design methods of Intelligent Systems and Intelligent Computing. Virtually all disciplines such as engineering, natural sciences, computer and information science, ICT, economics, business, e-commerce, environment, healthcare, life science are covered. The list of topics spans all the areas of modern intelligent systems and computing.

The publications within "Advances in Intelligent Systems and Computing" are primarily textbooks and proceedings of important conferences, symposia and congresses. They cover significant recent developments in the field, both of a foundational and applicable character. An important characteristic feature of the series is the short publication time and world-wide distribution. This permits a rapid and broad dissemination of research results.

More information about this series at http://www.springer.com/series/11156

Vadlamani Ravi · Bijaya Ketan Panigrahi
Swagatam Das · Ponnuthurai Nagaratnam
Suganthan
Editors

Proceedings of the Fifth International Conference on Fuzzy and Neuro Computing (FANCCO - 2015)

Editors
Vadlamani Ravi
IDRBT
Center of Excellence in Analytics
Hyderabad
India

Bijaya Ketan Panigrahi
Department of Electrical Engineering
IIT Delhi
Delhi
India

Swagatam Das
Electronics and Communication Sciences Unit
Indian Statistical Institute
Kolkata
India

Ponnuthurai Nagaratnam Suganthan
Division of Control and Instrumentation
Nanyang Technological University
Singapore
Singapore

ISSN 2194-5357 ISSN 2194-5365 (electronic)
Advances in Intelligent Systems and Computing
ISBN 978-3-319-27211-5 ISBN 978-3-319-27212-2 (eBook)
DOI 10.1007/978-3-319-27212-2

Library of Congress Control Number: 2015956362

Printed on acid-free paper

This Springer imprint is published by SpringerNature
The registered company is Springer International Publishing AG Switzerland

Preface

This volume presents the research papers presented in the 5th International Conference Fuzzy and Neuro Computing (FANCCO) held during 17–19 December 2015 and hosted by Institute for Development and Research in Banking Technology (IDRBT), Hyderabad, India. It brought together researchers from academia and industry to report the latest cutting-edge research made in the areas of fuzzy computing, neuro computing and hybrid neuro-fuzzy computing in the paradigm of soft computing or computational intelligence.

We are grateful to Dr. Y.V. Reddy, former Governor, Reserve Bank of India, for inaugurating the conference. Five internationally renowned experts from four countries, namely Prof. Nikola Kasabov, Auckland University of Technology, New Zealand; Prof. Vipin Kumar, University of Minnesota, USA; Prof. Tadihiko Murata, Osaka University, Japan; Prof. Dipti Srinivasan, National University of Singapore, Singapore and Prof. Raghuram Krishnapuram, Xerox Research India delivered keynotes on cutting-edge topics. Further, two pre-conference tutorials from internationally acclaimed Indian professors, Prof. B. Yagnanarayan and Prof. M. Narisimha Murty, respectively, from IIIT, Hyderabad and IISc, Bangalore were organized. We sincerely thank all the keynote speakers for having come from distant places to be present in the conference and sharing their invaluable knowledge and experience. We are also thankful to the tutorial speakers for readily accepting our request and delivering highly informative and thought-provoking tutorials.

After a rigorous academic review of the 68 submissions from all over the world, the International Programme Committee members and some additional reviewers selected 27 papers for presentation at the conference. We sincerely thank all the PC members and additional expert reviewers for their unwavering support and putting enormous efforts in reviewing and accepting the papers. Without their untiring efforts and the contributing authors, the conference would not have been possible. There are seven papers from outside India, including countries such as USA, Russia, Hungary, New Zealand, Palestine, Fiji Islands, etc.

The entire process of submission, refereeing, e-communication of decision-making of the PC for accepting the papers was done through the

EasyChair system. We thank the administrators and creators of EasyChair for providing a highly configurable conference management system.

The accepted papers have a good, balanced mix of theory and applications. The techniques ranged from fuzzy neural networks, decision trees, spiking neural networks, self-organizing feature map, support vector regression, adaptive neuro-fuzzy inference system, extreme learning machine, fuzzy multi-criteria decision-making, machine learning, web usage mining, Takagi–Sugeno Inference system, extended Kalman filter, Goedel-type logic, fuzzy formal concept analysis, biclustering, etc. The applications ranged from social network analysis, twitter sentiment analysis, cross-domain sentiment analysis, information security, education sector, e-learning, information management, climate studies, rainfall prediction, brain studies, bioinformatics, structural engineering, sewage water quality, movement of aerial vehicles, etc.

We are grateful to the Microsoft, India for being the Platinum sponsor and Teradata, India for being the Silver sponsor of the conference. We thank the IEEE Computer Society, Hyderabad Section and IEEE Computational Intelligence Society, Hyderabad Section for technically sponsoring the conference. We thank the Honorary Chair, Prof. Janusz Kacprzyk and general Chairs Prof. N.R. Pal, ISI Calcutta, Kolkata and Prof. Kalyanmoy Deb for their encouragement and support.

We thank the steering committee for providing IDRBT an opportunity to host the conference. We thank the local organizing committee at IDRBT which included Dr. B.M. Mehtre, Dr. Rajarshi Pal, Dr. P. Syam Kumar, Prof. Atul Negi, Dr. M. Naresh Kumar, G. Raghuraj, V. Belurgikar, P. Ratnakumar, S. Rashmi Dev, K.V.R. Murty, K. Srinivas, the administration, estate division, accounts division, library, publication division, housie keeping and security division for supporting and ensuring the successful organization of the conference. We profusely thank Dr. A.S. Ramasastri, Director, IDRBT for providing unflinching support to the overall conduct of the conference right from the concept stage through the successful organization without which it would not have been possible to conduct the conference.

Last but not least, we are thankful to Dr. Thomas Ditzinger from Springer for readily agreeing to publish the proceedings in the AISC series and also his team for preparing the proceedings very carefully well within time for the conference.

December 2015

Vadlamani Ravi
Bijaya Ketan Panigrahi
Swagatam Das
Ponnudurai Nagaratnam Suganthan

Organization

Honorary Chair

Janusz Kacprzyk, Poland

General Chairs

Kalyanmoy Deb, MSU, USA
Nikhil Pal, ISI Calcutta, Kolkata

Executive Chair

A.S. Ramasastri, Director, IDRBT, India

Programme Chairs

Vadlamani Ravi, IDRBT, Hyderabad, India
Bijaya Ketan Panigrahi, IIT Delhi, India
P.N. Suganthan, NTU, Singapore
Swagatam Das, ISI Calcutta, Kolkata, India

Organizing Committee

B.M. Mehtre, Associate Professor, IDRBT, India
Rajarshi Pal, Assistant Professor, IDRBT, India

P. Syam Kumar, Assistant Professor, IDRBT, India
S. Nagesh Bhattu, Assistant Professor, IDRBT, India
Atul Negi, Professor, University of Hyderabad, India
M. Naresh Kumar, NRSC, Hyderabad, India
G. Raghuraj, General Manager, Administration, IDRBT, India
V. Belurgikar, Assistant General Manager—Accounts, IDRBT, India
P. Ratnakumar, Assistant General Manager—Library, IDRBT, India
S. Rashmi Dev, Assistant General Manager—Human Resources & Publications, IDRBT, India
K.V.R. Murty, Systems Officer, IDRBT, India
K. Srinivas, Administrative Executive, IDRBT, India

International Advisory Committee/Programme Committee

C.A. Murthy, ISI Calcutta, Kolkata, India
M.K. Tiwari, IIT, Kharagpur, India
C. Chandra Sekhar, IIT Madras, Chennai, India
Suresh Sundaram, NTU, Singapore
Lipo Wang, NTU, Singapore
Amit Mitra, IIT, Kanpur, India
Aruna Tiwari, IIT, Indore, India
D. Nagesh Kumar, IISc, Bangalore, India
V. Sushila Devi, IISc, Bangalore, India
C. Hota, BITS Pilani, Hyderabad, India
Chilukuri Mohan, Syracuse University, Syracuse, USA
Debjani Chakraborty, IIT, Kharagpur, India
P.K. Kalra, IIT Kanpur, Kanpur, India
Vasant Pandian, University Putra Malaysia, Malaysia
Oscar Castillo, Tijuana Institute of Technology, Chula Vista CA, USA
Indranil Bose, IIM Calcutta, Kolkata, India
Sanghamitra Bandyopadhyay, ISI Calcutta, Kolkata, India
S. Bapi Raju, IIIT Hyderabad, India
Brijesh Verma, CQ University, Brisbane, Australia
C.R. Rao, UOHYD, Hyderabad, India
B.L. Deekshatulu, IDRBT, Hyderabad, India
Arun Agarwal, UOHYD, Hyderabad, India
Arnab Laha, IIM, Ahmedabad, India
Biplav Srivastava, IBM Research, New Delhi, India
B.K. Mohanty, IIM, Lucknow, India
M. Janga Reddy, IIT Bombay, Mumbai, India
M.C. Deo, IIT Bombay, Mumbai, India
Pankaj Dutta, IIT Bombay, Mumbai, India
Usha Anantha Kumar, IIT Bombay, Mumbai, India

Faiz Hamid, IIT, Kanpur, India
S. Chakraverty, NIT Rourkela, Rourkela
H. Fujita, Iwate Prefectural University, Iwate, Japan
Dries Benoit, Ghent University, Ghent, Belgium
S.A. Arul, Philips Electronics Singapore, Singapore
Pawan Lingars, Saint Mary's University Halifax, Canada
Samuelson Hong, Oriental Institute of Technology, Taiwan
Zhihua Cui, Taiyuan University of Science and Technology, Taiyuan, China
Balasubramaniam Jayaram, IIT, Hyderabad, India
K. Saman Halgamuge, The University of Melbourne, Melbourne, Australia
Nischal Verma, IIT, Kanpur, India
Laxmidhar Behera, IIT, Kanpur, India
YaoChu Jin, University of Surrey, Guildford, England
Vineeth Balasubramian, IIT, Hyderabad, India
Atul Negi, Professor, University of Hyderabad, India
M. Naresh Kumar, NRSC, Hyderabad, India

Contents

Tweet Sentiment Classification Using an Ensemble of Machine Learning Supervised Classifiers Employing Statistical Feature Selection Methods

K. Lakshmi Devi, P. Subathra and P.N. Kumar

Abstract Twitter is considered to be the most powerful tool of information dissemination among the micro-blogging websites. Everyday large user generated contents are being posted in Twitter and determining the sentiment of these contents can be useful to individuals, business companies, government organisations etc. Many Machine Learning approaches are being investigated for years and there is no consensus as to which method is most suitable for any particular application. Recent research has revealed the potential of ensemble learners to provide improved accuracy in sentiment classification. In this work, we conducted a performance comparison of ensemble learners like Bagging and Boosting with the baseline methods like Support Vector Machines, Naive Bayes and Maximum Entropy classifiers. As against the traditional method of using Bag of Words for feature selection, we have incorporated statistical methods of feature selection like Point wise Mutual Information and Chi-square methods, which resulted in improved accuracy. We performed the evaluation using Twitter dataset and the empirical results revealed that ensemble methods provided more accurate results than baseline classifiers.

Keywords Bagging · Boosting · Ensemble learners · Entropy · Naïve bayes · Sentiment classification · SVM

K.L. Devi · P. Subathra · P.N. Kumar (✉)
Department of CSE, Amrita Vishwa Vidyapeetham, Coimbatore 641112, India
e-mail: pn_kumar@cb.amrita.edu

K.L. Devi
e-mail: laksdevi115@gmail.com

P. Subathra
e-mail: p_subathra@cb.amrita.edu

V. Ravi et al. (eds.), *Proceedings of the Fifth International Conference on Fuzzy and Neuro Computing (FANCCO - 2015)*, Advances in Intelligent Systems and Computing 415, DOI 10.1007/978-3-319-27212-2_1

1 Introduction

Micro-blogging websites have become valuable source of information which varies from personal expressions to public opinions. People post status messages about their life, share opinions on their political and religious views, express support or protest against social issues, discuss about products in market, education, entertainment etc. Hence micro-blogging websites have become a powerful platform for millions of users to express their opinion. Among all micro-blogging websites, the most popular and powerful tool of information dissemination is Twitter launched in October 2006, which allows users to post free format textual messages called tweets. Owing to the free format of messages, users are relieved from concerning about grammar and spelling of the language. Hence Twitter allows short and fast posting of tweets where each tweet is of length 140 characters or less. These tweets are indicative of views, attitudes and traits of users and therefore they are rich sources of sentiments expressed by the people. Detecting those sentiments and characterizing them are very important for the users and product manufacturers to make informed decisions on products. Hence Twitter can be labeled as a powerful marketing tool. Recently, there has been a shift from blogging to micro-blogging. The reason for this shift is that micro-blogging provides faster mode of information dissemination when compared to blogging. Since there is restriction on the number of characters allowed, it reduces users' time consumption and effort for content generation. Another major difference is that the frequency of updates in micro blogging is very high when compared to blogging due to shorter posts [1].

Sentiment classification of social media data can be performed using many Machine Learning (ML) algorithms among which Support Vector Machine (SVM), Naive Bayes (NB), Maximum Entropy (Maxent) are widely used. These baseline methods use Bag Of Words (BOW) representation of words, where frequency of each word is used as a 'feature' which is fed to the classifier. These models ignore the word order but maintain the multiplicity. Ensemble techniques have gained huge prominence in the Natural Language Processing (NLP) field and they have been proved to provide improved accuracy. This is achieved by combining different classifiers which are trained using different subsets of data, resulting into a network of classifiers which can be further used to perform sentiment classification [2]. The major limitation associated with ensemble learners is that the training time is very huge since each of the classifiers must be individually trained. This increases the computational complexity as the dimensionality of the data increases. Therefore, the ensemble learning techniques are employed where the data has fewer number of dimensions and also when maximum possible accuracy of classification is a mandatory requirement [2].

In this work, we mainly focus on the sentiment classification of tweets using ensemble learners like Bagging and Boosting and performed a comparative analysis with base learners viz. SVM, NB and Maxent classifiers. We have employed Twitter dataset in this work and the results demonstrate that the ensemble methods outperformed baseline methods in terms of accuracy. We have adopted statistical feature selection methods like Point wise Mutual Information (PMI) and Chi-square methods for selecting the most informative features which are then fed to the classifier models. The rest of the paper is organized as follows: Sect. 2 discusses the related work. Section 3 describes the proposed system, feature selection methods, baseline methods and ensemble learners. Section 4 describes the implementation details and experiments. The results and analysis is given in Sect. 5. The conclusion and future enhancements of the paper are given in Sect. 6.

2 Related Work

Sentiment Analysis (SA) can be broadly classified into three: Sentence level SA, Document level SA and Aspect level SA [3]. The document level SA considers a single whole document as the fundamental unit assuming that the entire document talks about a single topic. The sentence level SA can be considered as a slight variation of document level SA where each sentence can be taken as a short document [4]. Each sentence can be subjective or objective and SA is performed on the subjective sentences to determine the polarity i.e., positive or negative. Aspect level SA [5] is performed when we have to consider different aspects of an entity. This is mostly applicable in the SA of product reviews, for example: "The battery life of the phone is too low but the camera quality is good". SA can be performed in other levels of granularity like clause, phrase and word level depending on the application under consideration. The sentiment classification approaches can be divided into three

- Machine Learning (ML) approach
- Lexicon Based approach
- Hybrid approach.

2.1 Machine Learning (ML) Approach

In machine learning approach, different learning algorithms like NB, SVM, Maxent etc., use linguistic features to perform the classification. The ML approaches can be supervised, where a labeled set of training data is provided, or unsupervised which is used where it is difficult to provide labeled training data. The unsupervised approach builds a sentiment lexicon in an unsupervised way and then evaluates the polarity of text using a function that involves positive, negative and neutral indicators [6]. The major goal aim of a semi-supervised learning approach is to produce more accurate results by using both labeled and unlabeled data [7]. Even though there exist unsupervised and semi-supervised techniques for sentiment classification, supervised techniques are considered to have more predictive power [6].

2.2 Lexicon Based Approach

This approach is based on the already available pre compiled collection of sentiment terms which is called as 'sentiment lexicon'. This approach aims at discovering the opinion lexicon and is classified into two: the dictionary based approach and corpus based approach. The dictionary-based approach works by considering some seed words which is a predefined dictionary that contains positive and negative words for which the polarity is already known. A dictionary search is then performed in order to pick up the synonyms and antonyms of these words and it then adopts word frequency or other measures in order to score all the opinions in the text data. The computational cost for performing automated sentiment analysis is very low when the dictionary is completely predefined [8]. The corpus based approach builds a seed list containing opinion words and uses semantics and statistics to find other opinion words from a huge corpus in order to find opinion words with respect to the underlying context [6].

2.3 Hybrid Approach

The third approach for performing sentiment classification is the hybrid approach which is a combination of machine learning and lexicon based approach. The hybrid methods have gained huge prominence in recent years. This approach aims at attaining the best of both approaches i.e. robustness and readability from a well-designed lexicon resource and improved accuracy from machine learning algorithms. The main aim of these approaches is to classify the text according to polarity with maximum attainable accuracy.

2.4 Ensemble Methods

Each of these approaches work well in different domains and there is no consensus as to which approach performs well in a particular domain. So, in order to mitigate these difficulties, 'an ensemble of many classifiers' can be adopted for achieving much more accurate and promising results. Recent research has revealed the potential of ensemble learners to provide improved accuracy in sentiment classification [9]. An ensemble should satisfy two important conditions: prediction diversity and accuracy [6]. Ensemble learning comprises of state of the art techniques like Bagging [10], Boosting [11] and Majority Voting [12]. Majority voting is the most prominent ensemble technique used in which there exists a set of experts which can classify a sentence and identify the polarity of the sentence by choosing the majority label prediction. This results in improved accuracy; however, this method does not address prediction diversity. Bagging and Boosting techniques are explicitly used to address the issue of prediction diversity. The bagging technique works by random sampling with replacement of training data i.e. every training subset drawn from the entire data (also called as bag) is provided to different baseline methods of learner of the same kind [9]. These randomly drawn subsets are known as bootstrapped replicas of entire training data. Boosting techniques generally constructs an ensemble incrementally in which a new model is trained with an emphasis on the instances which were misclassified by the previous models [10].

3 Proposed Work

In this paper, we have implemented Sentiment multi-class Classification of twitter dataset using Ensemble methods viz. Bagging and Boosting. A comparative analysis with baseline methods SVM, NB and Maxent has been done and their performance evaluated using evaluation metrics accuracy, precision, recall and f-measure. We have also employed statistical feature selection methods like PMI and Chi-square to generate n most informative features. The proposed system consists of:

- Tweet pre-processing.
- Feature selection using Statistical methods like PMI and Chi-square.
- Multi class Sentiment Classification into positive, negative and neutral.
- Performance Evaluation.

The system architecture is given in Fig. 1.

3.1 Tweet Pre-processing

Since Twitter allows free format of messages, people generally do not care about grammar and spelling. Hence the dataset must be pre-processed before it can be used for sentiment classification. The pre-processing steps include:

- Tokenizing.
- Removing non-English tweets, URLs, target (denoted using @), special characters and punctuations from hash tags (a hash tag is also called as summarizer of a tweet), numbers and stop words.
- Replacing negative mentions (words that end with 'nt' are replaced with 'not'), sequence of repeated characters (For example, 'wooooww' is replaced by 'wooow').

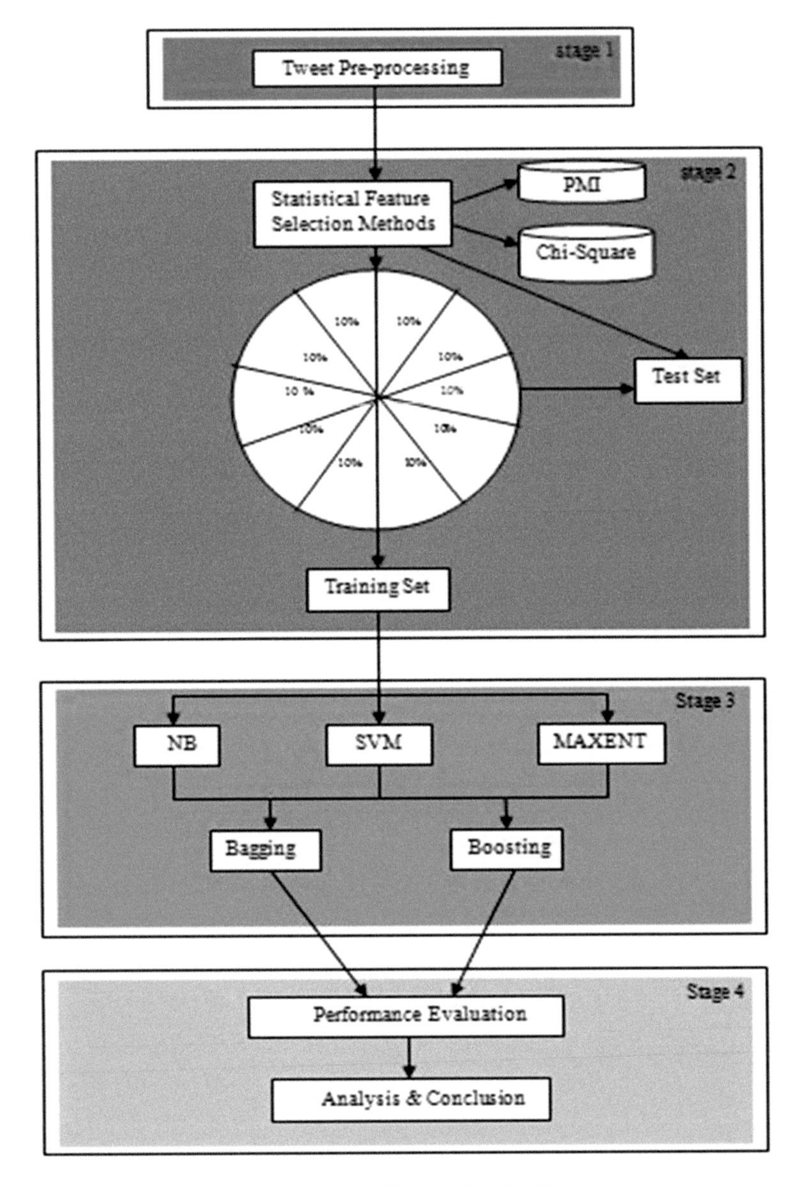

Fig. 1 Architecture diagram of sentiment multi-class classification

3.2 Feature Selection

Feature selection methods treat each sentence/document as a BOW or as a string that maintains the sequence of the words that constitute the sentence/document. BOW which is known for its simplicity is widely used by baseline methods for sentiment classification which often adopts stop word elimination, stemming etc. for feature selection. The feature selection methods can be broadly classified into two:

- Lexicon based: This method incrementally builds large lexicon using a set of 'seed words' and gradually bootstrapping the set by identifying synonyms and incorporating available online resources. The main limitation associated with this approach is that this method needs human annotation.
- Statistical method: This approach is fully automatic and they are widely adopted in sentiment analysis. We have employed two statistical methods of feature selection in this work.

3.3 Point Wise Mutual Information (PMI)

The mutual information measure indicates how much information one particular word provides about the other. In text classification, this method can be used to model the mutual information between the features selected from the text and the classes. The PMI between a given word w and a class c can be defined as the level or degree of co-occurrence between the given word w and the class c. The mutual information is defined as the proportion of expected co-occurrence of class c and word w based on mutual independence and the true co-occurrence which is given in the equation below:

$$M(w) = \log\left(\frac{F(w)\cdot p_i(w)}{F(w)\cdot p_i}\right) = \log\left(\frac{p_i(w)}{p_i}\right) \quad (1)$$

The value indicates how much a particular feature has influence on a class c. If the value is greater than zero, the word w is said to have positive correlation with class c and has negative correlation when the value is less than zero.

3.4 Chi-square Method

This method computes a score in order to determine whether a feature and a class are independent. This particular test is a statistic which determines the degree or the extent to which these two are independent. This method makes an initial assumption that the class and the feature are independent and then computes score; with large value indicating they are dependent. Let the number of given documents be denoted as n, $p_i(w)$ indicates the conditional probability of class i for the given documents which have word w, P_i and F(w) are the global fraction of documents that has class i and word w respectively. The chi-square statistic of the word between w and class i is given as

$$M\chi_i^2 = \frac{n.F(w)^2 \cdot (p_i(w) - P_i)^2}{F(w) \cdot (1 - F(w)) \cdot P_i \cdot (1 - P_i)} \tag{2}$$

χ_i^2 is considered to be better than PMI because χ_i^2 value is normalized and therefore are more comparable across terms in the same type [13].

3.5 *Multi-class Sentiment Classification*

3.5.1 Naive Bayes Classifier

The NB classifier is a simple and commonly used probabilistic classifier for text. This model calculates the posterior probability of a class by making use of word distribution in the given document. This is one of the baseline methods which uses BOW representation that ignores word order. The Bayes theorem is used in order to predict the probability that a given feature belongs to a particular label which is denoted using the equation below:

$$P(\text{label}|\text{features}) = \frac{P(\text{label})*P(\text{features}|\text{label})}{P(\text{features})} \tag{3}$$

P(*label*) is the prior probability of the label, P(features|label) is the prior probability that a given feature set is being classified as a label. P(features) is the prior probability that a given feature. Using the naive property which assumes that all features are independent, the above stated equation can be rewritten as below:

$$P(\text{label}|\text{features}) = \frac{P(\text{label})*P(\text{f1}|\text{label})*\cdots*P(\text{fn}|\text{label})}{P(\text{features})} \tag{4}$$

3.5.2 Maximum Entropy Classifier

The Maxent classifier which is also called as conditional exponential classifier is a probabilistic classifier that employs encoding in order to convert feature sets which are labeled into vector representation. The resultant encoded vector is used for computing weights for every feature which are subsequently combined to identify the most probable label for the given feature set. The model uses x weights as parameters which combine to form joint features generated using encoding function. The probability of every label is calculated using the equation below:

$$\text{P(fs|label)} = \frac{\text{dotprod(weights, encode(fs, label))}}{\text{sum(dotprod(weights, encode(fs, l))for l in labels)}} \tag{5}$$

3.5.3 Support Vector Machine

SVM is a linear classifier which is best suited for text data because text data is sparse in most cases. However they tend to correlate with each other and are mostly organized into 'linearly separable categories'. SVM maps the data to an inner product space in a non-linear manner thus building a 'nonlinear decision surface' in the real feature space. The hyper plane separates the classes linearly.

3.5.4 Ensemble Learners

Ensemble Learners came into existence with the realization that each of the machine learning approaches performs differently on different applications and there is no conformity on which approach is optimal for any given application. This uncertainty has led to the proposal of ensemble learners which exploit the capabilities and functionalities of different learners for determining the polarity of text. Each of the classifiers which are combined are considered to be independent and equally reliable. This assumption leads to biased decisions and this is considered to be a limitation which is to be addressed. The computational complexity associated with big data is very huge and the conventional state of the art approaches focus only on accuracy. Moreover, these approaches performed well on formal texts as well on noisy data. The major highlight of this paper is to evaluate the performance of ensemble learners in short and informal text such as tweets. We have implemented two 'instance partitioning methods' like Bagging and Boosting.

3.5.5 Bagging

Bagging also known as bootstrap aggregating [8] is one of the traditional ensemble techniques which is known for its simplicity and improved performance. This technique achieves diversity using bootstrapped replicas of training data [8] and each replica is used to train different base learners of same kind. The method of combination of different base learners is called as Majority Vote which can lessen variance when combined with baseline methods. Bagging is often employed in the cases where the dataset size is limited. The samples drawn are mostly of huge size in order to ensure the availability of sufficient instances in each sample. This causes significant overlapping of instances which can be avoided using an 'unstable base learner' that can attain varied decision boundaries [11].

3.5.6 Boosting

This technique applies weighting to the instances sequentially thereby creating different base learners and the misclassifications of the previous classifier is fed to the next sequential base classifier with higher weights than the previous round. The main aim of Boosting is to provide a base learner with modified data subsets repeatedly which results in a sequence of base learners. The algorithm begins by initializing all instances with uniform weights and then applying these weighted instances to the base learners in each iteration. The error value is calculated and all the misclassified instances and correctly classified instances are assigned higher and lower weights respectively. The final model will be a linear combination of the base learners in each iteration [9]. Boosting is mainly based on weak classifiers and we have employed AdaBoost, which works by calling a weak classifier several times.

4 Experiments

We have implemented this work in R platform. After the pre-processing steps, two feature selection methods viz. PMI and Chi-square were adopted and the most informative features were chosen. The Chi-square method seemed to produce better results than PMI because of the normalization scheme adopted in Chi-square method. Hence we chose Chi-square method as the feature selection strategy in this work and the features generated were fed to the classifiers. The experimental evaluation consists of three phases:

1. The supervised classifiers were trained using most informative features selected using PMI and Chi-square method.
2. Implementations of the state of the art methodologies like baseline methods viz. NB, SVM and Maxent and also ensemble learners like Bagging and Boosting.
3. Performance evaluation using 10 fold cross validation and calculation of precision, recall, f-measure and accuracy. The dataset was divided into ten equal sized subsets from which nine of them constituted training set and remaining one constituted the test set. This process iterated ten times so that every subset became test set once and the average accuracy is computed.

4.1 Baseline and Ensemble Methods

The baseline methods such as NB, SVM and Maxent were investigated and accuracy, precision, recall and f-measure of each of these models were calculated. We have employed 10-fold cross validation for evaluation. The results were compared with ensemble methods such as Bagging and Boosting.

4.2 Dataset

In this work, we have used the twitter dataset provided in SemEval 2013, Task 9 [14]. This dataset consists of tweet ids which are manually annotated with positive, negative and neutral labels. The training set contains 9635 tweets and the testing set contains 3005 tweets.

5 Results and Analysis

Figure 2 demonstrates the performance evaluation of baseline methods and ensemble methods using the evaluation metrics like precision, recall, f-measure and accuracy. The graph depicts the values of precision, recall, f-measure and accuracy for each of the positive, negative and neutral classes. The accuracy of sentiment multi class classification that include a neutral class using BOW approaches has not gone beyond 60 % in most cases [15]. The results demonstrate that the ensemble

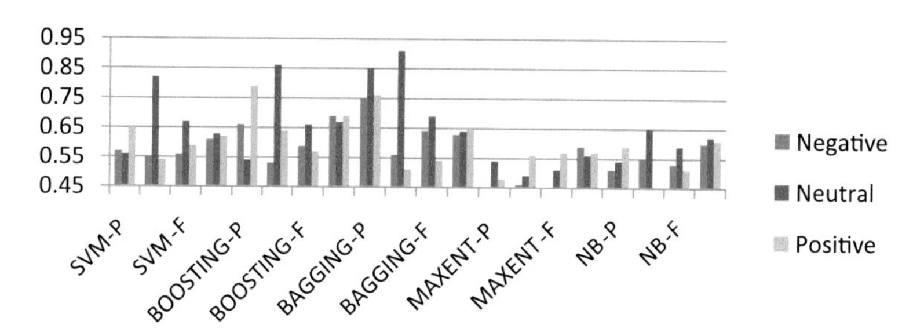

Fig. 2 Performance comparison of baseline methods (NB, SVM and MAXENT) and ensemble learners (bagging and boosting) using precision (P), recall (R), f-measure (F) and accuracy (ACC) for each of the three classes: positive, negative and neutral

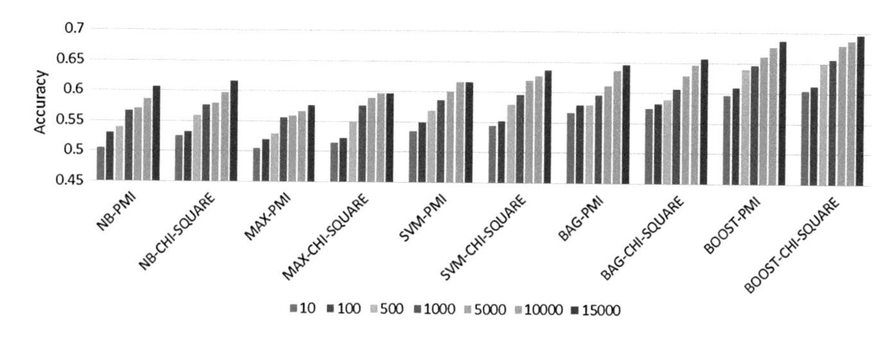

Fig. 3 Performance comparison of PMI and Chi-square feature selection methods for different number of features

methods like Bagging and Boosting has achieved significant performance improvement when compared to SVM, NB and Maxent classifiers.

Figure 3 demonstrates the comparison of PMI and Chi-square methods based on accuracy for each of the baseline classifiers and ensemble learners. The results demonstrate that Chi-square method provided improved accuracy than PMI.

6 Conclusion

Micro-blogging websites have become very popular and have fuelled significance of sentiment classification. In this work, we evaluated popular and widely used ensemble learners (Bagging and Boosting) for use in tweet sentiment classification into three classes: positive, negative and neutral. The Twitter dataset was used to perform the sentiment classification and the empirical results showed the effectiveness of the above stated Ensemble Learners by comparing with the baseline methods like NB, SVM and Maxent classifiers. The Ensemble Learners tend to produce improved accuracy than the baseline methods. We also incorporated statistical feature selection methods like PMI and chi-square for extracting most informative features instead of using traditional BOW features.

The future directions for this work include: bigger datasets must be used for validating the results obtained in our work because of the imbalanced nature of Twitter datasets. The high computational complexity and running time of the Ensemble Learners need to be tackled by using parallel computing. In addition, the knowledge learnt by the Ensemble techniques are often difficult to interpret by the humans and hence suitable methods must be incorporated in order to improve the interpretability.

References

1. Java, A., Song, X., Finin, T., Tseng, B.: Why we twitter: understanding microblogging usage and communities. In: Proceedings of the 9th WebKDD and 1st SNA-KDD 2007 Workshop on Web Mining and Social Network Analysis, pp. 56–65. ACM.2007
2. Whitehead, M., Yaeger, L.: Sentiment Mining Using Ensemble Classification Models: Innovations and Advances in Computer Sciences and Engineering, pp. 509–514. Springer, Netherlands (2010)
3. Medhat, W., Hassan, A., Korashy, H.: Sentiment analysis algorithms and applications: a survey. Ain Shams Eng. J. **5**(4), 1093–1113 (2014)
4. Liu, B.: Sentiment analysis & opinion mining. Synth. Lect. Hum. Lang. Technol. **5**(1), 1–167 (2012)
5. Lek, H.H, Poo, D.C.: Aspect-based Twitter sentiment classification. In: 2013 IEEE 25th International Conference on Tools with Artificial Intelligence (ICTAI), pp. 366–373. IEEE (2013)
6. Fersini, E., Messina, E., Pozzi, F.A.: Sentiment analysis: Bayesian ensemble learning. Decis. Support Syst. **68**, 26–38 (2014)

7. Rice, D.R, Zorn, C.: Corpus-based dictionaries for sentiment analysis of specialized vocabularies. In: Proceedings of NDATAD (2013)
8. Ortigosa-Hernández, J., Rodríguez, J.D., Alzate, L., Lucania, M., Inza, I., Lozano, J.A.: Approaching Sentiment Analysis by using semi-supervised learning of multi-dimensional classifiers. Neurocomputing **92**, 98–115 (2012)
9. Wang, G., Sun, J., Ma, J., Xu, K., Gu, J.: Sentiment classification: the contribution of ensemble learning. Decis. Support Syst. **57**, 77–93 (2014)
10. Breiman, L.: Bagging predictors. Mach. Learn. **24**(2), 123–140 (1996)
11. Schapire, R.E.: The strength of weak learnability. Mach. Learn. **5**(2), 197–227 (1990)
12. Dietterich, T.G: Ensemble Methods in Machine Learning. Multiple Classifier Systems, vol. 1, p. 15. Springer, Berlin (2000)
13. Aggarwal, C.C., Zhai, C.: Mining Text Data. Springer, Berlin (2012)
14. http://alt.qcri.org/semeval2014/task9/
15. Wang, H., Can, D., Kazemzadeh, A., Bar, F., Narayanan, S.: A system for real-time twitter sentiment analysis of 2012 US presidential election cycle: In: Proceedings of the ACL 2012 System Demonstrations. Association for Computational Linguistics, pp. 115–120 (2012)

Multiple Fuzzy Correlated Pattern Tree Mining with Minimum Item All-Confidence Thresholds

Radhakrishnan Anuradha, Nallaiah Rajkumar and Venkatachalam Sowmyaa

Abstract Rare item problem in association rule mining was solved by assigning multiple minimum supports for each item. In the same way rare item problem in correlated pattern mining with all-confidence as interesting measure was solved by assigning multiple minimum all-confidences for each items. In this paper multiple fuzzy correlated pattern tree (MFCP tree) for correlated pattern mining using quantitative transactions is proposed by assigning multiple item all-confidence (MIAC) value for each fuzzy items. As multiple fuzzy regions of a single item are considered, time taken for generating correlated patterns also increases. Difference in Scalar cardinality count for each fuzzy region is considered in calculating MIAC for fuzzy regions. The proposed approach first constructs a multiple frequent correlated pattern tree (MFCP) using MIAC values and generates correlated patterns using MFCP mining algorithm. Each node in MFCP tree serves as a linked list that stores fuzzy items membership value and the super—itemsets membership values of the same path. The outcome of experiments shows that the MFCP mining algorithm efficiently identifies rare patterns that are hidden in multiple fuzzy frequent pattern (MFFP) tree mining technique.

Keywords Fuzzy data mining · MIAC · MFCP tree · Fuzzy correlated patterns

R. Anuradha (✉) · N. Rajkumar · V. Sowmyaa
Department of Computer Science and Engineering, Sri Ramakrishna Engineering College, Coimbatore 641022, TN, India
e-mail: anuradha.r@srec.ac.in

N. Rajkumar
e-mail: nrk29@rediffmail.com

V. Sowmyaa
e-mail: sowmvenk@gmail.com

V. Ravi et al. (eds.), *Proceedings of the Fifth International Conference on Fuzzy and Neuro Computing (FANCCO - 2015)*, Advances in Intelligent Systems and Computing 415, DOI 10.1007/978-3-319-27212-2_2

1 Introduction

Data mining techniques have played an imperative role in deriving interesting patterns from a wide range of data [1]. Different mining approaches are divided based on the required necessity of knowledge like association rules [2, 3], classification rules [4] and clustering [5]. Most of the algorithms in association rule mining use level by level approach to generate and test candidate itemsets. One such algorithm is Apriori, which requires high computation cost for rescanning the whole database iteratively. To overcome the above drawback Han et al. [6] proposed the frequent pattern tree (FP Tree) approach which requires only two scans for processing the entire database and mining frequent patterns. The FP tree represents tree structure of a database which contains only the frequent items.

Fuzzy set theory [7] has been progressively used in intelligent systems for its easiness. Fuzzy learning algorithms for generating membership functions and inducing rules for a given dataset are proposed [8, 9, 10] and used in specific domains. In real world the transaction data mostly consists of quantifiable value. Hong et al. [6, 9] proposed a fuzzy mining algorithm for mining fuzzy association rules from quantitative data. Papadimitriou et al. [11] proposed an approach to mine fuzzy association rules based on FP trees. To mine frequent items from quantitative values, fuzzy mining has been proposed to derive the fuzzy rules using fuzzy FP-trees [12, 13, 14].

A correlation between two items becomes the key factor in detecting associations among items in market basket database. Correlated pattern mining was first introduced by Brin et al. [15] using contingency table that evaluates the relationship of two items. Many interesting measures like All-Confidence, Coherence and Cosine has been used in correlated pattern mining [16]. All confidence [17, 18] measure is used for mining interestingness of a pattern, which satisfies both anti-monotonic and null invariance properties. The above measure at higher threshold invokes rare item problem. Rare item problem in association rule mining was solved using multiple minimum supports by Liu et al. [19]. In the same way rare item problem in correlated pattern mining with all-confidence as interesting measure was solved using multiple minimum all-confidences by Rage et al. [20]. Fuzzy Correlated rule mining was first proposed by Lin et al. [21] using fuzzy correlation coefficient as the interesting measure.

Taking support and confidence as the only interesting measure would not produce rare correlated patterns. As fuzzy frequent and rare itemsets are mined by assigning multiple minimum supports to each fuzzy region, fuzzy correlated frequent and rare patterns are mined by assigning multiple item All-confidence values to each fuzzy regions. In this paper, a correlated pattern tree structure called multiple fuzzy correlated pattern tree (MFCP) is designed to mine frequent correlated patterns using Multiple Item all-confidence (MIAC). The proposed approach first constructs a MFCP tree using MIAC values and generates correlated patterns using MFCP mining algorithm. The amount of patterns generated is compared with the amount of patterns mined in multiple fuzzy frequent pattern rule mining algorithm.

1.1 Fuzzy FP-Tree Algorithm

Frequent pattern mining is one of the most significant research issues in data mining. Mining association rules was first given by Agarwal et al. [3] in the form of Apriori algorithm. The main drawback of the algorithm was repeated scanning of the database which proved to be too costly. To overcome the above problem Han et al. [22] introduced frequent pattern (FP) growth algorithm. FP growth algorithm uses a compressed data structure, called frequent pattern tree (FP-tree) which retains the quantitative information about frequent items. Frequent 1 itemsets will have nodes in the tree. The order of arrangement of tree nodes is that the more frequent nodes have better likelihood of sharing nodes when compared to less frequent nodes. FP-tree based pattern growth mining moves from frequent 1 itemset, scans the conditional pattern base, creates the conditional FP-tree and executes recursive mining till all its frequent itemset are scanned. Fuzzy FP Growth works in a same manner of FP-tree, with each node corresponding to a 1-itemset has a fuzzy membership function. The membership function for each 1-itemset is retrieved from the fuzzy dataset and the sum of all membership function values for the 1-itemset is its support. The support for a k-itemset (where k = 2) is calculated from the nodes corresponding to the itemset by using a suitable t-norm (Table 1).

Hong et al. [23] proposed mining fuzzy association rules using Aprioritid algorithm for quantitative transactions. Papadimitriou and Mavroudi [11] proposed an algorithm for mining fuzzy association rules based on FP trees. In their approach only local frequent 1-itemsets in each transaction were used for rule mining which was very primitive. Lin et al. [13] introduced a new fuzzy FP tree structure for mining quantitative data. In that approach the FP tree structure was huge as two transactions with identical fuzzy regions but different orders were put into two diverse paths. To overcome the above drawback Lin et al. [14] devised a compressed fuzzy frequent pattern tree. Here the items in the transactions, used for constructing the compressed fuzzy FP (CFFP) tree were sorted based on their occurrence frequencies. The above methodologies used only linguistic terms with maximum cardinality for mining process, making the number of fuzzy regions equal to the number of given items thus reducing the processing time.

1.2 Multiple Fuzzy FP-Tree Algorithm

Hong et al. [12] introduced a multiple fuzzy frequent pattern (MFFP) tree algorithm for deriving complete fuzzy frequent itemsets. Here a single itemset would have

Table 1 t-norms in fuzzy sets

$t-norm$
$T_M(x,y)=\min(x,y)$
$T_P(x,y)=xy$
$T_W(x,y)=\max(x+y-1,0)$

more than one fuzzy region which enables it to derive more fuzzy association rules than the previous models. A compressed multiple fuzzy frequent pattern tree (CMFFP) algorithm was proposed by Jerry Chun et al. [24]. This approach was an extension work of CFFP tree. CMFFP algorithm constructs tree structure similar to that of CFFP tree to store multiple fuzzy frequent itemsets for the subsequent mining process and each node contains additional array to keep track of the membership values of its prefix path. Hence the algorithm is efficiently designed for completely mine all the fuzzy frequent item sets.

Notation

n	Number of transactions in D
m	Number of items in D
D^i	The ith transaction datum, $i = 1$to n.
I_j	The jth item, $j = 1$ to m.
V_j^i	The quantity of an item I_j in. t ith transaction
R_{jk}	The K_th fuzzy region of item I_j.
$f_{jl}{}^{(i)}$	V_j^i's fuzzy membership value in the region R_{jl}
Sum_{jl}	The count of fuzzy region R_{jl} in D
max_Sum_{jl}	The maximum count value of the fuzzy region R_{jl} in D

2 The Multiple Fuzzy Correlated Pattern Tree Construction Algorithm

Input: A body of n quantitative transaction data, a set of membership functions, a minimum all confidence threshold minAllconf and MIAC (Minimum item all confidence) values for all frequent fuzzy regions.

Output: A multiple fuzzy correlated pattern tree (MFCP tree).

Step 1 Transform the quantitative value V_j^i of each transaction datum $D^i, i = 1$to n for each item $I_j, j = 1$to m, into a fuzzy set $f_j^{(i)}$ represented as $\left(\frac{f_{j1}^{(i)}}{R_{j1}} + \frac{f_{j2}^{(i)}}{R_{j2}} + \cdots + \frac{f_{jp}^{(i)}}{R_{jp}}\right)$ using the membership functions, where R_{jk} is the Kth fuzzy region of item I_j, $V_j^{(i)}$ is the quantity of item I_j in ith transaction, $f_{jl}^{(i)}$ is $V_j^{(i)}$'s fuzzy membership value in region R_{jl} and p is the number of fuzzy regions for I_j.

Step 2 Calculate the sum of membership (scalar cardinality) values of each fuzzy region R_{jl} in the transaction data as

$$Sum_{jl} = \sum_{i=1}^{n} f_{ijl} \tag{1}$$

Step 3 The sum of fuzzy regions for each item is checked against the predefined minimum support count (α). If the sum is equal to (or) greater than minimum support count, the corresponding fuzzy region is put in the set of frequent fuzzy region (F_1).

$$F_1 = \{R_{jl}\ Sum_{jl} > \alpha, 1 \leq j \leq m\} \quad (2)$$

Step 4 Build the Header table by arranging the frequent R_{jl} in F_1 in descending order of their MIAC values.

Step 5 Remove the item of R_{jl}'s not present in F_1 from the transactions and place the remaining R_{jl}'s in new transaction database (D_1).

Step 6 Sort the remaining R_{jl}'s in each transaction of (D_1) in descending order of their membership values.

Step 7 Set the root node of the MFCP tree as {null} and add the transactions of D_1 into the MFCP tree tuple by tuple.

While inserting, two cases can exist.

Sub step 7-1 If a fuzzy region R_{jl} in a transaction is at the corresponding branch of the MFCP tree for the transaction, add the membership value f_{ijl} of R_{jl} in the transaction to the node of R_{jl} in the branch.

Sub step 7-2 Else insert a node of R_{jl} at the end of the identical branch, fix the sum of the node as the membership value f_{ijl} of R_{jl}, and inset a link from the node of R_{jl} in the last branch to the current node. If there is no such branch with the node of R_{jl}, insert a link from the entry of R_{jl} in the header table of the added node.

2.1 An Example

This section illustrates the proposed multiple fuzzy correlated pattern mining algorithm in quantitative database using an example. Assume six transactions with five items $(A - E)$ as shown in Table 2. The amount given in the database for each items are denoted using three fuzzy regions *Low*, *Middle* and *High*. Thus, each item in a transaction gets three fuzzy membership values based on the predefined membership functions. The minimum support is set as 25 %. The membership functions in Fig. 1 are used for all the items in the database.

Table 2 Database with six transactions

TID	Items
1	(A:5) (C:10) (D:2) (E:9)
2	(A:8) (B:2) (C:3)
3	(B:3) (C:9)
4	(A:7) (C:9) (D:3)
5	(A:5) (B:2) (C:5)
6	(A:3) (C:10) (D:2) (E:2)

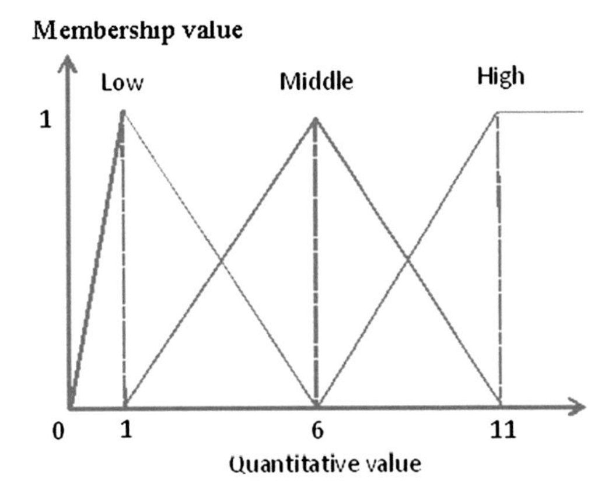

Fig. 1 The membership functions used in the example

Table 3 shows the transformation of quantitative values of the item to fuzzy sets. For example the amount "5" of item A in Transaction one is transformed into $\left(\frac{0.2}{A.low} + \frac{0.8}{A.middle}\right)$. The sum of membership values of each fuzzy region is then calculated. For example the sum of the membership value of *A.middle* is calculated as (0.8 + 0.6 + 0.8 + 0.8 + 0.4) which is 3.4. This step is repeated for all the regions and the results are displayed in Table 4.

Let minimum All-Confidence (minAllconf) threshold for the entire database is set to 70 % and the minimum support threshold is set as 24 %. The size of the dataset is 6 so the minimum support count is calculated as 6 * 24 % (=1.4). The sum of fuzzy regions that equals or larger than the minimum support count is kept in the set of frequent fuzzy regions (F_1) are used in successive mining. The frequent fuzzy regions (F_1) are displayed in Table 5.

Table 3 Fuzzy sets for Table 2 transactions

TID	Items
1	$\left(\frac{0.2}{A.low} + \frac{0.8}{A.middle}\right)\left(\frac{0.2}{C.middle} + \frac{0.8}{C.high}\right)$ $\left(\frac{0.8}{D.low} + \frac{0.2}{B.middle}\right)\left(\frac{0.4}{E.middle} + \frac{0.6}{E.High}\right)$
2	$\left(\frac{0.6}{A.middle} + \frac{0.4}{A.high}\right)\left(\frac{0.8}{B.low} + \frac{0.2}{B.middle}\right)\left(\frac{0.6}{C.low} + \frac{0.4}{C.middle}\right)$
3	$\left(\frac{0.6}{B.low} + \frac{0.4}{B.middle}\right)\left(\frac{0.4}{C.middle} + \frac{0.6}{C.high}\right)$
4	$\left(\frac{0.8}{A.middle} + \frac{0.2}{A.high}\right)\left(\frac{0.6}{D.low} + \frac{0.4}{D.middle}\right)$
5	$\left(\frac{0.2}{A.low} + \frac{0.8}{A.middle}\right)\left(\frac{0.8}{B.low} + \frac{0.2}{B.middle}\right)\left(\frac{0.2}{C.low} + \frac{0.8}{C.middle}\right)$
6	$\left(\frac{0.6}{A.low} + \frac{0.4}{A.middle}\right)\left(\frac{0.2}{C.middle} + \frac{0.8}{C.high}\right)$ $\left(\frac{0.8}{D.low} + \frac{0.2}{B.middle}\right)\left(\frac{0.8}{E.low} + \frac{0.2}{E.middle}\right)$

Table 4 Sum of membership values for the fuzzy regions given in Table 3

Item	Sum	Item	Sum
B.low	2.2	*A.low*	1.0
B.middle	0.8	*A.middle*	3.4
B.high	0.0	*A.high*	0.6
C.low	0.8	*D.low*	2.2
C.middle	2.4	*D.middle*	0.8
C.high	2.8	*D.high*	0.0
E.low	0.8	*E.high*	0.6
E.middle	0.6		

Table 5 The frequent fuzzy regions of set F_1

Frequent fuzzy regions	Sum
B.low	2.2
A.middle	3.4
C.middle	2.4
C.high	2.8
D.low	2.2

Mining frequent patterns and rare patterns requires Minimum item all confidence values (MIAC) values for all the fuzzy regions. To generate MIAC values for all fuzzy regions, it is necessary to calculate the difference in sum of the entire fuzzy region existing in the database.

Consider $\max\_Sum_{jl}$ as the maximum sum value in the database. (In this case it is 3.4). The difference in sum of entire fuzzy region (D^{sp}) is calculated using the formula.

$$\text{Sum Difference } (D^{sp}) = max\_Sum_{jl} - max\_Sum_{jl} * \text{ minAllconf} \quad (3)$$

$$= max\_Sum_{jl} * (1 - \text{ minAllconf}) \quad (4)$$

The MIAC value for each fuzzy region (R_{jl}) in the database is given by the formula.

$$\text{MIAC}(R_{jl}) = 1 - \left(\frac{D^{sp}}{Sum_{jl}}\right) \quad (5)$$

The MIAC value of any fuzzy region (R_{jl}) with maximum sum is the minAllconf value itself. For example the fuzzy region *A.middle* contains the maximum sum value of 3.4, the difference in sum value for the entire database is calculated as

$$\begin{aligned} D^{sp} &= \max\_Sum_{jl} * (1 - \text{ minAllconf}) \\ &= 3.4(1 - 0.70) \\ &= 1.02 \end{aligned}$$

Table 6 MIAC values for the frequent fuzzy items

Item	MIAC values (%)
B.low	54
A.middle	70
C.middle	57
C.high	63
D.low	54

Table 7 Header table

Fuzzy region	Sum
A.middle	3.4
C.high	2.8
C.middle	2.4
B.low	2.2
D.low	2.2

.

Substituting the above value in calculating MIAC (A.middle) we get

$$\text{MIAC (A.middle)} = 1 - \left(\frac{1.02}{3.4}\right) = 0.70$$

From this we come to a conclusion that the fuzzy regions of maximum sum would get minAllconf value as its MIAC value. In the same way MIAC values of other fuzzy regions are calculated and displayed in Table 6.

The MIAC values generated should be lesser or equal compared to the minAllconf of the entire database. So that rare fuzzy patterns are discovered.

The frequent fuzzy regions in F_1 are arranged in descending order of their MIAC value and are kept in the header table shown in Table 7.

The frequent fuzzy regions from each transaction are picked up from database (D) and placed into modified database (D_1). Sort the fuzzy regions in each transactions of (D_1) in descending order of their membership values. The results are displayed in Table 8. Build the MFCP tree by inserting fuzzy regions of each transactions tuple by tuple using the pattern tree. The resultant tree is given in Fig. 2.

Table 8 Ordered fuzzy regions for each transaction

TID	Items
1	$(\frac{0.8}{A.middle})(\frac{0.8}{C.high})(\frac{0.8}{D.low})(\frac{0.2}{C.middle})$
2	$(\frac{0.8}{B.low})(\frac{0.6}{A.middle})(\frac{0.4}{C.middle})$
3	$(\frac{0.6}{B.low})(\frac{0.6}{C.high})(\frac{0.4}{C.middle})$
4	$(\frac{0.8}{A.middle})(\frac{0.6}{C.high})(\frac{0.6}{D.low})(\frac{0.4}{C.middle})$
5	$(\frac{0.8}{A.middle})(\frac{0.8}{B.low})(\frac{0.8}{C.middle})$
6	$(\frac{0.8}{C.high}))(\frac{0.8}{D.low})(\frac{0.4}{A.middle})(\frac{0.2}{C.middle})$

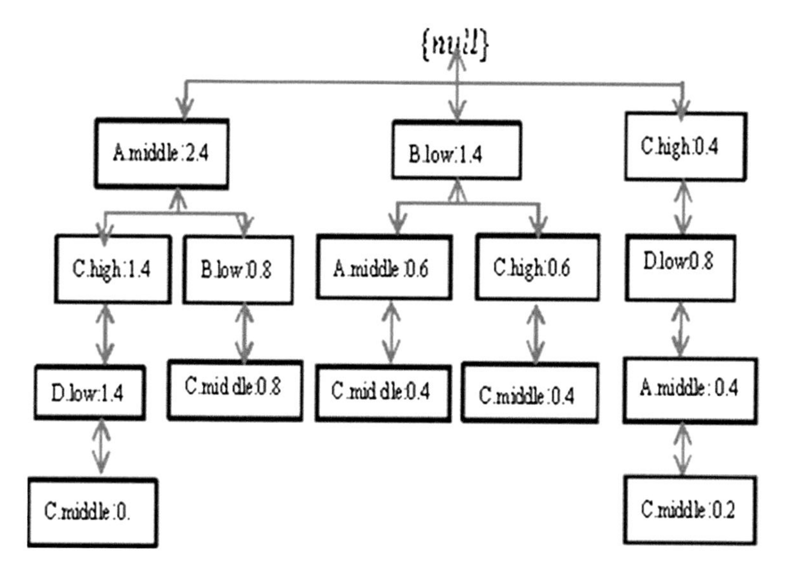

Fig. 2 MFCP tree after scanning all the transactions till tid = 6

3 Proposed Multiple Fuzzy Correlated Pattern Growth Mining Algorithm

The two measures for identifying frequent item set are support and confidence. Low level in these measures leads to the generation of too many patterns which are of least significance. To solve the above problem in quantitative databases All-Confidence measure is used which satisfies both anti-monotonic and null variance properties.

A fuzzy pattern using All-confidence measure for any two fuzzy regions F_x and F_y is defined as

$$\text{fAll} - \text{conf}(\{F_x, F_y\}) = \frac{\sum_{i=1}^{n} \min(f_j(t_i)|f_j \in \{F_x, F_y\})}{\max(fsupp\{F_x, F_y\})} \quad (6)$$

where $f_j(t_i)$ is the degree that a fuzzy region appears in transaction t_i and *fsupp* is the fuzzy support value of two fuzzy regions F_x, F_y. In both the cases the minimum operator is used for intersection. Using fuzzy All-confidence measure and generating MIAC values for each fuzzy region the proposed algorithm generates rare correlated patterns.

Input: The MFCP tree, header table and predefined MIAC values for frequent items.

Output: The fuzzy correlated patterns.

Step 1 The fuzzy regions in the header table, arranged in descending order of MIAC values are processed one by one from lowermost to the uppermost. Let the current fuzzy region is taken as R_{jl}.

Step 2 The items in the path of the fuzzy region R_{jl} $(\alpha.term)$ are extorted to form the conditional fuzzy patterns. The minimum operator is used for intersection. The fuzzy values are obtained by grouping of fuzzy itemsets related to R_{jl} using minimum operation, which excludes the items associated with the same I_j. (For example when considering the fuzzy region *C.middle*, all other frequent fuzzy region are associated with it except *C.High* as both regions are from same fuzzy item).

Step 3 Add the fuzzy values of the resultant fuzzy itemsets that are same and put as conditional pattern base of R_{jl}. Let $\beta.term$ be any fuzzy item in conditional pattern base, R_{jl}.

Step 4 The fuzzy correlated patterns are generated by checking it against the following conditions.

$$fsupp(\alpha.term, \beta.term) \geq minsup \tag{7}$$

and

$$\frac{fsupp(\beta.term)withR_{jl}}{sum(\beta.term)} \geq MIAC(\beta.term) \tag{8}$$

As $\beta.term$ would be region with maximum value compared with $\alpha.term$ the Eq. (6) can be rewritten in the form of Eq. (8) and checked against MIAC value of the fuzzy region.

Step 5 Repeat the steps 2–4 for subsequent fuzzy region until all the fuzzy regions in the header table are processed.

3.1 An Example

For the above generated MFCP tree, the items are processed from bottom to top in the order given by the header table $(D.low, B.low, C.middle, C.high, A.middle)$. To generate any correlated pattern of (α, β) where α is fuzzy region of smallest MIAC value compared to β and β is $\alpha' s$ conditional pattern base.

From the given MFCP tree, the conditional pattern base of $(D.low)$ consists of $(B.low)(C.middle)$ $(C.high)$ *and* $(A.middle)$. If $\beta.term$ is taken as $(C.high)$.

$$fsupp(D.low, C.high) = 1.4 \geq minsup$$

and

Table 9 Fuzzy correlated patterns

S.No	Fuzzy correlated Itemsets
1	$(C.middle, A.middle)$
2	$(D.low, C.high)$
3	$(B.low, C.middle)$

$$\frac{fsupp(D.low, C.high)}{sum(C.high)} = 79\% \geq MIAC(C.high)$$

Hence, $(D.low, C.high) is\, a\, correlated\, pattern$. The final lists of fuzzy correlated patterns are shown in the Table 9. For the above example database only 2-fuzzy correlated pattern are generated. There are no possibilities of mining 3-fuzzy correlated patterns from the given example.

4 Experimental Results

The testing environment consists of a 2.13 GHZ Intel® Core™ i3 processor with 2.00 GB of memory running a windows xp operating system. We took mushroom dataset and Retail dataset publicly available at the FIMI repository [25] is used for evaluation. The details of the datasets are shown in Table 10. Random values from 1 to 11 were assigned to the items in the dataset.

Minimum support value is set to 0.01. Experiments are conducted for a range of minimum All-Confidence values 100–30 %. MIAC values are calculated for the frequent-1 items (items in Header Table) with reference to the highest support item in the database. Figures 3 and 4 show the execution times taken for various minimum All-Confidence values for mushroom and retail datasets respectively.

Table 10 Details of databases

Database	No. of transactions	No. of items	Avg. length of transactions
Mushroom	8124	119	23
Retail	88,162	16,470	76

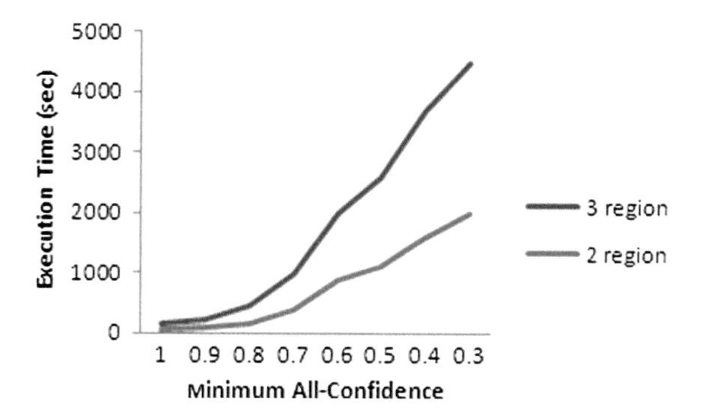

Fig. 3 Execution time at different MIAC values for mushroom dataset

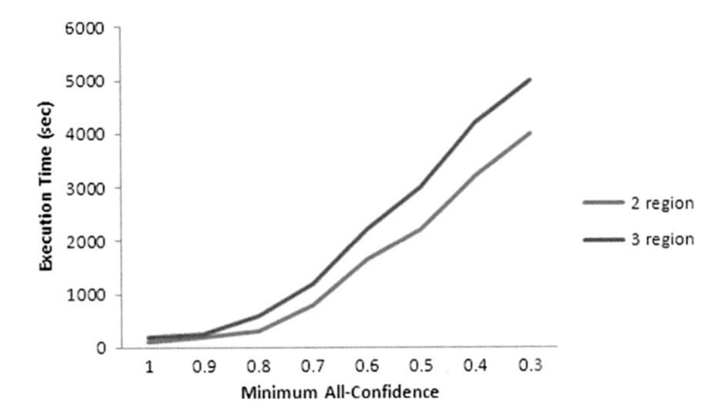

Fig. 4 Execution time at different MIAC values for retail dataset

Figures 3 and 4 show that the execution time is prolonged for three regions when compared to two regions for all minimum All-Confidence thresholds. This is due to the fact that three regions would produce more tree nodes than two fuzzy regions. The processing time also depends upon transformation of fuzzy regions using predefined membership functions and the quantitative values of items.

Figure 5 illustrates the number of correlated patterns generated for various minimum All-Confidence values for mushroom and retail dataset. The number of patterns obtained in the retail dataset is more when compared to the patterns generated from the mushroom dataset. This is because; increase in dataset size causes more patterns to be generated. The number of correlated patterns generated using the MFCP approach is more when compared with multiple fuzzy frequent pattern (MFFP) growth mining algorithm. The MFCP mining approach uses different MIAC values for each item based on its support difference, so at even high confidence levels rare items are generated. As support and confidence are the only interesting measures considered for MFFP growth mining algorithm, the time taken is slightly less when compared to the MFCP tree mining approach. Figure 7 displays the time taken for execution using the above methods using retail dataset. Fuzzy rare patterns are left out in MFFP growth mining algorithm as it works on single support and confidence measure for the entire database. The MFCP approach generates rare patterns along with frequent patterns. So Comparatively MFCP generates more number of patterns than MFFP approach. Using Retail dataset for

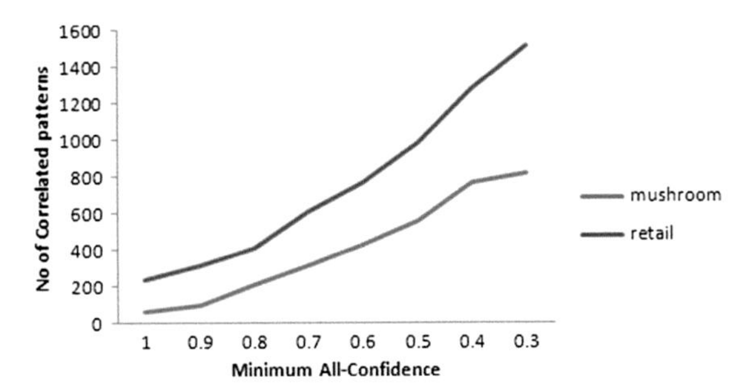

Fig. 5 Number of correlated patterns generated at different min-all-confidence

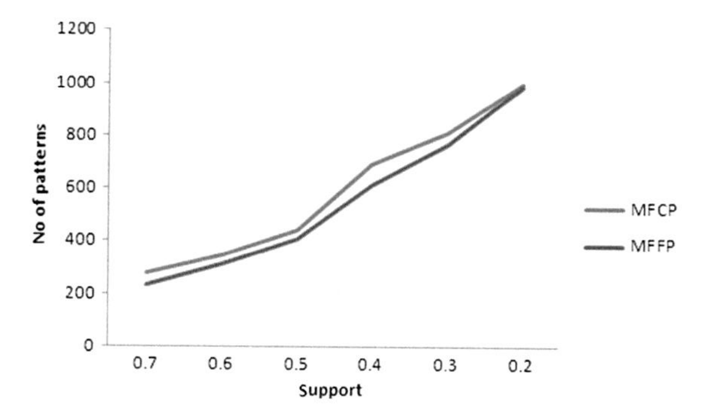

Fig. 6 Comparison on the number of patterns generated using MFFP and MFCP

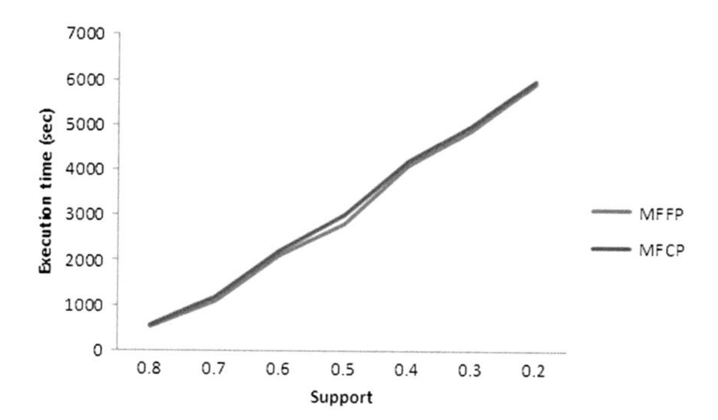

Fig. 7 Execution time of MFFP and MFCP

the given support measure, patterns are generated using the above mentioned approaches. The Fig. 6 displays the number of patterns generated using both MFCP and MFFP growth mining algorithms.

5 Conclusions and Future Work

In Traditional data mining approaches, maximum work is on binary databases to find significant information. The discovered knowledge is represented in a statistical or numerical way. In real world applications, the scope of using binary databases is less. Fuzzy set theory reveals human thinking and aids to take significant decisions. In this paper, the idea of Multiple Fuzzy Frequent Pattern tree [12] and Correlated pattern tree [20] are integrated to find fuzzy correlated patterns in a given quantitative data set. In the MFCP tree mining algorithm MIAC values are set for each item to find the fuzzy correlated patterns. The rational for using two confidence thresholds has been justified. An example is given to demonstrate that the MFCP tree mining algorithm can derive multiple frequent correlated patterns under multiple minimum item confidence in a simple and effective way.

In the future, the proposed approach is extended for dynamic databases and also investigates the extension of proposed work for generating fuzzy closed and maximal correlated patterns.

Acknowledgments The authors like to thank the anonymous referees for their notable comments and Sri Ramakrishna Engineering College for providing resources for our experiments.

References

1. Tan, P.N., Steinbach, M., Kumar, V.: Introduction to Data Mining. Pearson Addison Wesley, Boston (2006)
2. Agarwal, R., Imielinski, T., Swami, A.: Mining association rules between sets of items in large databases. The International Conference on Management of Data, pp. 207–216 (1993)
3. Agarwal, R., Srikant, R.: Fast algorithms for mining association rules in large databases. The International Conference on Very Large Databases, pp. 487–499 (1994)
4. Kim, E., Kim, W., Lee, Y.: Combination of multiple classifiers for the customer's purchase behavior prediction. Decis. Support Syst. (2003)
5. Berkhin, P.: A Survey of clustering data mining techniques. Technical Report, Accrue Software, San Jose, CA (2002)
6. Han, J., Pei, J., Yin, Y., Mao, R.: Mining frequent patterns without candidate generation: a frequent-pattern tree approach. Data Mining Knowl. Discov. (2004)
7. Zadeh, L.A.: Fuzzy set. Inf. Control **8**(3), 338–353 (1965)
8. de Campos, L.M., Moral, S.: Learning rules for a fuzzy inference model. Fuzzy Sets Syst. **59**, 247–257 (1993)
9. Hong, T.P., Chen, J.B.: Finding relevant attributes and membership functions. Fuzzy Sets Syst. **103**, 389–404 (1999)
10. Hong T.P., Lee, C.Y.: Induction of fuzzy rules and membership functions from training examples. Fuzzy Sets Syst. 33–47 (1996)
11. Papadimitriou, S., Mavroudi, S.: The fuzzy frequent pattern tree. In: The WSEAS International Conference on Computers (2005)
12. Hong, T.P., Lin C.W., Lin, T.C.: The MFFP-Tree fuzzy mining algorithm to discover complete linguistic frequent itemsets. Comput. Intell. **30**(1) (2014)
13. Lin, C.W., Hong, T.P., Lu, W.H.: Mining fuzzy association rules based on fuzzy fp-trees. In: The 16th National Conference on Fuzzy Theory and Its Applications, pp. 11–16 (2008)
14. Lin, C.W., Hong, T.P., Lu, W.H.: An efficient tree-based fuzzy data mining approach. Int. J. Fuzzy Syst. **12**(2) (2010)
15. Brin, S., Motwani, R., Silverstein, C.: Beyond market baskets: generalizing association rules to correlations. SIGMOD Rec. **26**, 265–276 (1997)
16. Kim, S., Barsky, M., Han, J.: Efficient mining of top correlated patterns based on null invariant measures. In: ECML PKDD. pp. 172–192 (2011)
17. Kim, W.Y., Lee, Y.K. Han, J.: Ccmine: efficient mining of confidence-closed correlated patterns. In: PAKDD, pp. 569–579 (2004)
18. Lee, Y.K., Kim, W.Y., Cao, D., Han, J.: CoMine: efficient mining of correlated patterns. In: ICDM, pp. 581–584 (2003)
19. Liu, B., Hsu, W., Ma, Y.: Mining association rules with multiple minimum supports. In: KDD, pp. 337–341 (1999)
20. Rage, U.K., Kitsuregawa, M.: Efficient discovery of correlated patterns using multiple minimum all-confidence thresholds. J. Intell. Inf. Syst. (2014)
21. Nancy P. Lin., Hao-En Chueh.: Fuzzy correlation rules mining. In: International Conference on Applied Computer Science, Hangzhou, China, 15–17 April (2007)

22. Han, J., Kamber, W.: Data Mining: Concepts and Techniques. Morgan Kaufman, San Francisco (2001)
23. Hong, T.P., Kuo, C.S., Wang, S.L.: A fuzzy aprioritid mining algorithm with reduced computational time. Appl. Soft Comput. **5**, 1–10 (2004)
24. Lin, J.C., Hong, T.P., Lin, T.C.: CMFFP-tree algorithm to mine complete multiple fuzzy frequent itemsets. Appl. Soft Comput. (2015)
25. Frequent Itemset Mining Implementations repository: http://fimi.helsinki.fi

Autonomous Visual Tracking with Extended Kalman Filter Estimator for Micro Aerial Vehicles

K. Narsimlu, T.V. Rajini Kanth and Devendra Rao Guntupalli

Abstract The objective of this paper is to estimate the Ground Moving Target position and track the Ground Moving Target continuously using Extended Kalman Filter estimator. Based on previous target positions in image sequences, this algorithm predicts the target next position in the image sequence. A Graphical User Interface based tool was developed for simulation and test the Autonomous Visual Tracking with Extended Kalman Filter estimator using MATLAB Graphical User Interface Development Environment tool.

Keywords Autonomous visual tracking system • Extended Kalman Filter Ground moving target • Ground stationary target • Micro aerial vehicles • Image tracking software

1 Introduction

Recently, the uses of Micro Aerial Vehicles (MAVs) have increased for various purposes such as civil aircraft applications, military aircraft applications [1, 2], crowd monitoring and control, environmental monitoring [3], forestry and fire monitoring [4], meteorology applications [5], aerial photography [6], agriculture [7],

K. Narsimlu (✉)
Department of Computer Science and Engineering, Jawaharlal Nehru Technological University, Hyderabad, India
e-mail: narsimlu@gmail.com

T.V. Rajini Kanth
Department of Computer Science and Engineering, Sreenidhi Institute of Science and Technology, Hyderabad, India
e-mail: rajinitv@gmail.com

D.R. Guntupalli
Information Systems, Cyient Limited, Hyderabad, India
e-mail: Devendra.Guntupalli@cyient.com

V. Ravi et al. (eds.), *Proceedings of the Fifth International Conference on Fuzzy and Neuro Computing (FANCCO - 2015)*, Advances in Intelligent Systems and Computing 415, DOI 10.1007/978-3-319-27212-2_3

security, surveillance and surveying applications. These MAVs can be controlled by on-board Autonomous Visual Tracking System (AVTS) [8, 9]. The main use of AVTS is to track the Ground Moving Target (GMT) [10–14] or Ground Stationary Target (GST) [15–19] from MAV.

On-board AVTS contains subsystems such as the Gimbaled Camera, Global Positioning System (GPS)/Inertial Navigation System (INS), Image Tracking Software [20–34], MAV Guidance, Gimbaled Camera Control and Autopilot. On-board AVTS receives the real-time video frame sequence from Gimbaled Camera and it computes the GMT position and GMT velocity in the real world based on the pixel position in image sequences. On-board INS/GPS sensor is used for measures the MAV present position and velocity. The Extended Kalman Filter (EKF) estimator is used for the estimation of GMT state parameters such as position and velocity.

In this paper, the EKF estimator is discussed along with algorithm usage. The main motivation of this paper is to estimate the GMT path in an image sequence and tracks the GMT continuously with the help of the Gimbaled Camera using EKF estimator.

A 3D geometry demonstration of AVTS configuration is shown in Fig. 1.

In this paper, the remaining sections are organized as follows: Section 2, Extended Kalman Filter Estimator. Section 3, Autonomous Visual Tracking System. Section 4, Simulation and Experimental Results. Section 5, Conclusions.

2 Extended Kalman Filter Estimator

This section discusses the Extended Kalman Filter (EKF) estimator process in detail. This EKF process being estimated can be accurately linearized at each point along the trajectory of the states. Hence, EKF is a very reliable state estimator for non-linear process [35–40]. The key feature of EKF estimator is the ability to track the GMT even when the target loss event occurs in the video frame sequence.

In order to start the GMT tracking, the EKF uses the initial state of GMT (position and velocity), which can be provided automatically by an object detector. The EKF predicts the GMT next state based on the GMT previous state from the previous frame sequence in the video frame sequence. The EKF measures the GMT state based on the pixel position, which was inherently a nonlinear discrete-time process. The EKF corrects the GMT state and estimate the trajectory of the GMT based on measured data. These predict, measure, correct steps are iteratively run in order to track GMT continuously.

The EKF estimator steps are: Step 1, Initialization: the EKF state vector (s_t) contains the GMT initial state (x_t , y_t), the search window size (w_t , l_t) and the GMT centroid (x_c , y_c), at time t, as follows:

$$s_t = (x_t , y_t , w_t , l_t , x_c , y_c) \quad (1)$$

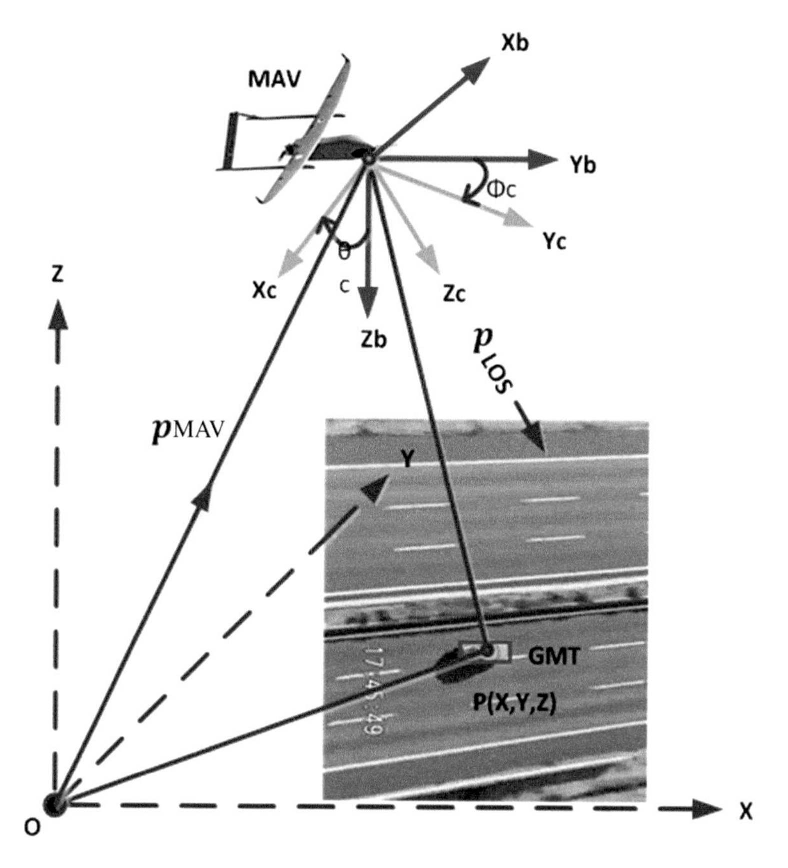

Fig. 1 3D Demonstration of Autonomous Visual Tracking System

The EKF measurement vector (z_t) at time t, as follows:

$$z_t = (x_t\, , \, y_t\, , \, w_t\, , \, l_t) \tag{2}$$

The EKF estimation process updates the GMT state.
The EKF state process (s_t) at time t, as follows:

$$s_t = A^* s_{t-1} + w \tag{3}$$

where, the EKF Jacobian matrix (A), at the time difference (dt), as follows:

$$A = \begin{bmatrix} 1 & 0 & dt & 0 & 0 & 0 \\ 0 & 1 & 0 & dt & 0 & 0 \\ 0 & 0 & 1 & 0 & 0 & 0 \\ 0 & 0 & 0 & 1 & 0 & 0 \\ 0 & 0 & 0 & 0 & 1 & 0 \\ 0 & 0 & 0 & 0 & 0 & 1 \end{bmatrix} \tag{4}$$

The EKF process noise (w) at time t, as follows:

$$w = \begin{bmatrix} 1 \\ 1 \\ 1 \\ 1 \\ 1 \\ 1 \end{bmatrix} \tag{5}$$

The EKF measurement model (z_t) at time t, as follows:

$$z_t = H^* s_t + v \tag{6}$$

where, the EKF Jacobian matrix (H) at time t, as follows:

$$H = \begin{bmatrix} 1 & 0 & 0 & 0 & 0 & 0 \\ 0 & 1 & 0 & 0 & 0 & 0 \\ 0 & 0 & 1 & 0 & 0 & 0 \\ 0 & 0 & 0 & 1 & 0 & 0 \end{bmatrix} \tag{7}$$

The EKF measurement noise (v) at time t, as follows:

$$v = \begin{bmatrix} 0.1 \\ 0.1 \\ 0 \\ 0 \end{bmatrix} \tag{8}$$

The EKF process noise covariance (Q) at time t, as follows:

$$Q = \begin{bmatrix} 1 & 0 & 0 & 0 & 0 & 0 \\ 0 & 1 & 0 & 0 & 0 & 0 \\ 0 & 0 & 1 & 0 & 0 & 0 \\ 0 & 0 & 0 & 1 & 0 & 0 \\ 0 & 0 & 0 & 0 & 1 & 0 \\ 0 & 0 & 0 & 0 & 0 & 1 \end{bmatrix} \tag{9}$$

The EKF measurement noise covariance (R) at time t, as follows:

$$R = \begin{bmatrix} 0.1 & 0 & 0 & 0 \\ 0 & 0.1 & 0 & 0 \\ 0 & 0 & 1 & 0 \\ 0 & 0 & 0 & 1 \end{bmatrix} \tag{10}$$

Step 2, Prediction: the EKF predicts the GMT next state (s_t^-) based on GMT previous state (s_{t-1}):

$$s_t^- = A * s_{t-1} \tag{11}$$

The EKF update the covariance (P_t^-) based on previous covariance (P_{t-1}) and EKF process noise covariance (Q):

$$P_t^- = A * P_{t-1} * A^T + Q \tag{12}$$

Step 3, Correction: the EKF corrects the GMT state and estimates the GMT trajectory based on the measured data.

The EKF Kalman gain (K_t) specifies the measurement:

$$K_t = P_t^- * H^T (H * P_t^- * H^T + R)^{-1} \tag{13}$$

Based on the EKF Kalman gain (K_t) measurement, the corrected state (s_t) is:

$$s_t = s_t^- + K_t (z_t - H * s_t^-) \tag{14}$$

Based on the EKF Kalman gain (K_t) measurement, the corrected covariance (P_t):

$$P_t = P_t^- (I - K_t{}^* H) \tag{15}$$

where, H is the measurement Jacobians at time t, and R is the measurement noise covariance at time t.

These predict, measure, correct steps are iteratively run in order to track GMT continuously and estimate the GMT state (position and velocity) accurately.

The EKF estimator cycle is shown in Fig. 2.

3 Autonomous Visual Tracking System

On-board Autonomous Visual Tracking System (AVTS) contains subsystems such as Gimbaled Camera, INS/GPS, Image Tracking Software, MAV Guidance Law, Gimbaled Camera Control Law and Autopilot [41–55].

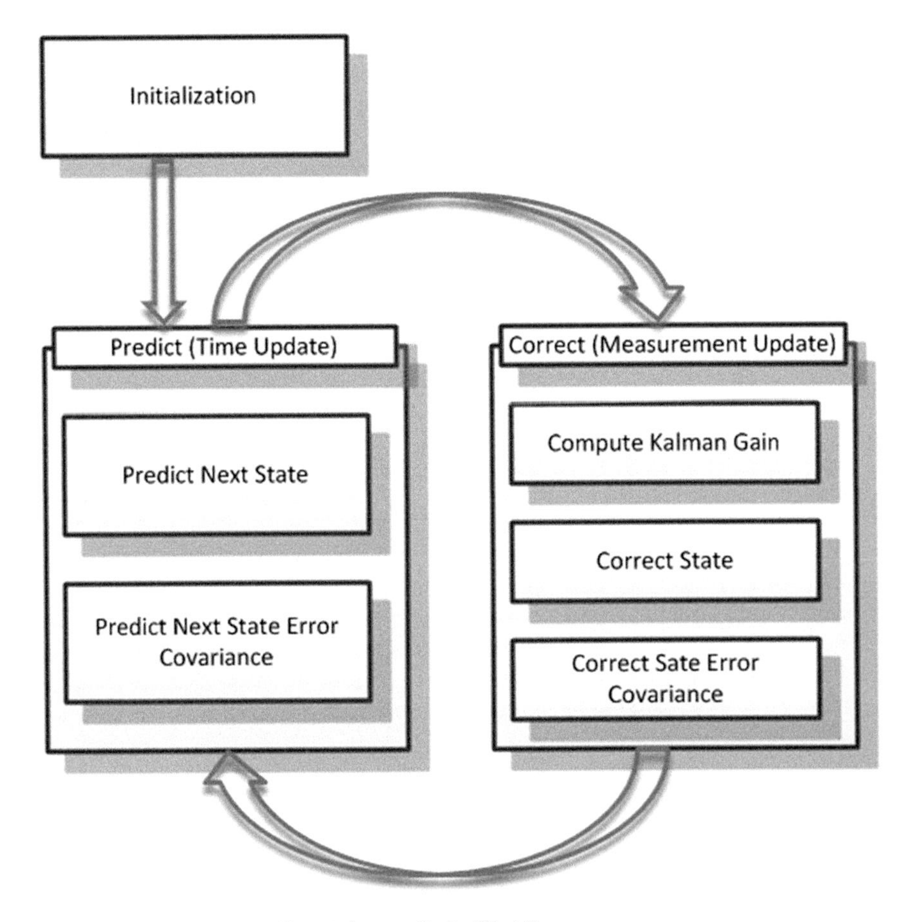

Fig. 2 Extended Kalman Filter Estimator Cycle [35–40]

A Simulation of AVTS is developed using MATLAB GUIDE R2011a, is shown in Fig. 3.

The main purpose of AVTS simulation is to estimate and track the GMT continuously from MAV. We implemented EKF estimator in MATLAB. The implemented EKF estimator is included in AVTS for the simulation.

4 Simulation and Experimental Results

We have considered the video of aerial tracking [56] as an input video in AVTS for the GMT tracking simulation purpose.

The GMT tracking by the EKF estimator is shown in Fig. 4.

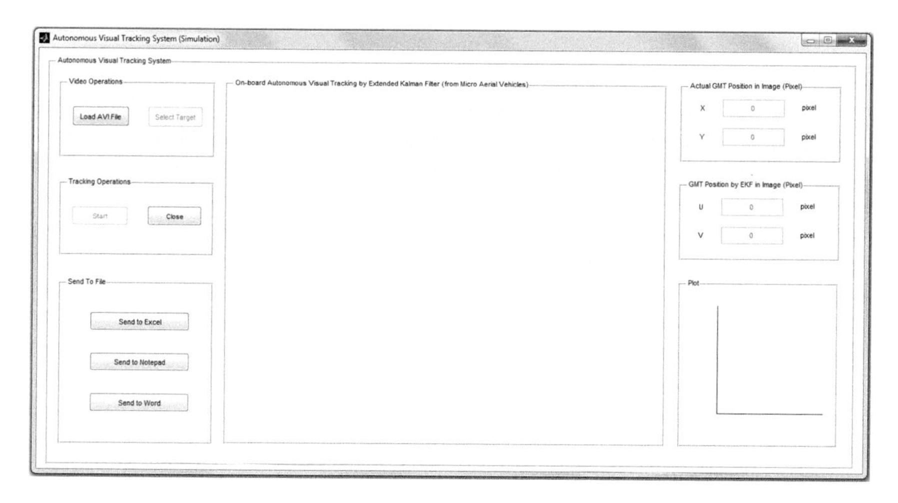

Fig. 3 Autonomous Visual Tracking System (Simulation)

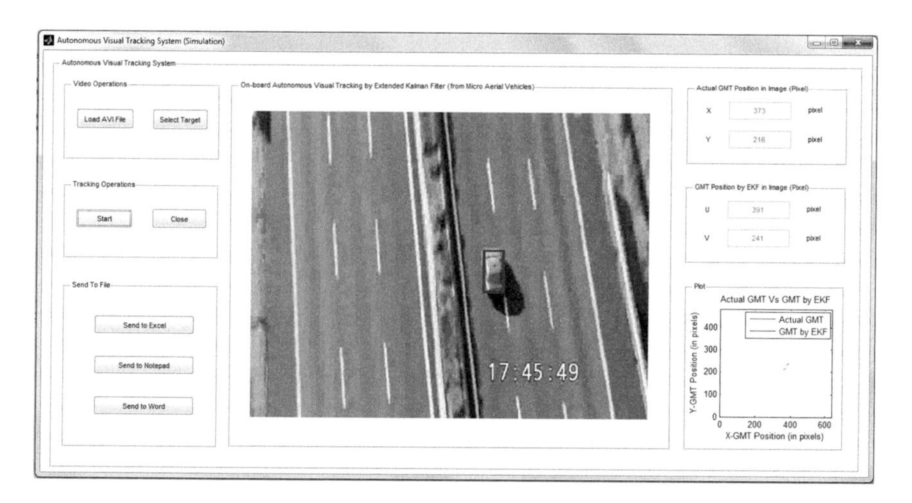

Fig. 4 GMT Tracking by Extended Kalman Filter

The captured results of GMT tracking by the EKF estimator (Frame-by-Frame), is shown in Fig. 5.

We considered the input video frame size (X, Y) as (640, 480) pixels, whereas input video frame resolution is 640 × 480 pixels. We have given 12 input video frame sequences in AVTS to experimental results analysis. In AVTS, the actual GMT position in the video frame is marked with red "+" and stored the actual GMT position (X, Y) pixels. In AVTS, the computed the GMT position of the EKF in the video frame is marked with blue rectangle and stored GMT position by the EKF estimator as (U, V) pixels, is shown in a Table 1.

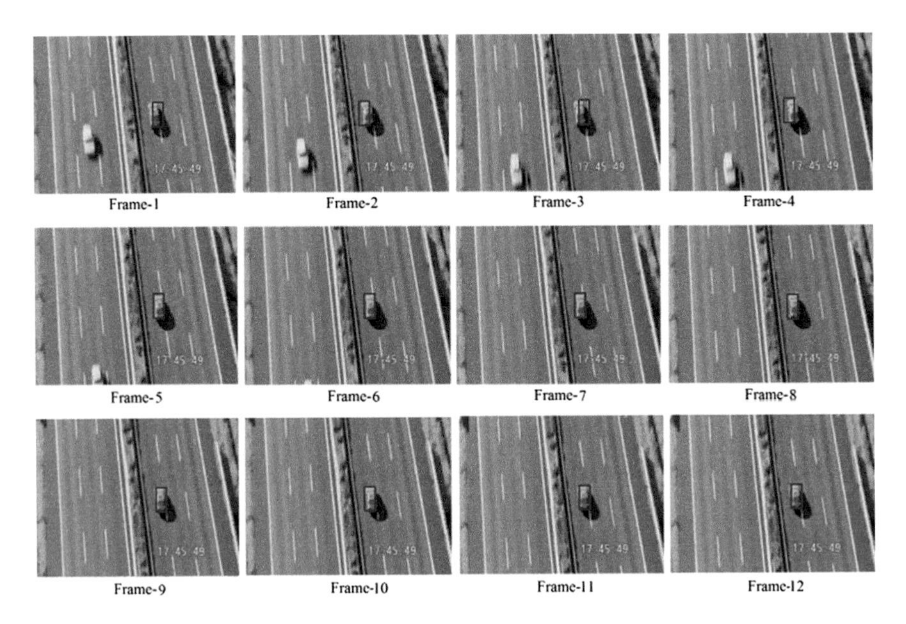

Fig. 5 GMT Tracking by Extended Kalman Filter (Frame-by-Frame)

Table 1 Actual GMT position (X, Y) versus GMT position by the EKF (U, V) (in pixels)

Frame number	Actual GMT position (X, Y)	GMT position by EKF (U, V)	Error (δX, δY) = (Actual GMT position−GMT position by EKF)
1st-Frame	(378, 234)	(367, 215)	(11, 19)
2nd-Frame	(381, 232)	(371, 211)	(10, 21)
3rd-Frame	(387, 233)	(376, 214)	(11, 19)
4th-Frame	(390, 234)	(379, 218)	(11, 16)
5th-Frame	(390, 234)	(377, 218)	(13, 16)
6th-Frame	(392, 236)	(380, 219)	(12, 17)
7th-Frame	(392, 237)	(379, 221)	(13, 16)
8th-Frame	(393, 237)	(380, 221)	(13, 16)
9th-Frame	(393, 237)	(389, 222)	(4, 15)
10th-Frame	(391, 239)	(390, 225)	(1, 14)
11th-Frame	(391, 239)	(389, 224)	(2, 15)
12th-Frame	(391, 241)	(389, 226)	(2, 15)

Using AVTS, we can export the actual GMT position data and the GMT position of the EKF estimator data to Microsoft Word, Microsoft Excel and Microsoft Notepad for off-line analysis. We calculated the error in the GMT position (δX, δY) between the actual GMT position (X, Y) and the GMT position by the EKF estimator (U, V) in pixels.

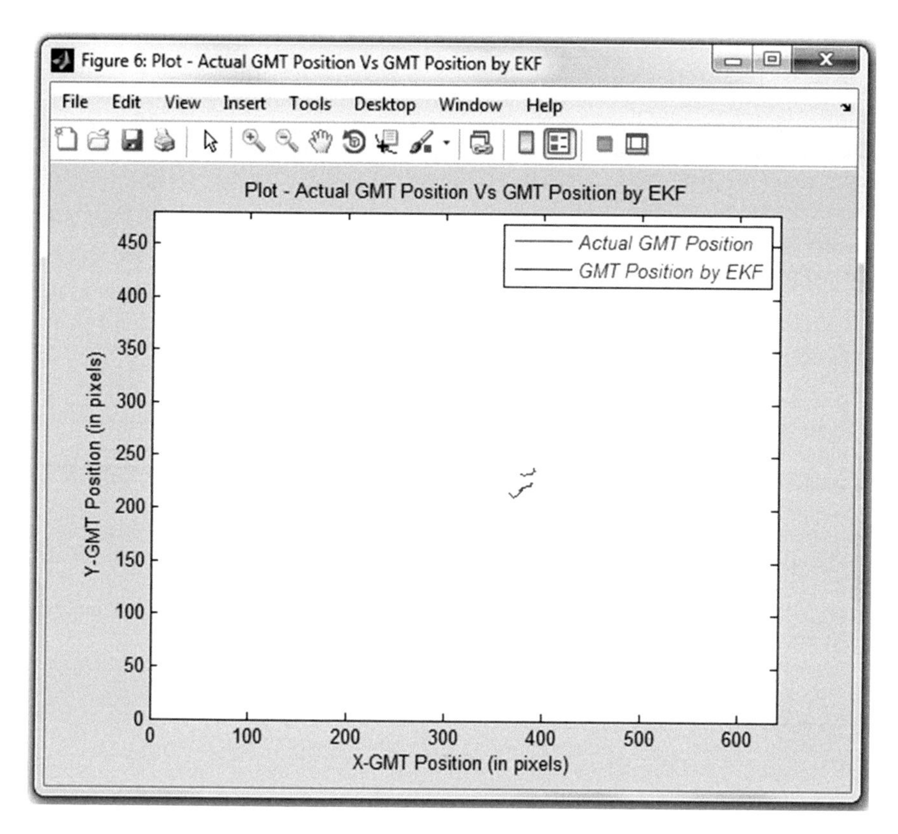

Fig. 6 Plot—Actual GMT Position Vs GMT Position by the EKF

In AVTS, the experimental results of actual GMT position and computed GMT position by the EKF are plotted on a graph, as shown in Fig. 6.

On the graph, we considered the GMT position (X, Y) as (640, 480) pixels. The red trajectory of the graph indicates the actual GMT position and the blue trajectory GMT position by the EKF estimator.

The experimental results show that the error between the actual GMT position and the GMT position by the EKF estimator is very less. This EKF estimate the GMT state very accurately.

5 Conclusions

The Extended Kalman Filter is explained along with algorithm usage. A Simulation of Autonomous Visual Tracking System is developed using MATLAB. The Extended Kalman Filter is implemented in MATLAB and included in Autonomous Visual Tracking System for simulation purpose. The Autonomous Visual Tracking

System is tested on the input video of aerial tracking and observed the Extended Kalman Filter estimator performance. We observed that the error between the actual Ground Moving Target position and Ground Moving Target position by the Extended Kalman Filter estimator is very less. This Extended Kalman Filter estimator predicts, measures, corrects iteratively in order to track Ground Moving Target continuously and estimate the Ground Moving Target state (position and velocity) accurately.

References

1. Narsimlu, K., Rajinikanth, T.V., Guntupalli, D.R.: An experimental study of the autonomous visual target tracking algorithms for small unmanned aerial vehicles. In: 1st International Conference on Rough Sets and Knowledge Technologies (ICRSKT-2014), pp. 80–84. Elsevier Publications (2014)
2. Narsimlu, K., Rajinikanth, T.V., Guntupalli, D.R.: A comparative study on image fusion algorithms for avionics applications, Int. J. Adv. Eng. Glob. Technol. **2**(4), 616–621 (2014)
3. Rajinikanth, T.V., Rao, T., Rajasekhar, N.: An efficient approach for weather forecasting using support vector machines. In: International Conference on Intelligent Network and Computing (ICINC-2012), vol. 47, issue 39, pp. 208–212 (2012)
4. Rajinikanth, T.V., Rao T.: A hybrid random forest based support vector machine classification supplemented by boosting, Int. Res. J. Publ. **14**(1), 43–53 (2014) (Global Journals Inc. (USA))
5. Rajinikanth, T.V., Rao T.: Supervised classification of remote sensed data using support vector machine, Int. Res. J. Publ. **14**(14), 71–76 (2014). Global Journals Inc. (USA)
6. Rajinikanth, T.V., Nagendra Kumar, Y.J.: Managing satellite imagery using geo processing tools and sensors—mosaic data sets. In: 1st International Conference on Rough Sets and Knowledge Technologies (ICRSKT-2014), pp. 52–59. Elsevier Publications (2014)
7. Kiranmayee, B.V., Nagini, S., Rajinikanth T.V.: A recent survey on the usage of data mining techniques for agricultural datasets analysis. In: 1st International Conference on Rough Sets and Knowledge Technologies (ICRSKT-2014), pp. 38–44. Elsevier Publications (2014)
8. Zhang, M., Liu, H.H.: Vision-based tracking and estimation of ground moving target using unmanned aerial vehicle. In: IEEE, American Control Conference, pp. 6968–6973 (2010)
9. Dobrokhodov, V.N., Kaminer, I., Jones, K.D., Ghabcheloo, R.: Vision-based tracking and motion estimation for moving targets using small UAVs. In: IEEE, American Control Conference (2006)
10. El-Kalubi, A.A., Rui, Z., Haibo, S.: Vision-based real time guidance of UAV. In: IEEE, Management and Service Science (MASS), pp. 1–4 (2011)
11. Xin, Z., Fang, Y., Xian, B.: An on-board pan-tilt controller for ground target tracking systems. In: IEEE, Control Applications (CCA), pp. 147–152 (2011)
12. Li, Z., Hovakimyan, N., Dobrokhodov, V., Kaminer, I.: Vision based target tracking and motion estimation using a small UAV. In: IEEE, Decision and Control (CDC), pp. 2505–2510 (2010)
13. Watanabe, Y., Calise, A.J. Johnson, E.N.: Vision-based obstacle avoidance for UAVs. In: AIAA, Guidance, Navigation and Control Conference and Exhibit (2007)
14. Theodorakopoulos, P., Lacroix, S.: A strategy for tracking a ground target with a UAV. In: IEEE, Intelligent Robots and Systems (IROS 2008), pp. 1254–1259 (2008)
15. Peliti, P., Rosa, L., Oriolo, G., Vendittelli, M.: Vision-based loitering over a target for a fixed-wing UAV. In: 10th IFAC Symposium on Robot Control, International Federation of Automatic Control, Dubrovnik, Croatia (2012)

16. Barber, D.B., Redding, J.D., Mclain, T.W., Beard, R.W., Taylor, C.N.: Vision-based target geo-location using a fixed-wing miniature air vehicle. J. Intell. Robot. Syst. **47**(4), 361–382 (2006)
17. Johnson, E.N., Schrage, D.P.: The Georgia Tech Unmanned Aerial Research Vehicle: GTMax. School of Aerospace Engineering, Georgia Institute of Technology, Atlanta (2003)
18. Cohen, I., Medioni, G.: Detecting and tracking moving objects in video from and airborne observer. In: IEEE Image Understanding Workshop, pp. 217–222 (1998)
19. Yau, W.G.; Fu, L-C., Liu, D.: Design and implementation of visual servoing system for realistic air target tracking. In: Proceedings of the IEEE International Conference on Robotics and Automation—ICRA, vol. 1, pp. 229–234 (2001)
20. Yilmaz, A., Javed, O., Shah, M.: Object tracking: a survey. ACM Comput. Surv. **38**(4) (2006)
21. Joshi, K.A., Thakore, D.G.: A survey on moving object detection and tracking in video surveillance system. Int. J. Soft Comput. Eng. (IJSCE) **2**(3) (2012)
22. Badgujar, C.D., Sapkal, D.P.: A Survey on object detect, track and identify using video surveillance. IOSR J. Eng. (IOSRJEN) **2**(10), 71–76 (2012)
23. Deori, B., Thounaojam, D.M.: A survey on moving object tracking in video. Int. J. Inf. Theory (IJIT) **3**(3) (2014)
24. Hu, W., Tan, T., Wang, L., Maybank, S.: A survey on visual surveillance of object motion and behaviors. IEEE Trans. Syst. Man Cybern. Part C, 334–352 (2004)
25. Comaniciu, D., Meer, P.: Mean shift: a robust approach toward feature space analysis. IEEE Trans. PAMI **24**(5) (2002)
26. Comaniciu, D., Ramesh, V., Meer, P.: Real-time tracking of non-rigid objects using mean shift. In: Proceedings of the Conference on Computer Vision and Pattern Recognition, vol. 2, pp. 142–149 (2000)
27. Comaniciu, D., Meer, P.: Mean shift analysis and applications. In: Proceedings of the IEEE International Conference on Computer Vision (ICCV), pp. 1197–1203 (1999)
28. Comaniciu, D., Ramesh, V., Meer, P.: Kernel-based object tracking. IEEE Trans. Pattern Anal. Mach. Intell. **25**(5), 564–577 (2003)
29. Collins, R.T.: Mean-shift blob tracking through scale space. In: Proceedings of IEEE Conference on Computer Vision and Pattern Recognition, pp. 234–240 (2003)
30. Leung, A., Gong, S.: Mean-shift tracking with random sampling. In: BMVC, pp. 729–738 (2006)
31. Intel Corporation: OpenCV Reference Manual v2.1, (2010)
32. Allen, J.G., Xu, R.Y.D., Jin, J.S.: Object tracking using camshift algorithm and multiple quantized feature spaces, In: Proceedings of the Pan-Sydney Area Workshop on Visual Information Processing, pp. 3–7 (2004)
33. Stolkin, R., Florescu, I., Kamberov, G.: An adaptive background model for camshift tracking with a moving camera, In: Proceedings of the 6th International Conference on Advances in Pattern Recognition, (2007)
34. Emami, E, Fathy, M.: Object tracking using improved camshift algorithm combined with motion segmentation (2011)
35. Kalman, R.E.: A new approach to linear filtering and prediction problems. Trans. ASME: J. Basic Eng. **82**, 35–45 (1960)
36. Kalman, R.E., Bucy, R.S.: New results in linear filtering and prediction theory. Trans. ASME: J. Basic Eng. **83**, 95–107 (1961)
37. Welch, G., Bishop, G.: An introduction to the Kalman filter. In: Proceedings of SIGGRAPH, pp. 19–24 (2001)
38. Janabi, F., Marey, M.: A Kalman filter based method for pose estimation in visual servoing. IEEE Trans. Rob. **26**(5), 939–947 (2010)
39. Torkaman, B., Farrokhi, M.: A Kalman-filter-based method for real-time visual tracking of a moving object using pan and tilt platform. Int. J. Sci. Eng. Res. **3**(8) (2012)
40. Salhi, A., Jammoussi A.Y.: Object tracking system using camshift, meanshift and Kalman filter. World Acad. Sci. Eng. Technol. (2012)

41. Raja, A.S., Dwivedi, A., Tiwari, H.: Vision based tracking for unmanned aerial vehicle. Adv. Aerosp. Sci. Appl. **4**(1), 59–64 (2014)
42. Wang, X., Zhu, H., Zhang, D., Zhou, D., Wang, X.: Vision-based detection and tracking of a mobile ground target using a fixed-wing UAV. Int. J. Adv. Rob. Syst. (2014)
43. Qadir, A., Neubert, J., Semke, W.: On-board visual tracking with unmanned aircraft system. In: AIAA Infotech@Aerospace Conference, St. Louis, MO (2011)
44. Redding, J.D., McLain, T.W., Beard, R.W., Taylor, C.N: Vision-based target localization from a fixed-wing miniature air vehicle. In: Proceedings of the 2006 American Control Conference, Minneapolis, Minnesota, USA (2006)
45. Al-Radaideh, A., Al-Jarrah, M.A., Jhemi, A., Dhaouadi, R.: ARF60 AUS-UAV modeling, system identification, guidance and control: validation through hardware in the loop simulation. In: 6th International Symposium on Mechatronics and its Applications (ISMA09), Sharjah, UAE. (2009)
46. Chui, C.K., Chen, G.: Kalman Filtering with Real-Time Applications, 3rd edn. Springer, Berlin (1986)
47. Grewal, M.S., Andrews, A.P.: Kalman Filtering: Theory and Practice Using MATLAB. Wiley, New York (2001)
48. Brown, R.G., Hwang, P.Y.C.: Introduction to Random Signals and Applied Kalman Filtering with MATLAB Exercises and Solutions. Wiley, New York (1997)
49. Prince, RA.: Autonomous visual tracking of stationary targets using small unmanned aerial vehicles. M.Sc. Dissertation, Department of Mechanical and Astronautical Engineering, Naval Postgraduate School, Monterey, California (2004)
50. Trago, T.M.: Performance analysis for a vision-based target tracking system of a small unmanned aerial vehicle. M.Sc. Dissertation, Department of Mechanical and Astronautical Engineering, Naval Postgraduate School, Monterey, California (2005)
51. Brashear, T.J.: Analysis of dead time and implementation of smith predictor compensation in tracking servo systems for small unmanned aerial vehicles. M.Sc. Dissertation, Department of Mechanical and Astronautical Engineering, Naval Postgraduate School, Monterey, California (2005)
52. Bernard, S.M.L.: Hardware in the loop implementation of adaptive vision based guidance law for ground target tracking. M.Sc. Dissertation, Department of Mechanical and Astronautical Engineering, Naval Postgraduate School, Monterey, California (2008)
53. Homulle, H., Zimmerling, J: UAV camera system for object searching and tracking. Bachelor Thesis, Faculty of Electrical Engineering, Mathematics and Computer Science, Delft University of Technology (2012)
54. Garcia, R.A.: Gimbal control. M.Sc. Thesis, School of Engineering, Cranfield University (2012)
55. Amer, A., Al-Radaideh, K.H.: Guidance, control and trajectory tracking of small fixed wing unmanned aerial vehicles (UAV's). M.Sc. Thesis, American University of Sharjah (2009)
56. UAV Vision Pty Ltd. http://www.uavvision.com (2014)

Identification of Uncertain Mass and Stiffness Matrices of Multi-Storey Shear Buildings Using Fuzzy Neural Network Modelling

S. Chakraverty and Deepti Moyi Sahoo

Abstract In this paper a fuzzy neural network based modelling has been presented for the identification of structural parameters of uncertain multi-storey shear buildings. Here the method is developed to identify uncertain structural mass and stiffness matrices from the dynamic responses of the structure. In this method, the related equations of motion are modified using relative responses of stories in such a way that the new set of equations can be implemented in a series of Fuzzy Neural Networks (FNNs).

Keywords System identification · Mass · Stiffness · Damping · Fuzzy neural network

1 Introduction

In system identification problems, a set of inputs and resulting outputs for a system is known and we develop a mathematical description or model of the system. In system identification for structural problems we identify or estimate structural parameters such as stiffness, mode shapes, damping ratios and structural response. System identification techniques have become an increasingly interesting research topic for the purpose of structural health monitoring, damage assessment and safety evaluation of existing engineering structures.

As regards [1–5] gave various methodologies for different type of problems in system identification. A procedure which systematically modifies and identifies the structural parameters by using the prior known estimates of the parameters with the corresponding vibration characteristics and the known dynamic data is given by [1]. [2] presented a novel inverse scheme based on consistent mass Transfer Matrix

---

S. Chakraverty (✉) · D.M. Sahoo
Department of Mathematics, National Institute of Technology Rourkela,
Rourkela 769008, Odisha, India
e-mail: sne_chak@yahoo.com

V. Ravi et al. (eds.), *Proceedings of the Fifth International Conference on Fuzzy and Neuro Computing (FANCCO - 2015)*, Advances in Intelligent Systems and Computing 415, DOI 10.1007/978-3-319-27212-2_4

(TM) to identify the stiffness parameters of structural members. They used a non-classical heuristic Particle Swarm Optimization Algorithm (PSO). [3] discussed selective sensitivity analysis and used this method to solve system identification problems. [4] used modal analysis procedure, NExT/ERA and frequency domain decomposition (or IPP) to study ambient vibration in tall buildings. [5] used Holzer criteria along with some other numerical methods to estimate the global mass and stiffness matrices of the structure from modal test data. Physical parameter system identification methods to determine the stiffness and damping matrices of shear storey buildings have been proposed by [9]. A number of studies [6–10] have used ANN for solving structural identification problems. [6] proposed a neural network-based substructural identification for the estimation of the stiffness parameters of a complex structural system, particularly for the case with noisy and incomplete measurement of the modal data. In particular, [7] identified the dynamic characteristics of a steel frame using the back-propagation neural network. Neural network based strategy was also developed by [8] for direct identification of structural parameters from the time domain dynamic responses of a structure without any eigen value analysis. A neural network based method to determine the modal parameters of structures from field measurement data was given by [9]. A procedure for identification of structural parameters of two storey shear buildings by an iterative training of neural networks was proposed by [10]. System identification of an actively controlled structure using frequency response functions with the help of artificial neural networks (ANN) has also been studied by [11]. It reveals from above literature survey that Artificial Neural Networks (ANNs) gives a different approach to system identification problems. They are successfully applied for various identification and control of dynamic systems because of the excellent learning capacity.

The above literature review shows that various types of ANN are developed to deal with data in exact or crisp form. But experimental data obtained from equipments may be with errors that may be due to human or equipment errors giving uncertain form of the data. On the other hand we may also use probabilistic methods to handle these problems. But the probabilistic method requires large quantity of data which may not be easy or feasible in these problems. In view of the above, various research works are being done by using fuzzy neural networks in different fields. [12] developed architecture for neural networks where the input vectors are in terms of fuzzy numbers. A methodology for fuzzy neural networks where the weights and biases are taken as fuzzy numbers and the input vectors as real numbers has been proposed [13]. A fuzzy neural network with trapezoidal fuzzy weights was also presented by [14]. They have developed the methodology in such a way that it can handle fuzzy inputs as well as real inputs. In this respect [15] derived a general algorithm for training a fuzzified feed-forward neural network that has fuzzy inputs, fuzzy targets and fuzzy connection weights. The derived algorithms are also applicable to the learning of fuzzy connection weights with various shapes such as triangular and trapezoidal. Another new algorithm for learning fuzzified neural networks has also been developed by [16].

In this paper structural parameters such as mass and stiffness matrices of shear buildings have been identified using single layer neural network in fuzzified form.

To identify the physical parameters in fuzzified form the related equations of motion are used systematically in a series of fuzzy neural networks.

2 Fuzzy Set Theory and Preliminaries

The following Fuzzy arithmetic operations are used in this paper for defining fuzzified neural network in [17].

A. Let X be an universal set. Then the fuzzy subset A of X is defined by its membership function

$$\mu_A : X \rightarrow [0, 1]$$

which assign a real number $\mu_A(x)$ in the interval [0, 1], to each element $x \in X$, where the value of $\mu_A(x)$ at x shows the grade of membership of x in A.

B. Given a fuzzy set A in X and any real number $\alpha \in [0, \ 1]$, then the $\alpha -$ cut or $\alpha -$ level or cut worthy set of A, denoted by A_α is the crisp set

$$A_\alpha = \{x \in X | \ \mu_A(x) \geq \alpha\}$$

C. A Triangular Fuzzy Number (TFN) A is defined as a triplet $[\underline{a}, \ ac, \ \overline{a}]$. Its membership function is defined as

$$\mu_A(x) = \begin{cases} 0, & x < \underline{a} \\ \dfrac{x - \underline{a}}{ac - \overline{a}}, & \underline{a} \leq x \leq ac \\ \dfrac{\overline{a} - x}{\overline{a} - ac}, & ac \leq x \leq \overline{a} \\ 0, & x > \overline{a} \end{cases}$$

Above TFN may be transformed to an interval form A_α by α—cut as

$$A_\alpha = \left[\underline{a}^{(\alpha)}, \overline{a}^{(\alpha)}\right] = [(ac - \underline{a})\alpha + \underline{a}, \ -(\overline{a} - ac)\alpha + \overline{a}]$$

where $\underline{a}^{(\alpha)}$ and $\overline{a}^{(\alpha)}$ are the lower and upper limits of the α—level set A_α.

D. A Trapezoidal Fuzzy Number (TRFN) A is defined as $A = (a_1, \ a_2, \ a_3, \ a_4)$. Its membership function is defined as

$$\mu_A(x) = \begin{cases} 0, & x < a_1 \\ \dfrac{x - a_1}{a_2 - a_1}, & a_1 \leq x \leq a_2 \\ 1, & a_2 \leq x \leq a_3 \\ \dfrac{a_4 - x}{a_4 - a_3}, & a_3 \leq x \leq a_4 \\ 0, & x > a_4 \end{cases}$$

Above TRFN may be transformed to an interval form A_α by α—cut as

$$A_\alpha = [(a_2 - a_1)\alpha + a_1, \; -(a_4 - a_3)\alpha + a_4]$$

3 Learning Algorithm for Single Layer Fuzzy Neural Network

A Neural Network is said to be a Fuzzy Neural Network if at least one of its input, output or weight have values in fuzzified form. In Fuzzy Neural Networks, neurons are connected as they are connected in traditional Neural Networks. Following are the steps in FNN using the fuzzy computation defined above.

Step 1: Initialize input weights $(\tilde{W}_{ji})$ and bias weights $\tilde{\theta}_i$ in fuzzified form.

Step 2: Present the training pairs in the form $\{\tilde{Z}_1, \tilde{d}_1; \tilde{Z}_2, \tilde{d}_2; \ldots \tilde{Z}_I, \tilde{d}_I\}$ where $\tilde{O}_I = \tilde{Z}_I = ([\underline{z}_1, z_1c, \overline{z}_1], [\underline{z}_2, z_2c, \overline{z}_2], \ldots, [\underline{z}_n, z_nc, \overline{z}_n])$ are inputs and $\tilde{d}_I = ([\underline{d}_1, d_1c, \overline{d}_1], [\underline{d}_2, d_2c, \overline{d}_2], \ldots, [\underline{d}_n, d_nc, \overline{d}_n])$ are desired values for the given inputs in fuzzified form.

Step 3: Calculate the output of the network for the input $\tilde{Z}_I$

$$\tilde{O}_J = \tilde{f}(Net_j)$$

where $(Net_j) = \sum_{j=1}^{J} \tilde{W}_{ji}\tilde{O}_I + \tilde{\theta}_i$

and $\tilde{f}$ is the fuzzy unipolar activation function defined by

$$\tilde{f}(net) = 1/[1 + \exp(-\gamma \, net)]$$

Step 4: The error value is then computed as

$$\tilde{E} = \frac{\alpha}{2}\left[\left(\underline{d}_J^\alpha - \underline{O}_J^\alpha\right)^2 + \left(\overline{d}_J^\alpha - \overline{O}_J^\alpha\right)^2\right]$$

Step 5: The weight is modified as

$$\tilde{W}_{ji}^\alpha(New) = \left[\underline{W}_{ji}^\alpha(New), \overline{W}_{ji}^\alpha(New)\right] = \left[\underline{W}_{ji}^\alpha(Old), \overline{W}_{ji}^\alpha(Old)\right] + \left[\Delta\underline{W}_{ji}^\alpha, \Delta\overline{W}_{ji}^\alpha\right]$$

where
change in weights are calculated as

$$\Delta \tilde{W}_{ji}^{\alpha} = \left[\Delta \underline{W}_{ji}^{\alpha}, \Delta \overline{W}_{ji}^{\alpha} \right] = \left[-\eta \frac{\partial \tilde{E}}{\partial \underline{W}_{ji}^{\alpha}}, -\eta \frac{\partial \tilde{E}}{\partial \overline{W}_{ji}^{\alpha}} \right]$$

In the similar fashion the bias weights are also updated.

4 System Identification of Structural Parameter in Fuzzified Form

Let us consider a three storey shear building with structural system governed by the following set of linear differential equations in fuzzified form as

$$[\tilde{M}]\left\{\ddot{\tilde{X}}\right\}_3 + [\tilde{K}]\{\tilde{X}\}_3 = \{\tilde{F}\}_3 \tag{1}$$

where $\left\{\ddot{\tilde{X}}\right\}_3$ and $\{\tilde{X}\}_3$ indicate known acceleration and displacement vectors in fuzzified form respectively. Moreover, $\{\tilde{M}\} = [\underline{M}, Mc, \overline{M}]$ is a 3 × 3 mass matrix of the structure in fuzzified form and is given by

$$\{\tilde{M}\} = \begin{bmatrix} [\underline{m}_1, m_1c, \overline{m}_1] & 0 & 0 \\ 0 & [\underline{m}_2, m_2c, \overline{m}_2] & 0 \\ 0 & 0 & [\underline{m}_3, m_3c, \overline{m}_3] \end{bmatrix}$$

and $\{\tilde{K}\} = [\underline{K}, Kc, \overline{K}]$ is *a* 3 × 3 stiffness matrix of the structure in fuzzified form which may be obtained as

$$\{\tilde{M}\} = \begin{bmatrix} [\underline{k}_1, k_1c, \overline{k}_1] + [\underline{k}_2, k_2c, \overline{k}_2] & -[\underline{k}_2, k_2c, \overline{k}_2] & 0 \\ -[\underline{k}_2, k_2c, \overline{k}_2] & [\underline{k}_2, k_2c, \overline{k}_2] + [\underline{k}_3, k_3c, \overline{k}_3] & -[\underline{k}_3, k_3c, \overline{k}_3] \\ 0 & -[\underline{k}_3, k_3c, \overline{k}_3] & [\underline{k}_3, k_3c, \overline{k}_3] \end{bmatrix}$$

Solution of Eq. (1) for ambient gives the corresponding fuzzy eigenvalues and eigenvectors. These are denoted respectively by $\tilde{\lambda}_3$ and $\{\tilde{A}\}_3 = \{\underline{A}, Ac, \overline{A}\}_3$ where $\tilde{\omega}_3^2\ (=\tilde{\lambda}_3)$ designate the natural frequency in fuzzified form. It may be noted that the free vibration equation will be a fuzzy eigen value problem. The fuzzy eigen value and vector are obtained by considering different sets of lower, centre and upper stiffness and mass values.

After finding the solution, Eq. (1) is rewritten to get the following set of equations in fuzzified form

$$\tilde{m}_1\ddot{\tilde{x}}_1 + \left(\tilde{k}_1 + \tilde{k}_2\right)\tilde{x}_1 - \tilde{k}_2\tilde{x}_2 = \tilde{f}_1 \tag{2}$$

$$\tilde{m}_2\ddot{\tilde{x}}_2 - \tilde{k}_2\tilde{x}_1 + \left(\tilde{k}_2 + \tilde{k}_3\right)\tilde{x}_2 - \overleftrightarrow{k}_3\tilde{x}_3 = \tilde{f}_2 \tag{3}$$

$$\tilde{m}_3\ddot{\tilde{x}}_3 - \tilde{k}_3\, x_2 + \tilde{k}_3\, \tilde{x}_3 = \tilde{f}_3 \tag{4}$$

Equation (4) is then written as

$$\tilde{m}_3\ddot{\tilde{x}}_3 + \tilde{k}_3\left[\tilde{x}_3 - \tilde{x}_2\right] = \tilde{f}_3 \tag{5}$$

The above equation may now be presented as

$$\tilde{m}_3\ddot{\tilde{x}}_3 + \tilde{k}_3\tilde{d}_3 = \tilde{f}_3 \tag{6}$$

where $\tilde{d}_3 = \left[\tilde{x}_3 - \tilde{x}_2\right]$

Here $\tilde{d}_3$ indicate the known relative displacement in fuzzified form for 3rd storey. Using the single layer fuzzy neural network, Eq. (6) is solved. Inputs for this fuzzy neural network are taken as structural acceleration and relative displacement for 3rd storey. Output of the network is taken as the applied force at time *t*. To solve Eq. (6) a continuous training process with *n* training patterns are done using fuzzy neural network and the converged weight matrix of the neural network is thus obtained. From this weight matrix the corresponding physical parameters such as $\tilde{m}_3$ and $\tilde{k}_3$ in fuzzified form are obtained. Identified parameters of 3rd storey are then used to identify the parameters of 2nd storey using Eq. (5). Finally the unknown parameters of the first storey are obtained using Eq. (2). In the similar fashion one may model multistory structure with two or more than three stories. The cluster of fuzzy neural network diagram for three storey structure is shown in Fig. 1.

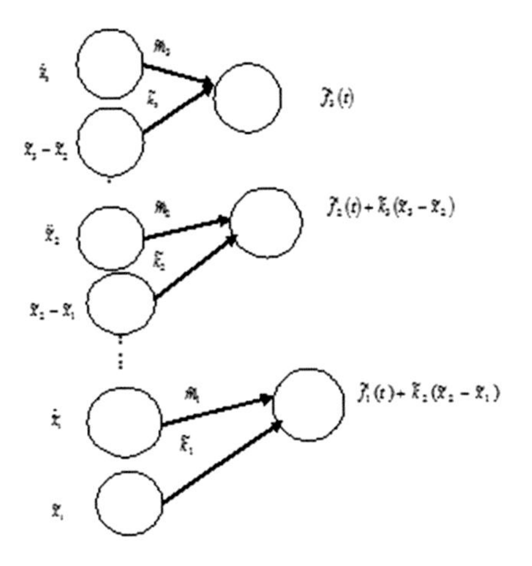

Fig. 1 Proposed cluster fuzzy neural network model for three storey structure

5 Results and Discussion

The above developed method has been used for different storey shear structures. Fuzzy neural network training is done till a desired accuracy is achieved. The methodology has been discussed by giving the results for following two problems.

Here two problems have been considered with different type of fuzzy numbers as below.

Two storeys (with TRFN)
Three storeys (with TFN)

(a) Two storey shear building:
Structural parameter of shear building has been identified using the direct method where the data are considered to be in fuzzified form. To generate the data initially the theoretical values of the structural parameters are taken. These generated data are used first to train the neural network for n training patterns thus by establishing the converged weight matrix of the neural network. Corresponding component of the converged weight matrix gives the unknown or present structural parameters. Then the trained and theoretical data has been compared to show the efficiency of the proposed method. These ideas have been applied in all the cases. The initial structural parameters in trapezoidal fuzzified form are taken as: storey masses $\tilde{m}_1 = [2.5,\ 2.8,\ 3.2,\ 3.5]$, $\tilde{m}_2 = [2.5,\ 2.8,\ 3.2,\ 3.5]$, kNs^2 m^{-1} and the storey stiffnesses $\tilde{k}_1 = [1190, 1195, 1205, 1210]$, $\tilde{k}_2 = [790, 795, 805, 810]$ kN m^{-1}. The harmonic forces exerted in the shear building are assumed in fuzzified form as $\tilde{f}_1(t) = [90\sin(1.6\pi t) + \pi,\ 95\sin(1.6\pi t) + \pi,\ 105\ \sin(1.6\pi t) + \pi,\ 110\sin(1.6\pi t) + \pi]$ and $\tilde{f}_2(t) = [90\sin(1.6\pi t), 95\ \sin(1.6\pi t),\ 105\sin(1.6\pi t),\ 110\sin(1.6\pi t)]$ kN. The Trapezoidal Fuzzy Number (TRFN) plot of mass and stiffness for two storeys have been plotted in Figs. 2,3,4,5 to show the comparison between the identified and theoretical structural parameters.

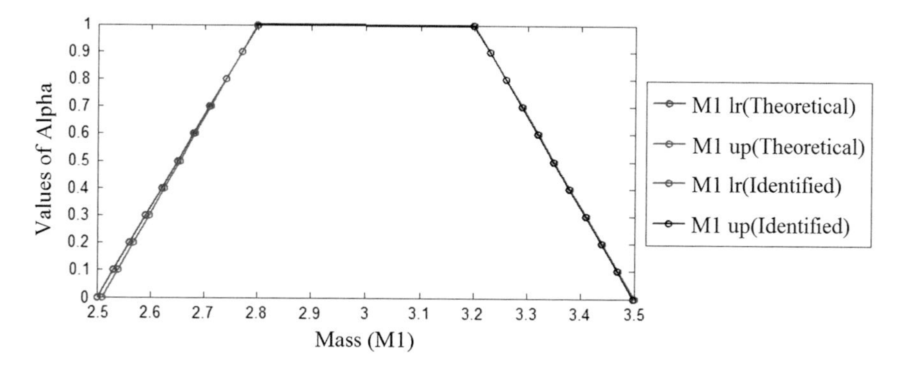

Fig. 2 TRFN of identified and theoretical mass (M_1) of two-storey building under the forced vibration test

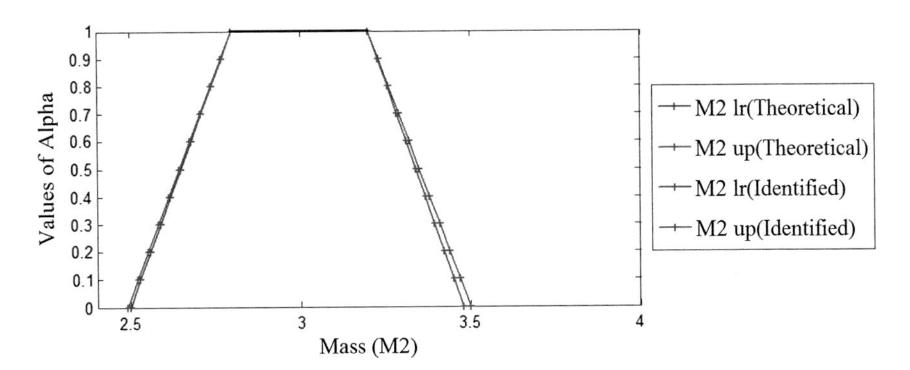

Fig. 3 TRFN of identified and theoretical mass (M_2) of two-storey building under the forced-vibration test

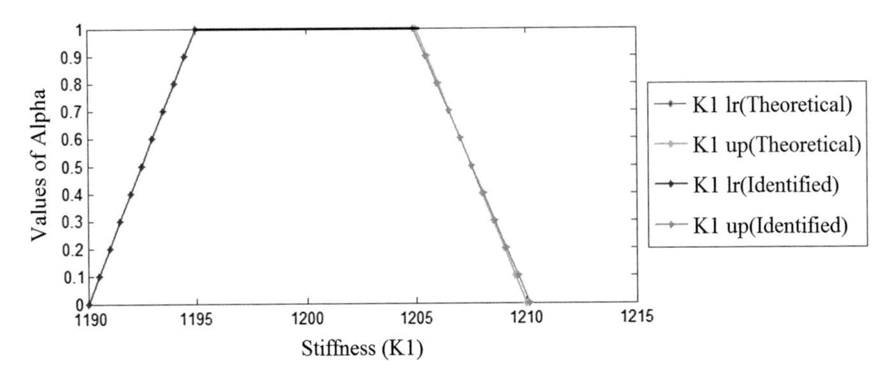

Fig. 4 TRFN of identified and theoretical stiffness (K_1) of two-storey building under the forced-vibration test

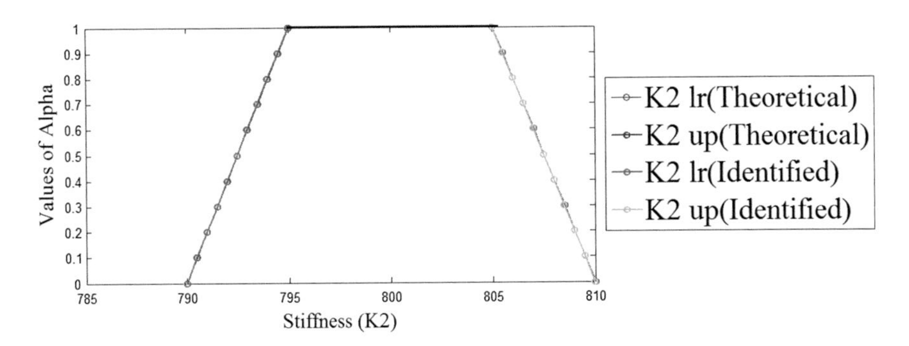

Fig. 5 TRFN of identified and theoretical stiffness (K_2) of two-storey building under the forced-vibration test

Table 1 Identified mass and stiffness parameters of three-storey building in fuzzified form under the forced-vibration test

Parameter	Storey	Theoretical	Identified
Mass (kN s^2 m^{-1})	M_1	[3.5,4,4.5]	[3.466,3.986,4.494]
	M_2	[2.5,3,3.5]	[1.045,1.686,2.308]
	M_3	[1.5,2,2.5]	[1.496,1.898,2.499]
Stiffness (kN m^{-1})	K_1	[990,1000,1010]	[989.991,999.994,1009.997]
	K_2	[790,800,810]	[789.597,799.189,809.529]
	K_3	[590,600,610]	[589.976,599.289,609.897]

(b) Three storey shear building:
In this case structural parameters in fuzzified form (TFN) are taken as: the storey masses $\tilde{m}_1 = [3.5,\ 4,\ 4.5]$, $\tilde{m}_2 = [2.5, 3, 3.5]$, $\tilde{m}_3 = [1.5,\ 2,\ 2.5]$ kNs^{2}m^{-1} and the storey stiffnesses $\tilde{k}_1 = [990, 1000, 1010]$, $\tilde{k}_2 = [790, 800,\ 810]$, $\tilde{k}_3 = [590, 600, 610]$ kNm^{-1}. The harmonic forces exerted in the shear building in fuzzified form are considered as $\tilde{f}_1(t) = [90\sin(1.6\pi t) + \pi,$ $100\sin(1.6\pi t) + \pi, 110\sin(1.6\pi t) + \pi]$, $\tilde{f}_2(t) = [90\sin(1.6\pi t),\ 100\sin(1.6\pi t),$ $110\sin(1.6\pi t)]$ and $\tilde{f}_3(t) = [0.5\sin(3.2\pi t), 1.0\sin(3.2\pi t), 1.5\sin(3.2\pi t)]$ kN. Comparisons between the identified and theoretical structural parameters are incorporated in Table 1.

6 Concluding Remarks

This paper uses the powerful soft computing technique viz. single layer Fuzzy Neural Network (FNN) for identification of uncertain structural parameters. Fuzzy neural network has been developed so that it can handle uncertain or fuzzified data. Here, direct method for system identification has been proposed in fuzzified form. Related equations of motion are modified using relative responses of stories and are implemented in a series of fuzzy neural network models. Two example problems are investigated and corresponding results are given to show the powerfulness of the proposed system identification method.

References

1. Chakraverty, S.: Modelling for identification of stiffness parameters of multi-storey Structure from dynamic data. J. Sci. Ind. Res. **63**(2), 142–148 (2004)
2. Nandakumar, P., Shankar, K.: Identification of structural parameters using consistent mass transfer matrix. Inverse Prob. Sci. Eng. **22**(3), 436–457 (2013)
3. Billmaier, M., Bucher, C.: System identification based on selective sensitivity analysis: A case-study. J. Sound Vib. **332**(11), 2627–2642 (2013)

4. Brownjohn, J.M.W.: Ambient vibration studies for system identification of tall Buildings. Earthquake Eng. Struct. Dynam. **32**(1), 71–96 (2003)
5. Chakraverty, S.: Identification of structural parameters of multistorey shear buildings from modal data. Earthquake Eng. Struct. Dynam. **34**(6), 543–554 (2005)
6. Yun, C.B., Bahng, E.Y.: Substructural identification using neural networks. Comput. Struct. **77**(1), 41–52 (2000)
7. Huang, C.S., Hung, S.L., Wen, C.M., Tu, T.T.: A neural network approach for structural identification and diagnosis of a building from seismic response data. Earthquake Eng. Struct. Dynam. **32**(2), 187–206 (2003)
8. Xu, B., Wu, Z., Chen, G., Yokoyama, K.: Direct identification of structural parameters from dynamic responses with neural networks. Eng. Appl. Artif. Intell. **17**(8), 931–943 (2004)
9. Chen, C.H.: Structural identification from field measurement data using a neural network. Smart Mater. Struct. **14**(3), 104–115 (2005)
10. Chakraverty, S.: Identification of Structural Parameters of two-storey shear buildings by the iterative training of neural networks. Architectural Sci. Rev. **50**(4), 380–384 (2007)
11. Pizano, D.G.: Comparison of frequency response and neural network techniques for system identification of an actively controlled structure. Dyna **78**(170), 79–89 (2011)
12. Ishibuchi, H., Fujioka, R., Tanaka, H.: An architecture of neural networks for input vectors of fuzzy numbers. Fuzzy Systems, IEEE International Conference, San Diego, CA, March 8–12, pp. 1293–1300 (1992)
13. Ishibuchi, H., Tanaka, H., Okada, H.: Fuzzy neural networks with fuzzy weights and fuzzy biases. Neural Networks, IEEE International Conference, San Francisco, CA, March 28–April 1, 3, pp. 1650–1655 (1993)
14. Ishibuchi, H., Morioka, K., Tanaka, H.: A fuzzy neural network with trapezoid fuzzy weights. Fuzzy Systems, IEEE World Congress on Computational Intelligence Proceedings of the third IEEE Conference, Orlando, FL, June 26–29, 1, pp. 228–233 (1994)
15. Ishibuchi, H., Kwon, K., Tanaka, H.: A learning algorithm of fuzzy neural networks with triangular fuzzy weights. Fuzzy Sets Syst. **71**(3), 277–293 (1995)
16. Ishibuchi, H., Morioka, K., Turksen, I.B.: Learning by Fuzzified Neural Networks. Int. J. Approximate Reasoning **13**(4), 327–358 (1995)
17. Lee, K.H.: First course on fuzzy theory and applications, Springer International Edition, pp. 1–333 (2009)

Fuzzy Approach to Rank Global Climate Models

K. Srinivasa Raju and D. Nagesh Kumar

Abstract Eleven coupled model intercomparison project 3 based global climate models are evaluated for the case study of Upper Malaprabha catchment, India for precipitation rate. Correlation coefficient, normalised root mean square deviation, and skill score are considered as performance indicators for evaluation in fuzzy environment and assumed to have equal impact on the global climate models. Fuzzy technique for order preference by similarity to an ideal solution is used to rank global climate models. Top three positions are occupied by MIROC3, GFDL2.1 and GISS with relative closeness of 0.7867, 0.7070, and 0.7068. IPSL-CM4, NCAR-PCMI occupied the tenth and eleventh positions with relative closeness of 0.4959 and 0.4562.

Keyword Global climate models · India · Rank · Performance indicators · Fuzzy · TOPSIS

1 Introduction

Global climate models are becoming familiar to simulate future climate changes due to their adaptability and ability to consider impacts on water resources planning. This necessitates evaluating the skill of global climate models for their suitability. In addition, ability of fuzzy logic for its imprecise data handling flexibility is an added advantage in complex climate change problems [1].

K. Srinivasa Raju
Department of Civil Engineering, Birla Institute of Technology and Science, Pilani, Hyderabad, India
e-mail: ksraju@hyderabad.bits-pilani.ac.in

D. Nagesh Kumar (✉)
Department of Civil Engineering, Indian Institute of Science, Bangalore, India
e-mail: nagesh@civil.iisc.ernet.in

V. Ravi et al. (eds.), *Proceedings of the Fifth International Conference on Fuzzy and Neuro Computing (FANCCO - 2015)*, Advances in Intelligent Systems and Computing 415, DOI 10.1007/978-3-319-27212-2_5

Extensive study of global climate models was made to understand their simulating capabilities. Walsh et al. [2] evaluated the performance of 15 global climate models for a case study of Greenland and Alaska. Kundzewicz and Stakhiv [3] reviewed the climate models regarding their applicability for water resources applications. Weigel et al. [4] analysed the effects of model weighting related to climate change projections and concluded that single models are not performing well compared to equally weighted multi models.

Radić and Clarke [5] assessed the 22 global climate models skill over North America and its western sub region. It was concluded that choice of climate variable play a major role in the ranking of global climate models. Johnson et al. [6] explored wavelet dependent skill measures to analyse global climate models performance for representing interannual variability.

Su et al. [7] evaluated 24 coupled model intercomparison project 5 based global climate models for the case study of Eastern Tibetan Plateau by comparing ground observations and outputs of the model for variables precipitation and temperature. It was emanated that most of the global climate models simulating ability is satisfactory. Perez et al. [8] evaluated the global climate models skill for the north-east Atlantic Ocean region to reproduce inter-annual variability in historical perspective and synoptic situations. Northrop and Chandler [9] quantified relative impact of uncertainties from climate model selection, emission scenario, internal variability. Sun et al. [10] analysed the skill of global climate models for daily precipitation and temperature for China. They also made extensive comparative analysis. Similar studies were reported by McMahon et al. [11]. Gulizia and Camilloni [12] evaluated the global climate models to simulate precipitation patterns in South America south of the equator and sub-regions.

Raju and Nagesh Kumar [13] explored eleven global climate models for upper Malaprabha catchment for precipitation rate and considered five indicators. Importance of indicators was computed using entropy method. They employed outranking based decision making method, PROMETHEE and found that MIROC3, GFDL2.1 and GISS occupied the first, second and third positions respectively. Equal importance of indicators is also assumed and in this case MIROC3, GISS and GFDL2.1 occupied the first, second and third positions. Raju and Nagesh Kumar [14] made similar studies considering both temperature and precipitation using skill score indicator. They employed TOPSIS for ranking global climate models. GFDL2.0, GFDL2.1 and MIROC3 occupied first, second and third positions respectively.

It is inferred from the literature review that no fuzzy based decision making approach was used till now for prioritizing the global climate models. In this regard, objectives proposed are: (1) selection of performance indicators for precipitation rate (2) applicability of fuzzy technique for order preference by similarity to an ideal solution (F-TOPSIS) for the case study of Upper Malaprabha catchment, India. Eleven coupled model intercomparison project 3 based global climate models, UKMO-HADGEM1, GISS, GFDL2.0, BCCR-BCCM2.0, IPSL-CM4, UKMO-HADCM3, GFDL2.1, INGV-ECHAM4, MIROC3, MRI-CGCM2 and NCAR-PCMI are selected for evaluation and details of the same are presented in Table 1 [13–15].

Table 1 Details of chosen coupled model intercomparison project 3 based GCMs

S.No	Organisation involved in the development	Name of global climate model
1	UK Met Office, UK	UKMO-HADGEM1
2	Goddard Institute for Space Studies, USA	GISS
3	Geophysical Fluid Dynamic Laboratory, USA	GFDL2.0
4	Bjerknes Centre for Climate Research, Norway	BCCR-BCCM 2.0
5	Institut Pierre Simon Laplace, France	IPSL-CM 4
6	UK Met Office, UK	UKMO-HADCM3
7	Geophysical Fluid Dynamic Laboratory, USA	GFDL2.1
8	Istituto Nazionale Di Geofisica E Vulcanologia, Italy	INGV-ECHAM 4
9	Centre for Climate Research, Japan	MIROC3
10	Meteorological Research Institute, Japan	MRI-CGCM2
11	Parallel Climate Models, NCAR, USA	NCAR-PCMI

2 Performance Indicators

Three performance indicators, skill score (SS), correlation coefficient (CC), normalised root mean square deviation (NRMSD) are considered for evaluation of global climate models and they respectively measures linear strength relationship, difference, similarity between observed and the computed values [16]. Imprecision in indicator values that may arise due to interpolation and averaging procedures, approximations are tackled through fuzzy logic approach which in our opinion is necessary. Triangular membership function is explored for handling impreciseness due to its simplicity (Fig. 1) and its compatibility while applying with the decision making technique F-TOPSIS [1].

3 F-TOPSIS

Fuzzy technique for order preference by similarity to an ideal solution is based on distance of each indicator for each global climate model from the ideal solution, D_i^+ (Eq. 1) and distance of each indicator for each global climate model from the anti-ideal solution, D_i^- (Eq. 2) [1, 17]. Relative closeness C_i (Eq. 3) is based on D_i^-, D_i^+ and higher relative closeness based global climate model is preferred. Flow chart of F-TOPSIS methodology is presented in Fig. 2.

$$D_i^+ = \sum_{j=1}^{J} d(\tilde{Y}_{ij}, \tilde{Y}_j^*) = \sqrt{\frac{[(p_{ij} - p_j^*)^2 + (q_{ij} - q_j^*)^2 + (r_{ij} - r_j^*)^2]}{3}} \quad (1)$$

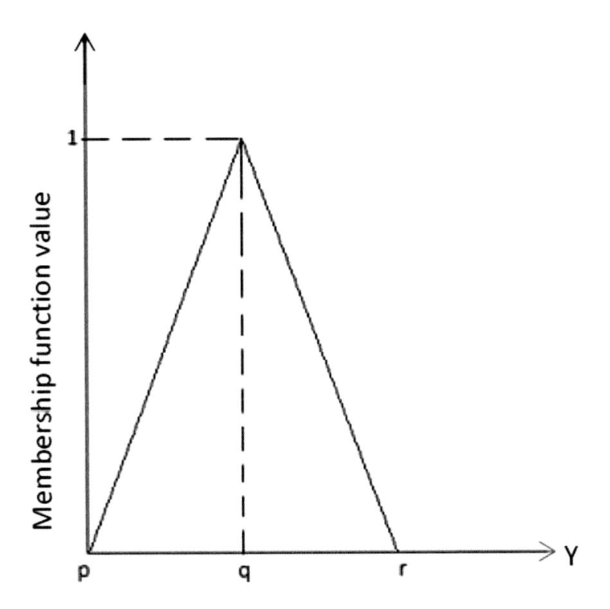

Fig. 1 Typical triangular membership function

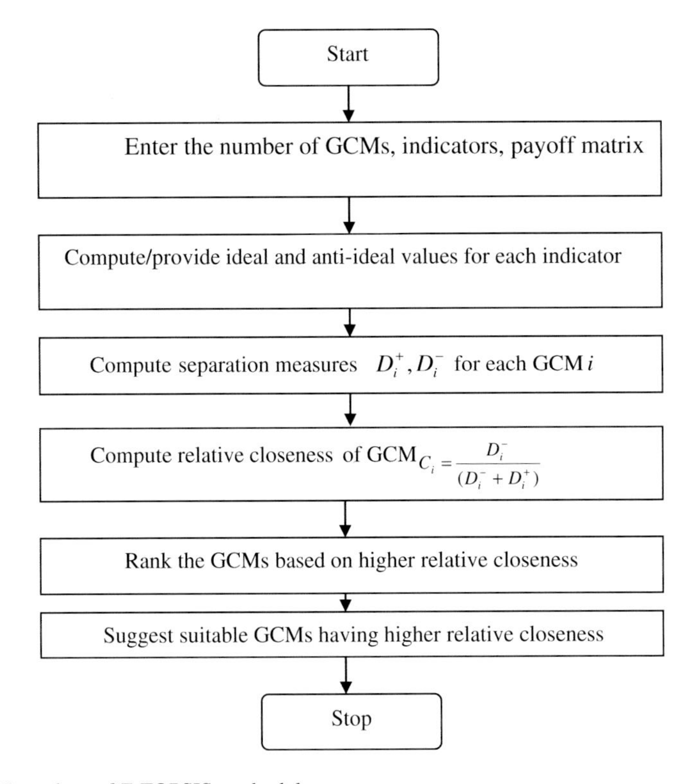

Fig. 2 Flow chart of F-TOPSIS methodology

$$D_i^- = \sum_{j=1}^{J} d(\tilde{Y}_{ij}, \tilde{Y}_j^{**}) = \sqrt{\frac{[(p_{ij} - p_j^{**})^2 + (q_{ij} - q_j^{**})^2 + (r_{ij} - r_j^{**})^2]}{3}} \quad (2)$$

$$C_i = \frac{D_i^-}{(D_i^- + D_i^+)} \quad (3)$$

here triangular fuzzy numbers $\tilde{Y}_{ij}$ (given data), $\tilde{Y}_j^*$(ideal), $\tilde{Y}_j^{**}$(anti-ideal) represents with elements (p_{ij}, q_{ij}, r_{ij}), (p_j^*, q_j^*, r_j^*), (p_j^{**}, q_j^{**}, r_j^{**}) respectively; p, q, r represents lower, middle and upper values. Here i and j are indices representing GCMs and indicators.

4 Results and Discussion

Upper Malaprabha catchment is selected as the case study for demonstration of the methodology. The National Centre for Environmental Prediction-National Centre for Atmospheric Research monthly data for the period of 1950–1999 was used at $2.5^{o} \times 2.5^{o}$ grid resolution and the outputs of all the global climate models are interpolated accordingly.

All the three indicators are assumed to have equal impact on the ranking of global climate models. Table 2 presents the payoff matrix obtained for the three indicators—CC, NRMSD, SS for the eleven global climate models in triangular membership environment. Three values for each indicator represent the range that

Table 2 Normalised performance indicators obtained for 11 global climate models

Model	CC			NRMSD			SS		
	p_{ij}	q_{ij}	r_{ij}	p_{ij}	q_{ij}	r_{ij}	p_{ij}	q_{ij}	r_{ij}
UKMO-HAD GEM1	0.649	0.806	0.964	0.390	0.466	0.578	0.714	0.788	0.863
GISS	0.670	0.828	0.985	0.436	0.534	0.687	0.704	0.778	0.853
GFDL2.0	0.629	0.787	0.945	0.433	0.529	0.680	0.741	0.815	0.889
BCCR-BCCM 2.0	0.617	0.775	0.933	0.448	0.551	0.717	0.697	0.772	0.846
IPSL-CM4	0.316	0.474	0.632	0.305	0.350	0.410	0.584	0.658	0.733
UKMO-HADCM3	0.644	0.802	0.960	0.413	0.499	0.631	0.736	0.810	0.884
GFDL2.1	0.582	0.740	0.897	0.452	0.557	0.727	0.761	0.835	0.909
INGV-ECHAM 4	0.629	0.787	0.944	0.466	0.579	0.765	0.609	0.683	0.758
MIROC3	0.684	0.842	0.999	0.544	0.705	1.000	0.782	0.857	0.931
MRI-CGCM2	0.613	0.771	0.929	0.391	0.467	0.581	0.681	0.755	0.829
NCAR- PCMI	0.198	0.355	0.513	0.322	0.372	0.441	0.554	0.628	0.703

can be handled by the fuzzy approach. It is found that evaluation of performance indicators in fuzzy environment for each global climate model is found to be helpful. Ideal (p_j^*, q_j^*, r_j^*) and anti-ideal ($p_j^{**}, q_j^{**}, r_j^{**}$) values for all the indicators are chosen as (1, 1, 1) and (0, 0, 0).

Table 3 presents D_i^+, D_i^-, C_i and ranking pattern for eleven global climate models. It is observed from Table 3 that highest and lowest (a) D_i^+ values observed are 1.6576 and 0.6732 respectively for NCAR-PCMI and MIROC3 with a range of 0.9844 (b) D_i^- values observed are 2.4833 and 1.3906 respectively for MIROC3 and NCAR-PCMI with a range of 1.0927 (c) C_i values observed are 0.7867 and 0.4562 respectively for MIROC3 and NCAR-PCMI with a range of 0.3305. MIROC3 occupied first position with D_i^+, D_i^-, C_i values, 0.6732, 2.4833, 0.7867 followed by GFDL 2.1 with D_i^+, D_i^-, C_i values, 0.9024, 2.1776, 0.7070. Third position is occupied by GISS with D_i^+, D_i^-, C_i values, 0.9043, 2.1802, 0.7068. Note that relative closeness is almost same with slight difference of 0.0002 for GFDL 2.1 and GISS for both second and third positions.

Similarly, UKMO-HADGEM1 and INGV-ECHAM4 occupied seven and eighth positions with relative closeness of 0.6808 and 0.6802 with slight difference of 0.0006. Tenth and eleventh positions are occupied by IPSL-CM4, NCAR-PCMI with relative closeness of 0.4959 and 0.4562. No single global climate model can be recommended due to insignificant relative closeness difference between the top rated global climate models. Sample calculation for computation of D_i^+, D_i^-, C_i for global climate model GISS is presented in Appendix.

Keeping this in view, ensemble of GFDL2.1, GISS, MIROC3 is suggested for further applications such as downscaling and hydrological modeling. Model ensemble can be performed by adopting arithmetic average or suitable weighting procedure on the corresponding historic climate simulations.

Table 3 Ranking pattern of global climate models

Model	D_i^+	D_i^-	C_i	Rank
UKMO-HADGEM1	0.9804	2.0914	0.6808	7
GISS	0.9043	2.1802	0.7068	3
GFDL2.0	0.9076	2.1714	0.7052	4
BCCR-BCCM 2.0	0.9378	2.1424	0.6955	6
IPSL-CM 4	1.5351	1.5100	0.4959	10
UKMO-HADCM3	0.9296	2.1466	0.6978	5
GFDL2.1	0.9024	2.1776	0.7070	2
INGV-ECHAM 4	0.9869	2.0989	0.6802	8
MIROC3	0.6732	2.4833	0.7867	1
MRI-CGCM2	1.0413	2.0251	0.6604	9
NCAR-PCMI	1.6576	1.3906	0.4562	11

5 Summary and Conclusions

Correlation coefficient, normalised root mean square deviation, and skill score are employed to rank eleven global climate models using F-TOPSIS for the case study of Upper Malaprabha catchment. It is considered to be the first multicriterion decision making effort in fuzzy perspective for ranking global climate models and conclusions emanated from the study are:

- Top three positions are occupied by MIROC3, GFDL2.1 and GISS with relative closeness of 0.7867, 0.7070, and 0.7068.
- IPSL-CM4, NCAR-PCMI occupied the tenth and eleventh positions with relative closeness of 0.4959 and 0.4562.
- Global climate models, namely, GISS, MIROC3, GFDL2.1 may be explored as ensemble for further downscaling and hydrological modeling applications.

Further work can be explored with different membership functions such as trapezoidal, Gaussian, hyperbolic and exponential to understand their applicability. Some of the other possible extensions that can be explored are computation of weights of indicators and application of other multicriterion decision making methods.

Acknowledgments Present research work is supported by Council of Scientific and Industrial Research, New Delhi vide project number 23(0023)/12/EMR-II dated 15.10.2012.

Appendix

Sample Calculation for Computation Of $\mathbf{D_i^+}$, $\mathbf{D_i^-}$, $\mathbf{C_i}$

Chosen global climate model: GISS
Values of indicators in triangular membership environment:
Correlation coefficient (CC) = (0.670, 0.828, 0.985)
Normalised root mean square deviation (NRMSD) = (0.436, 0.534, 0.687)
Skill Score (SS) = (0.704, 0.778, 0.853)
Ideal values of indicators = (1, 1, 1)
Anti-ideal values of indicators = (0, 0, 0)

(i) Separation measure of GISS from ideal solution (Eq. 1):

$D_{GISS}^{+} = \sqrt{\frac{[(p_{ij}-p_j^*)^2+(q_{ij}-q_j^*)^2+(r_{ij}-r_j^*)^2]}{3}} = \sqrt{\frac{[(0.670-1)^2+(0.828-1)^2+(0.985-1)^2]}{3}}$ for correlation coefficient + $\sqrt{\frac{[(0.436-1)^2+(0.534-1)^2+(0.687-1)^2]}{3}}$ for normalised root mean square deviation + $\sqrt{\frac{[(0.704-1)^2+(0.778-1)^2+(0.853-1)^2]}{3}}$ for skill score
$= 0.2150 + 0.4594 + 0.2299 = 0.9043$

(ii) Separation measure of GISS from anti-ideal solution (Eq. 2):

$$D^-_{Giss} = \sqrt{\frac{\left[(p_{ij} - p_j^{**})^2 + (q_{ij} - q_j^{**})^2 + (r_{ij} - r_j^{**})^2\right]}{3}} = \sqrt{\frac{\left[(0.670 - 0)^2 + (0.828 - 0)^2 + (0.985 - 0)^2\right]}{3}}$$

for correlation coefficient + $\sqrt{\frac{\left[(0.436 - 0)^2 + (0.534 - 0)^2 + (0.687 - 0)^2\right]}{3}}$ for normalised root mean square deviation + $\sqrt{\frac{\left[(0.704 - 0)^2 + (0.778 - 0)^2 + (0.853 - 0)^2\right]}{3}}$ for skill score $= 0.8376 + 0.5619 + 0.7807 = 2.1802$

(iii) Relative closeness of GISS with reference to anti-ideal measure (Eq. 3):

$$C_{GISS} = \frac{D^-_{GISS}}{(D^-_{GISS} + D^+_{GISS})} = \frac{2.1802}{(2.1802 + 0.9043)} = 0.7068$$

References

1. Raju, K.S., Nagesh Kumar, D.: Multicriterion Analysis in Engineering and Management. Prentice Hall of India, New Delhi (2014)
2. Walsh, J.E., Chapman, W.L., Romanovsky, V., Christensen, J.H., Stendel, M.: Global climate model performance over Alaska and Greenland. J. Clim. **21**, 6156–6174 (2008)
3. Kundzewicz, Z.W., Stakhiv, E.Z.: Are climate models "ready for prime time" in water resources management applications, or is more research needed? Hydrol. Sci. J. **55**, 1085–1089 (2010)
4. Weigel, A.P., Knutti, R., Liniger, M.A., Appenzeller, C.: Risks of model weighting in multimodel climate projections. J. Clim. **23**, 4175–4191 (2010)
5. Radić, V., Clarke, G.K.C.: Evaluation of IPCC models' performance in simulating late-twentieth-century climatologies and weather patterns over North America. J. Clim. **24**, 5257–5274 (2011)
6. Johnson, F., Westra, S., Sharma, A., Pitman, A.J.: An assessment of GCM skill in simulating persistence across multiple time scales. J. Clim. **24**, 3609–3623 (2011)
7. Su, F., Duan, X., Chen, D., Hao, Z., Cuo, L.: Evaluation of the global climate models in the CMIP5 over the Tibetan Plateau. J. Clim. **26**, 3187–3208 (2013)
8. Perez, J., Menendez, M., Mendez, F.J., Losada, I.J.: Evaluating the performance of CMIP3 and CMIP5 global climate models over the north-east Atlantic region. Clim. Dyn. **43**, 2663–2680 (2014)
9. Northrop, P.J., Chandler, R.E.: Quantifying sources of uncertainty in projections of future climate. J. Clim. **27**, 8793–8808 (2014)
10. Sun, Q., Miao, C., Duan, Q.: Comparative analysis of CMIP3 and CMIP5 global climate models for simulating the daily mean, maximum, and minimum temperatures and daily precipitation over China. J. Geophys. Res. Atmos. 120, 4806–4824 (2015)
11. McMahon, T.A., Peel, M.C., Karoly, D.J.: Assessment of precipitation and temperature data from CMIP3 global climate models for hydrologic simulation. Hydrol. Earth Syst. Sci. **19**, 361–377 (2015)
12. Gulizia, C., Camilloni, I.: Comparative analysis of the ability of a set of CMIP3 and CMIP5 global climate models to represent precipitation in South America. Int. J. Climatol. **35**, 583–595 (2015)
13. Raju, K.S.: Nagesh Kumar D.: Ranking of global climatic models for India using multicriterion analysis. Clim. Res. **60**, 103–117 (2014)
14. Raju, K.S., Nagesh Kumar, D.: Ranking general circulation models for India using TOPSIS. J. Wat. Clim. Chan. 6:288–299 (2015)

15. Meehl, G.A., Stocker, T.F., Collins, W.D., Friedlingstein, P., Gaye, A.T., Gregory, J.M., Kitoh, A., Knutti, R., Murphy, J.M., Noda, A., Raper, S.C.B., Watterson, I.G., Weaver, A.J., Zhao, Z.C.: Global climate projections. Climate change: The physical science basis. contribution of working group I to the fourth assessment report of the intergovernmental panel on climate change. In: Solomon, S., Qin, D., Manning, M., Chen, Z., Marquis, M., Averyt, K.B., Tignor, M., Miller, H.L. (eds.) Cambridge University Press, Cambridge, 747–846 (2007)
16. Perkins, S.E., Pitman, A.J., Holbrook, N.J., McAveney, J.: Evaluation of the AR4 climate models' simulated daily maximum temperature, minimum temperature and precipitation over Australia using probability density functions. J. Clim. **20**, 4356–4376 (2007)
17. Opricovic, S., Tzeng, G.H.: Compromise solution by MCDM methods: A comparative analysis of VIKOR and TOPSIS. Eur. J. Oper. Res. **156**, 445–455 (2004)

Comparative Analysis of ANFIS and SVR Model Performance for Rainfall Prediction

Akash Dutt Dubey

Abstract In this paper, a comparative study of adaptive neural fuzzy inference system and support vector machine regression models for the purpose of monthly rainfall prediction has been done. The models were trained, validated and tested using 50 years (1960–2010) of historical climatic data of Varanasi, a district in Uttar Pradesh state of India. The data used as the predictors in these models were monthly relative humidity, atmospheric pressure, average temperature and wind speed of Varanasi. The adaptive neural fuzzy inference system (ANFIS) model used in this study is generated using grid partitioning method and its performance was analyzed. The v-support vector regression (v-SVR) model is also developed for the comparative analysis of the rainfall prediction and the kernel used for this model is Radial Basis Function. The performance criteria of these models were Mean Square Error (MSE), Root Mean Square Error (RMSE), Mean Absolute Error (MAE), Nash-Sutcliffe model efficiency coefficient (E) and correlation coefficient R. In this study, it is observed that the performance of adaptive neural fuzzy inference system (ANFIS) model which is optimized using hybrid method has better performance than v-SVR.

Keywords Artificial neural network · ANFIS · Support vector regression · Rainfall prediction

1 Introduction

Rainfall prediction has been one of the most prominent issues that have been encountered in the catchment management applications. It also plays an important role for the countries having agro-based economy. Rainfall forecasting is often considered to be one of the most difficult tasks keeping in mind that the variability of the

A.D. Dubey (✉)
Department of Computer Science and Information Systems, SMCE-CEST,
Fiji National University, Suva, Fiji
e-mail: akash.dubey@fnu.ac.fj

V. Ravi et al. (eds.), *Proceedings of the Fifth International Conference on Fuzzy and Neuro Computing (FANCCO - 2015)*, Advances in Intelligent Systems and Computing 415, DOI 10.1007/978-3-319-27212-2_6

spatial and temporal factors which affect rainfall is very high. Moreover, the rainfall prediction methods can be used promptly by the flood management systems to increase the lead time for early warning. Over the years, several rainfall prediction methods have been applied to predict the rainfall accurately. These methods have been primarily divided into two major categories. The first category of the rainfall prediction methods uses the conventional methods, modeling the physical laws in order to predict the rainfall. One of the major drawbacks of this method is that it may cease to be feasible since the rainfall is a product which comprises several complex atmospheric processes which vary both in spatial and temporal forms. This method also involves relatively huge calculations which limit the models' capacities.

The second method for rainfall prediction involves the use of intelligent computing methods. Generally in this category, attempts are made to establish a pattern of the rainfall in order to get accurate rainfall prediction. For this purpose data are collected over years to analyze and establish a pattern among the factors affecting the rainfall. The intelligent computing methods used for this study may involve artificial neural networks, support vector machine and adaptive neural fuzzy inference systems. In this paper, two intelligent models have been developed for rainfall prediction i.e. support vector machines and ANFIS. The models developed in this paper use the major factors affecting rainfall and attempt to establish a relation between these factors and the rain-fall of the region. This paper also compares the performances of these two models.

2 Related Works

The learning and predicting ability of the ANN involving missing meteorological values were evaluated by Dawson and Wilby [1] and the results proved to be more accurate than the conventional methods of prediction. In 2001, Toth et al. [2] found out that the ANNs predicted the short term rainfall with better accuracy as compared to the conventional rainfall prediction methods. A comparison of three different neural networks (multilayer feed forward network, Elman partial recurrent neural network and time delay neural network) was done by Luk et al. [3].

The comparative study of the predictive ability of Radial Basis Function Neural Network (RBFNN) and Hopfield Model (HFM) revealed that the RBFNN had better accuracy (Maqsood et al.) [4]. However, it has also been suggested that the neural networks should be re-initialized several times in order to ensure the best solution (Lekkas et al.) [5]. Several studies have suggested that the neural networks have better prediction accuracy than the regressive models [6, 7]. Yu et al. [8] applied Support Vector Regression for the estimation of real time flood stage.

Hung et al. [9] developed a forecasting model that could forecast rainfall at 75 different rain gauge stations 1–6 h ahead for flood management. Ramesan et al. [10] used Neural Network Auto Regressive with exogenous input (NNARX) and adaptive neuro-fuzzy inference system (ANFIS) for rainfall predictions and found out that both the methods provide high accuracy and reliability.

For the daily rainfall prediction, neuro-fuzzy approach was adopted by Luenam et al. [11]. Prognostic models were developed by Nastos et al. [12] to predict the rainfall for next 4 months with satisfactory results. A clustering method to cluster the average extreme rainfall of a year and then corresponding neural networks were chosen to predict the rainfall [13]. A support vector machine method was developed which used genetic programming during training and validation period to predict the hourly typhoon rain-fall [14].

A study of various soft computing methods was done using SVR, ANN, moving average (MA) and singular spectrum analysis (SSA) for the time series forecasting of rainfall which stated the modular models (ANN-SVR for daily forecasting and MSVR for monthly forecasting) gave better results as compared to ANN or SVR techniques [15]. Datta et al. [16] used artificial neural networks to establish the relation between the rainfalls of a given region to that of the rainfall of the neighboring regions using the temporal as well as spatial data for increasing the accuracy of the model.

Awan and Bae [17] modeled the ANFIS model using data from six different dams categorically and concluded that this method was more accurate for the long term inflow as compared to simple ANFIS model. SVM technique was coupled with single spectrum analysis (SSA) since single spectrum analysis was helpful in the preprocessing of the data Simões et al. [18]. The results obtained by this method were compared to the results of SVM model only and it was found that the results improved significantly by this method.

Li et al. [19] used multi objective genetic algorithm with support vector machines to decide the optimal inputs and then predict 1–6-h-ahead typhoon rainfall in Tsengwan river basin. Dubey [20] studied the performance of three different neural network models (back propagation algorithm, layer recurrent and distributed time delay model) for rainfall prediction and concluded that the prediction of feed forward time delay network model was better as compared to the other models. Wu et al. [21] proposed a Hybrid Particle Swarm Optimization Genetic Algorithm (HPSOGA) method which was applied to the RBF-NN design for rainfall prediction. Their study proved that this method can be effectively used for higher generalization ability as well as better forecasting accuracy.

3 Methodology

3.1 Adaptive Neural Fuzzy Inference System (ANFIS)

Adaptive neural fuzzy inference system are adaptive neural networks which combine the fuzzy inference system with the learning, adaptability and the approximate reasoning features of the artificial neural networks. ANFIS is a hybrid system that is used to solve complex problems with the use of intelligent systems and have the ability to deal with different issues like uncertainty and unknown variations effectively, thereby improving the robustness of the system.

The Fuzzy Inference system has the ability to employ rules and knowledge that can be expressed by linguistic expressions. The expression technique is one of the strengths of the FIS as compared to the ANNs because ANNs are difficult to express qualitative information such as "the temperature was high". Therefore, the main contribution of the FIS towards ANFIS is the modeling using if-then-else rules. With the help of the learning ability of the neural networks, the fuzzy rule sets are derived and the parameters are optimized.

Fuzzy logic is a flexible mathematical tool which is used in information modeling. Fuzzy logic is different with the conventional logic in the sense that while the conventional logic sets strict boundaries on the membership of a variable to a definite set, fuzzy logic provides membership ranks within two sets and provides the solution which is based on this dual.

The sugeno model in Fuzzy logic provides a systematic approach for creating a collection of fuzzy rules according to the given input-output set. Sugeno model is based on the if-then-else form as given below

$$\text{if } x_1 \text{ is A and } x_2 \text{ is B then } y = f(x_1, x_2) \tag{1}$$

given A and B are the fuzzy sets and $y = f(x_1, x_2)$ is defined as the function associated with the facts of x_1 and x_2 which correspond to A and B respectively.

ANFIS Architecture:

Figure 1 illustrates the basic architecture of the ANFIS models [22]. The basic architecture of is divided into five layers which can be described as follows:

Layer 1 The nodes in this layer are the adaptive nodes associated with a node function. The input to this layer is x to node i and the labels A_i are the linguistic labels associated with them. The output of the ith node in layer 1 will be represented as $O_{1,i}$ and will be the membership rank of a fuzzy set A(= A_1, A_1, B_2 and B_2). This output indicates the degree to which the inputs are satisfied by quantifier A. The Gaussian 2 parameter used in this study will be:

$$\mu_a(x) = exp\left(-\frac{1}{2}\left(\frac{x - c_i}{a_i}\right)^2\right) \tag{2}$$

where c_i and a_i are known as the premise or antecedent parameters.

Layer 2 All the nodes in this layer are fixed and are used to multiply the incoming signals and outputs the products. Each output of this layer will illustrate the firing strength of a rule. W_i is the firing strength of the ith node.

$$O_{2,i} = w_i = \mu_{A_i}(x)\mu_{B_i}(y) \tag{3}$$

Layer 3 The nodes in this layer are also fixed and the ith node calculates the ratio of the firing strength of the ith rule to the sum of the entire rule's firing strengths.

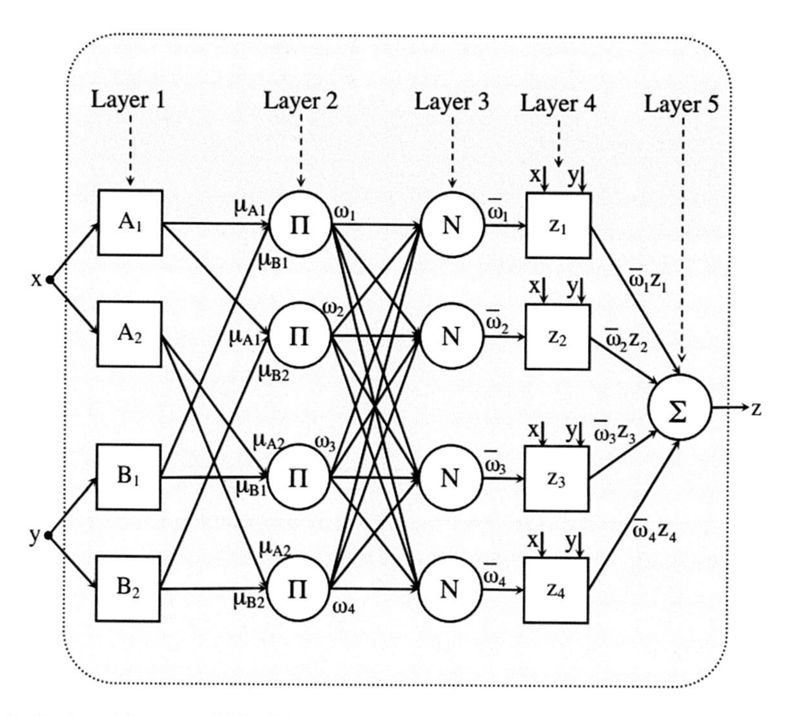

Fig. 1 Basic architecture of ANFIS

$$O_{3,i} = \overline{w}_i = \frac{w_i}{w_1 + w_2} \tag{4}$$

where $\overline{w}_i$ is the normalized firing strength.

Layer 4 This layer comprises of an adaptive node with the following node function:

$$O_{4,i} = \overline{w}_i f = \overline{w}_i (p_i x + q_i y + r_i) \tag{5}$$

The parameters p_i, q_i and r_i are the parameters of the nodes.

Layer 5 The total output is calculated as the sum of all the incoming signals using the single fixed node in this layer. The calculation can be done as below:

$$overall\,output = O_{5,i} = \sum_i \overline{w}_i f_i = \frac{\sum_i \overline{w}_i f_i}{\sum_i \overline{w}_i} \tag{6}$$

More details about ANFIS can be found in [23, 24]. Various identification methods such as Grid Partitioning (GP) and Subtractive Clustering (SC) can be used to develop the ANFIS Sugeno models. Grid partitioning method in ANFIS used axis paralleled partitions which is based on a defined number of membership functions, to divide the input space into a number of local fuzzy regions [25].

ANFIS GP works better if the number of input functions for the models is less than six along with a sufficient number of data samples. Therefore, in this research work, the number of data samples used is quite large and each input has been assigned four membership functions.

3.2 Support Vector Regression

In 1992, Cortes and Vapnik [26] developed classification and regression procedures based on statistical learning known as Support Vector Machines. The Support Vector Machines work on the principle of mapping the original data X into a high-dimensional feature space using the kernel function and then constructing an optimal hyper plane in the new space which separates the classes. The advantage of the SVM over conventional learning algorithms is that it avoids the local minima and the risk of over-fitting during training and its solution is always globally optimal. The support vector regression (SVR) describes the regression with SVM. In regression estimation with SVR, attempts are made to evaluate the dependency between the set of input points and the target points.

For a given set of data $x_i \subset R^n$, where $1 \leq i \leq l$, l being the size of the training data, and $y_i = \pm 1$ class labels, support vector machines find a hyper plane direction w and an offset scalar b such that $f(x) = w * \Phi(x) + b$ which will be more than zero for positive examples and less than 0 for negative examples. Subsequently, the main role of support vector regression is to find a function that can approximate the future values correctly. The support vector regression function can be represented as follows:

$$f(x) = w * \Phi(x) + b \tag{7}$$

where $w \subset R^n$, $b \subset R$ and Φ denotes the nonlinear transformation from R^n to high-dimensional space. The main goal in this equation is to compute the values of w and b in such a way that there is minimum regression risk in determination of the values of x.

$$R_{reg}(f) = C \sum_{i=0}^{l} \tau(f(x_i) - y_i) + \frac{1}{2}||w||^2 \tag{8}$$

τ being the cost function and C is a constant. The vector w can be described in the terms of data points as

$$w = \sum_{i=1}^{l} \left(\alpha_i - \alpha_i^*\right) \Phi(x_i) \tag{9}$$

Table 1 Kernels used in the support vector machine

Kernel	Function
Linear	x*y
Polynomial	$[(x*x_i) + 1]^d$
Radial basis function (RBF)	$\mathrm{Exp}\{-\gamma\lvert x - x_i\rvert^2\}$

On substituting the value of w in the Eq. 7,

$$\begin{aligned} f(x) &= \sum_{i=1}^{l} \left(\alpha_i - \alpha_i^*\right)\left(\Phi(x_i).\Phi(x)\right) + b \\ &= \sum_{i=1}^{l} \left(\alpha_i - \alpha_i^*\right) k(x_i, x) + b \end{aligned} \tag{10}$$

The term $\Phi(x_i).\Phi(x)$ in the equation can be substituted with the kernel function since it facilitates the evaluation of the dot product in high dimensional space using the low dimensional space data input without any knowledge of Φ. The kernel function used in this study is the RBF (Radial Basis Function). Some of the common kernels are given in Table 1.

The most common cost function that is widely used is the ε-insensitive loss function which is defined as follows:

$$\tau(f(x) - y) = \begin{cases} f(x) - y| - \varepsilon, & \text{for } f(x) - y| \geq \varepsilon \\ 0 & otherwise \end{cases} \tag{11}$$

The constant C used in the Eq. (8) is the penalties which are assigned for the estimation error. Higher value of C indicates that the penalties for the error in estimation is high and the regression is trained to minimize error and lesser generalization while a low value of C indicates greater generalization of the regression. As the value of C grows towards infinity, the errors minimize and more complex models are trained. As the value of C decreases to 0, simpler models are designed with high error tolerance. The support vector machines and the support vector regression can be used without the knowledge of transformation. For the better performance of the models, experiment with the kernel functions, values of C, the penalty function and the radius of the ε tube is needed so that the data inside it can be ignored during regression.

4 Area of Study and Data Set

The area under study for rainfall prediction is Varanasi is located at 25°0′ to 25°16′ N latitude and 82 5′ to 83°1′E longitude, at an elevation of 80 m. above sea level, along the banks of river Ganga. Varanasi has humid subtropical climate with a large difference between the summer and winter temperatures. The temperature in winter can drop up to 0° due to the Himalayan waves while the summers witness a temperature of up to 47°. The average annual rainfall in Varanasi is 1100 mm. The rainfall in Varanasi is mainly due to the South-Western monsoon which takes about 90 % of the total rainfall. The rainfall in Varanasi is prominent from the month of June to September, August being the wettest month of the year.

The data samples collected for the study included four predictors which were monthly average temperature, relative humidity, atmospheric pressure and wind speed. These four predictors were used together to predict the average monthly rainfall of the next month. Since these predictors have different units, it would be difficult to establish correlation between them using soft computing methods. Therefore, min-max normalization method was used to limit the value of these four predictors between 0 and 1. The formula for the normalization can be given as:

$$Normalized(z_i) = \frac{z_i - z_{min}}{z_{max} - z_{min}} \tag{12}$$

where $z_i \varepsilon \{z_1, z_2, z_3 \ldots z_n\}$, z_{min} is the minimum value and z_{max} is the maximum value from the data sample z.

5 Evaluation Criteria of the Performance of the Models

For the evaluation of the model performance in this study, different statistical evaluation criteria were applied. These criteria included Mean Square Error (MSE), Root Mean Square Error (RMSE), Mean Absolute Error (MAE), Nash-Sutcliffe model efficiency coefficient (E) and correlation coefficient R. These statistical criteria can be de- fined as follows:

$$MSE = \frac{\sum_{i=1}^{n} (y_i - \hat{y}_i)^2}{n} \tag{13}$$

$$RMSE = \sqrt{\frac{\sum_{i=1}^{n} (y_i - \hat{y}_i)^2}{n}} \tag{14}$$

$$MAE = \frac{\sum_{i=1}^{n} |y_i - \hat{y}_i|}{n} \tag{15}$$

$$E = 1 - \frac{\sum_{i=1}^{n} (y_i - \hat{y}_i)^2}{\sum_{i=1}^{n} (y_i - \bar{y}_0)^2} \tag{16}$$

$$R = \frac{n(\sum y_i \hat{y}_0) - \sum y_i \sum \hat{y}_0}{\sqrt{\left[n \sum y_i^2 - (\sum y_i)^2\right]\left[n \sum \hat{y}_0^2 - (\sum \hat{y}_0)^2\right]}} \quad (17)$$

where n signifies the number of months, y_i is the observed rainfall for the given month i, $\hat{y}_i$ is the estimated rainfall for the given month i and $\bar{y}_0$ is the mean rainfall over a given period of time.

6 Results and Discussion

In the applications, the ANFIS and v-SVR models were trained, validated and tested according to the given specifications. In this study, grid partition (GP) method has been used to construct the model. For ANFIS-GP model, different number of MFs were trained and tested to check the accuracy of the models. The MFs simulated for ANFIS models were triangular (trimf), two Gaussian (gauss2mf) and generalized bell (gbellmf) and the results obtained by gauss2mf were found to be better than the two other methods. The equation for the gauss2mf method has been given in Eq. (2).

The iteration number for training of all the models in ANFIS was set to 100 since any value more than 100 resulted in overtraining of the model. The node function used in this model is linear and Eq. (5) is the representation of the method. For the purpose of weight optimization, hybrid optimization method has been used which is a combination of least squared and back propagation approaches.

For the development of v-SVR model, two main steps that were involved were selection of kernel function and identification of C and v. The kernel used in this work was radial basis function (RBF) since it maps the data samples non-linearly into a higher dimension space. The advantage of using RBF kernel is that it can also work efficiently in the cases of The RBF kernel can also handle the case where the class labels and attributes are discrete.

As observed from the Table 2, it was concluded that the performance of ANFIS-Grid Partition model was better than the support vector regression model.

Table 2 Performance of the prediction models

	ANFIS-GP		v-SVR	
	Training	Validation	Training	Validation
MSE	0.00315	0.00418	0.00543	0.06251
RMSE	0.05620	0.06465	0.07371	0.07905
MAE	0.02517	0.03158	0.04710	0.05218
E	0.92642	0.90154	0.87345	0.86597
R	0.96323	0.94151	0.93609	0.92589

The regression coefficient R obtained in training and validation phases of the ANFIS model was found to be 0.96 and 0.94 respectively while those obtained for the support vector regression model were 0.93 and 0.92 respectively. The Nash-Sutcliffe model efficiency coefficient obtained for the ANFIS model 0.926 while for the support vector regression model was found to be 0.873. The results suggest that the performance of the ANFIS-GP model outperformed SVR model with a good margin, suggesting that the prediction power of ANFIS model was much better than the SVR model in the case of rainfall prediction using the four aforementioned predictors.

The Figs. 2 and 3 represent the regression plots for the performance during the validation of the ANFIS and SVR models. The performance of the prediction models of SVR and ANFIS has been illustrated in the Figs. 4 and 5 respectively which depict the performance of ANFIS is better than the SVR model, thus consolidating the results.

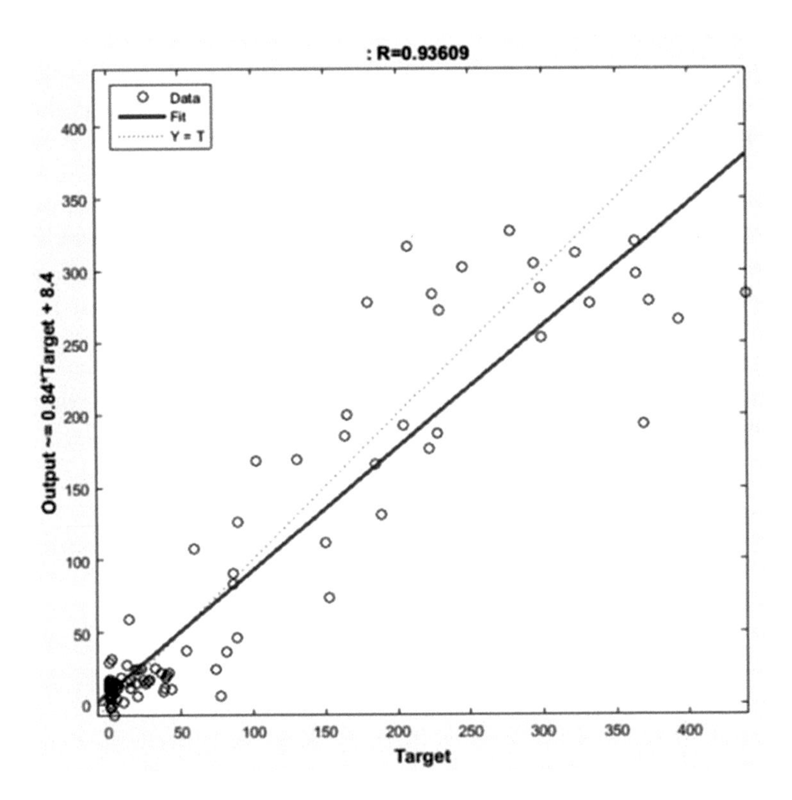

Fig. 2 Regression plot for support vector regression

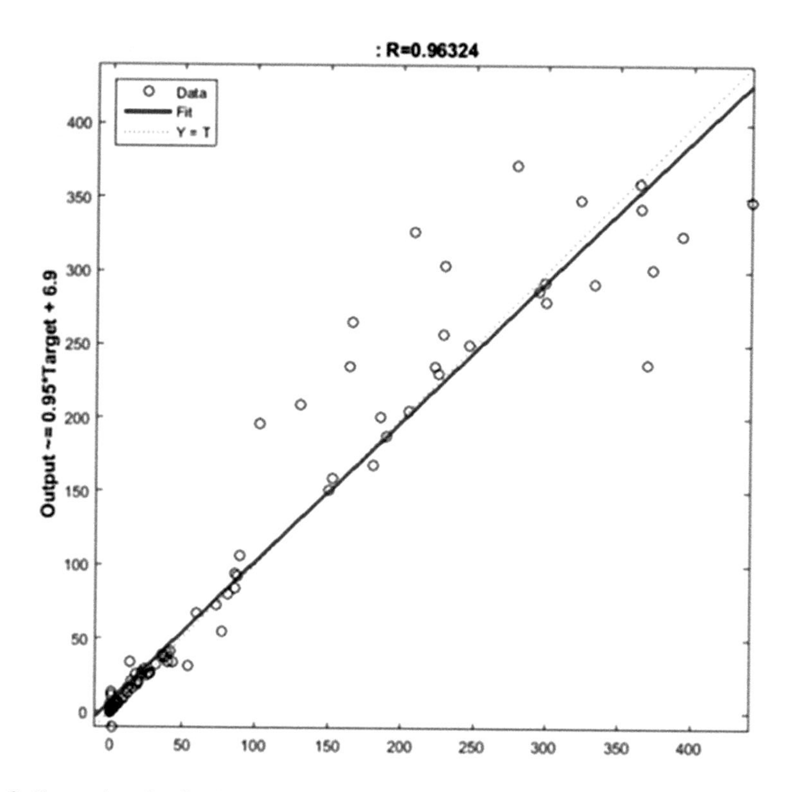

Fig. 3 Regression plot for ANFIS

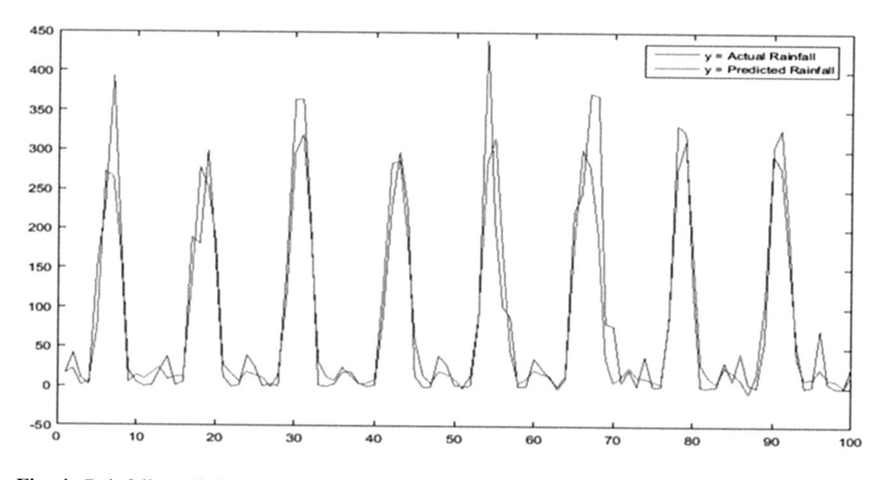

Fig. 4 Rainfall prediction using support vector regression

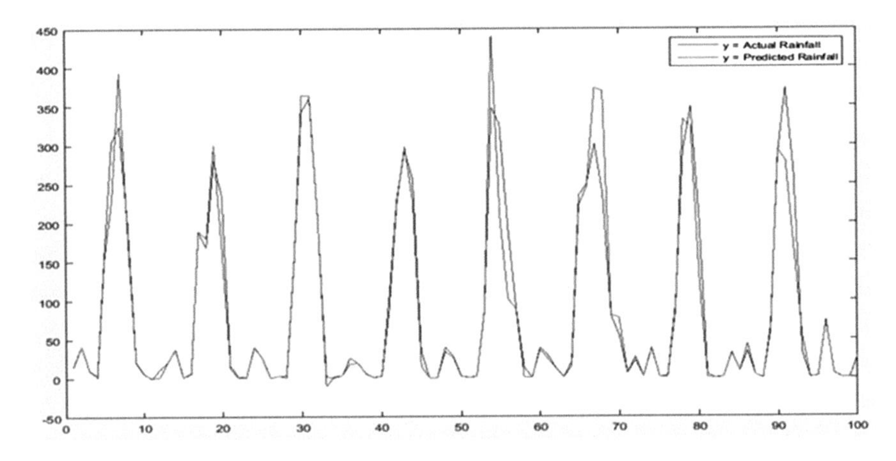

Fig. 5 Rainfall prediction using ANFIS model

7 Conclusions

In this research work, monthly rainfall prediction models using adaptive neural fuzzy inference system and support vector regression have been developed. The predictors used for these models were monthly average temperature, relative humidity, atmospheric pressure and wind speed. The performance of these two models were evaluated and compared to each other. It was concluded that the ANFIS model outperformed the support vector regression model. For the future studies, these adaptive neural fuzzy inference system and support vector regression models can be modified and optimized for better performance and better rainfall accuracy. These models can also be mixed with different soft computing methods thus generating hybrid models for better rainfall accuracy.

References

1. Dawson, C., Wilby, R.: A comparison of artificial neural networks used for river forecasting. Hydrol. Earth Syst. Sci. **3**, 529–540 (1999)
2. Toth, E., Brath, A., Montanari, A.: Comparison of short-term rainfall prediction models for real-time flood forecasting. J. Hydrol. **239**, 132–147 (2000)
3. Luk, K., Ball, J., Sharma, A.: An application of artificial neural networks for rainfall forecasting. Math. Comput. Model. **33**, 683–693 (2001)
4. Lekkas, D.F., Onof, C., Lee, M.J., Baltas, E.A.: Application of artificial neural networks for flood forecasting. Glob. Nest J. **6**, 205–211 (2004)
5. Maqsood, I., Khan, M., Abraham, A.: An ensemble of neural networks for weather forecasting. Neural Comput. Appl. 13, (2004)
6. Kisi, O.: Daily river flow forecasting using artificial neural networks and autoregressive models. Turk. J. Eng. Environ. Sci. **29**, 9–20 (2005)

7. Modelling and prediction of rainfall using artificial neural network and ARIMA techniques. J. Indian Geophys. Union **10**, 141–151 (2006)
8. Yu, P., Chen, S., Chang, I.: Support vector regression for real-time flood stage forecasting. J. Hydrol. **328**, 704–716 (2006)
9. Hung, N., Babel, M., Weesakul, S., Tripathi, N.: An artificial neural network model for rainfall forecasting in Bangkok. Thai. Hydrol. Earth Syst. Sci. **13**, 1413–1425 (2009)
10. Maheswaran, R., Khosa, R.: Long term forecasting of groundwater levels with evidence of non-stationary and nonlinear characteristics. Comput. Geosci. **52**, 422–436 (2013)
11. Nastos, P., Moustris, K., Larissi, I., Paliatsos, A.: Rain intensity forecast using artificial neural networks in Athens. Greece Atmos. Res. **119**, 153–160 (2013)
12. Luenam, P., Ingsriswang, S., Ingsrisawang, L., Aungsuratana, P., Khantiyanan, W.: A neuro-fuzzy approach for daily rainfall prediction over the central region of Thailand. In: International Multiconference of Engineers and Computers Scientists, Hong Kong (2010)
13. Wan, D., Wang, Y., Gu, N., Yu, Y.: A novel approach to extreme rainfall prediction based on data mining. In: Proceedings of 2012 2nd International Conference on Computer Science and Network Technology (2012)
14. Lin, K., Pai, P., Lu, Y., Chang, P.: Revenue forecasting using a least-squares support vector regression model in a fuzzy environment. Inf. Sci. **220**, 196–209 (2013)
15. Wu, C., Chau, K.: Prediction of rainfall time series using modular soft computing methods. Eng. Appl. Artif. Intell. **26**, 997–1007 (2013)
16. Datta, B., Mitra, S., Pal, S.: Estimation of average monthly rainfall with neighbourhood values: comparative study between soft computing and statistical approach. IJAISC **4**, 302 (2014)
17. Awan, J., Bae, D.: Improving ANFIS based model for long-term dam inflow prediction by incorporating monthly rainfall forecasts. Water Resour. Manage. **28**, 1185–1199 (2014)
18. Simões, N., Wang, L., Ochoa-Rodriguez, S., Leitão, J.P., Pina, R., Onof, C., Maksimovic, C.: A coupled SSA-SVM technique for stochastic short-term rainfall forecasting. In: 12th International Conference on Urban Drainage, Porto Alegre, Brazil (2011)
19. Li, L., Li, W., Tang, Q., Zhang, P., Liu, Y.: Warm season heavy rainfall events over the Huaihe River Valley and their linkage with wintertime thermal condition of the tropical oceans. Clim. Dyn. (2015)
20. Dubey, A.D.: Artificial neural network models for rainfall prediction in Pondicherry. Int. J. Comput. Appl. **120**, 30–35 (2015)
21. Wu, J., Long, J., Liu, M.: Evolving RBF neural networks for rainfall prediction using hybrid particle swarm optimization and genetic algorithm. Neurocomputing **148**, 136–142 (2015)
22. Sarikaya, N., Guney, K., Yildiz, C.: Adaptive neuro-fuzzy inference system for the computation of the characteristic impedance and the effective permittivity of the micro-coplanar strip line. Prog. Electromagnet. Res. B. **6**, 225–237 (2008)
23. Jang, J.: ANFIS: adaptive-network-based fuzzy inference system. IEEE Trans. Syst. Man Cybern. **23**, 665–685 (1993)
24. Jang, J., Sun, C., Mizutani, E.: Neuro-fuzzy and soft computing. Prentice Hall, Upper Saddle River, NJ (1997)
25. Abonyi, J., Andersen, H., Nagy, L., Szeifert, F.: Inverse fuzzy-process-model based direct adaptive control. Math. Comput. Simul. **51**, 119–132 (1999)
26. Cortes, C., Vapnik, V.: Support-vector networks. Mach. Learn. **20**, 273–297 (1995)

Cross Domain Sentiment Analysis Using Different Machine Learning Techniques

S. Mahalakshmi and E. Sivasankar

Abstract Sentiment analysis is the field of study that focuses on finding effectively the conduct of subjective text by analyzing people's opinions, sentiments, evaluations, attitudes and emotions towards entities. The analysis of data and extracting the opinion word from the data is a challenging task especially when it involves reviews from completely different domains. We perform cross domain sentiment analysis on Amazon product reviews (books, dvd, kitchen appliances, electronics) and TripAdvisor hotel reviews, effectively classify the reviews to positive and negative polarities by applying various preprocessing techniques like Tokenization, POS Tagging, Lemmatization and Stemming which can enhance the performance of sentiment analysis in terms of accuracy and time to train the classifier. Various methods proposed for document-level sentiment classification like Naive Bayes, k-Nearest Neighbor, Support Vector Machines and Decision Tree are analysed in this work. Cross domain sentiment classification is useful because many times we might not have training corpus of specific domains for which we need to classify the data and also cross domain is favoured by lower computation cost and time. Despite poor performance in accuracy, the time consumed for sentiment classification when multiple testing datasets of different domains are present is far less in case of cross domain as compared to single domain. This work aims to define methods to overcome the problem of lower accuracy in cross-domain sentiment classification using different techniques and taking the benefit of being a faster method.

Keywords Natural language processing · Machine learning · Cross domain

S. Mahalakshmi (✉) · E. Sivasankar
Department of Computer Science and Engineering, National Institute of Technology, Tiruchirappalli, India
e-mail: maha.81193@gmail.com

E. Sivasankar
e-mail: sivasankar@nitt.edu

V. Ravi et al. (eds.), *Proceedings of the Fifth International Conference on Fuzzy and Neuro Computing (FANCCO - 2015)*, Advances in Intelligent Systems and Computing 415, DOI 10.1007/978-3-319-27212-2_7

1 Introduction

With the evolution of web, people express or exchange their perspectives, feedbacks and opinions through online social media sites, such as user forums, blogs, discussion boards and review sites, etc. [1]. User views are considered to be valuable sources for improving the caliber of the service provided [2]. The user reviews are usually in unstructured format. Analyzing and gaining information from them can be a difficult task. Sentiment Classification (SC) is the technique which provides a solution to this problem. It focuses on identifying sentiment words and their polarity from user's review, which are then classified as positive and negative. This will help the user to arrive at a correct decision. Cross Domain Sentiment Classification (CDSC) is a ranging field of research in Machine Learning (ML). A fraction of the dataset is used for training the classifier and remaining fraction is used for testing in single domain whereas in cross domain, two completely different domains are considered and the classifier is trained with one domain dataset and tested using the other. The challenge in SC of user reviews is to improve the computational efficiency and also to ensure real time response for processing huge amounts of data with the feasibility of the technology to work across heterogeneous commercial domain [3]. Thus the computational complexity of the proposed technology is crucial.

Major research in sentiment analysis focuses on identifying people's emotion or attitude from text. In heterogeneous domain, the group is split into clusters and various techniques can be adopted for improving the results like spotting the keyword, lexical affinity, statistical methods and concept based approach [3]. Each method help in certain way to the problem formulation. Main reason for considering heterogeneous domains is the crucial problem faced by people in finding the sentiment of the texts or reviews extracted from the web accurately in an automated way irrespective of the domain. This paper is structured as follows: Sect. 2 describes the related work on CDSC. Different ML methods are elaborated in Sect. 3. Proposed frameworks are detailed in Sect. 4. Section 5 gives the experimental result obtained for the proposed models and performance of the CDSC for all classifiers are analyzed. Finally Sect. 6 concludes the work.

2 Literature Review

Bisio et al. [3] have done research on the classification across heterogeneous domains. In their research, they have defined a feature space for reviews using an integrated strategy which combines semantic network and contextual valence shifters. They have used empirical learning to map the reviews and its sentiment labels. Augmented k-Nearest Neighbor (kNN) [4] aided their model by improving the computational efficiency. Li et al. [5] proposed a sentiment transfer mechanism based on constrained non-negative matrix tri-factorizations of term-document matrices in the source and target domains.

Bollegala et al. [6] proposed a technique for CDSC that uses an automatically extracted Sentiment Sensitive Thesaurus (SST). They have used labeled data from multiple source domains and unlabeled data from source and target domains to construct SST. This constructed thesaurus is used for training and testing the binary classifier. A relevant subset of the features is selected using L1 regularization. Jambhulkar and Nirkhi [7] have compared and analyzed three techniques of cross domain sentiment analysis viz SST, spectral feature alignment, structural correspondence learning and discussed the challenges. This gave us an idea about the existing techniques and scope. We propose our model which uses techniques from the above literatures and the effectiveness is talked about in the following sections.

The performance of our proposed approach is evaluated experimentally by considering two large heterogeneous datasets of reviews : the first one is product reviews from AmazonTM [6] and the other one is hotel reviews from TripAdvisorTM [8]. The chosen datasets are available online to reproduce the showed results. The experimental results show that the proposed framework attains satisfactory performance in terms of both accuracy and computational efficiency.

The experimental datasets used for the analysis are two heterogeneous domains: Amazon Dataset (books, dvd, kitchen appliances, electronics) [6] and hotel reviews from TripAdvisor [8] which are quite distant commercial areas. The complications in SC of user reviews are majorly due to Linguistic nuances such as the quality of lexicon and incongruity in syntax that make it hard to apply any classical text mining approaches [6]. Drawing separating boundaries can be extremely tricky for classifiers especially in sparse scenarios. Thus the efficiency of classification is most dependent on the review patterns [9]. Apart from these impediments, cognitive issues also occurs for text to sentiment mapping of an established model. To facilitate these issues a holistic approach is required that integrates lexical and semantic information with ease [13].

3 Methodology

This part describes various ML techniques used to categorize the review as positive or negative. The reviews are preprocessed to improve the efficiency. Classifiers are trained using one dataset to predict the sentiment of the reviews from another dataset in cross domain. We measure the exactness of the result using confusion matrix.

For analysis we have used the following classifiers, K-Nearest Neighbour, Naive Bayes-Gaussian, Multinomial, Bernoulli, Support Vector Machine using Linear kernel and Radial Basis Function (RBF) kernel, Decision Tree.

K-Nearest Neighbour The kNN algorithm is a non-parametric method [4] which takes training set with class label as input for classification. Output is the class membership. This algorithm classifies the sample based on majority vote of its neighbors, the output is the most common class among its kNN [3]. K is a fixed constant which is generally small and an odd number is often chosen to avoid the neutrality case while doing the majority voting.

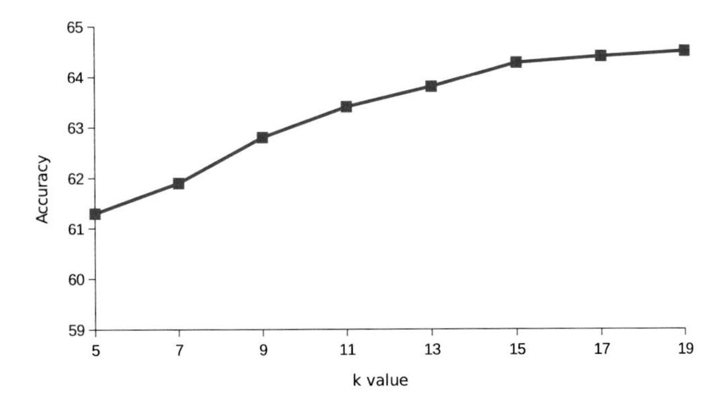

Fig. 1 Accuracy of KNN for different K value

The training set given as input to the classifier consist of vectors in multi-dimensional feature space, each with its class label. While training the classifier, the algorithm stores only the feature vectors and corresponding class labels for each sample. A user defined constant is chosen for k. An unlabeled feature vector is given as input for the classifier to identify the class. While classifying, k nearest feature vectors to the test data are identified by opting a distance metric. Most trivial and efficient distance metric is Euclidean Distance. Majority voting is done for the k identified neighbors and the class is assigned to the test vector.

An experiment is carried out on finding the accuracy of the k-NN classifier by training the Amazon dataset (pair 1) and testing with TripAdvisor dataset for Model 1. The Fig. 1 illustrates the experiment and it shows the increase in accuracy for increasing k and after k = 15, the increase in accuracy is not considerable and so we fixed the value of k as 15.

Naive Bayes NB classifiers [9] are simple probabilistic classifiers based on applying Bayes' theorem with strong (naive) independence assumptions between the features. In Gaussian NB, the assumption is that the continuous values associated with each class are distributed according to a Gaussian distribution.

$$p(x = v \mid c) = \frac{1}{\sqrt{2\pi\sigma_c^2}} e^{\frac{(v-\mu_c)^2}{2\mu_c^2}} \tag{1}$$

where x is a continuous attribute, c is a class, σ_c^2 is the variance and μ is the mean of the values in x associated with class c. With a multinomial event model, samples (feature vectors) represent the frequencies with which certain events are generated by a multinomial. Most document classification uses this model, with events representing the occurrence of a word in a single document. The likelihood of observing a histogram X is given by,

$$p(x \mid C_k) = \frac{(\sum_i x_i)!}{\prod_i x_i!} \prod_i p_{k_i}^{x_i} \tag{2}$$

where p_i is the probability that event i occurs. A feature vector $X = (x_1, ...x_n)$ is then a histogram, with X_i counting the number of times event i was observed in a particular instance.

In the multivariate Bernoulli event model, features are independent booleans (binary variables) describing inputs. If X_i is a boolean denoting the occurrence or absence of the $i'th$ term from the vocabulary, then the likelihood of a document given a class C_k is given by,

$$p(x \mid C_k) = \prod_{i=1}^{n} p_{k_i}^{x_i}(1 - p_{ki})^{(1-x_i)} \tag{3}$$

where p_{ki} is the probability of class C_k generating the term w_i. Same like Multinomial NB, this is also most suited for document classification [10]. Small amount of training data is sufficient to train the NB model to efficiently classify the test data. It is a simple technique for constructing classifiers.

Support Vector Machine SVMs or Support vector networks are supervised learning models. The data are analyzed and a pattern is recognized using learning algorithms that are part of the model. These patterns are used for analysis [9]. A set of training sample, each belonging to one of the two categories, is given as input to the SVM training algorithm. The algorithm constructs a model that will classify the new samples into one of the categories. Thus SVM is a non-probabilistic binary linear classifier. SVM model represents the training sample as points in space. Analysis of the points for pattern recognition results in identifying a clear gap which separates the two categories. When a test data is given to the model, it maps the data to the space and categorizes it based on which side of the gap it falls on [10]. Linear svm are generally applied on linearly separable data. If the data is not linearly separable, you'll need to map your samples into another dimensional space, using kernels like RBF.

Decision Tree DT is used in this learning technique. It as a predictive model which maps the details of the data to its class. This approach is used in statistics, data mining and ML [9]. Classification trees are models that accept the class to take finite set of values. In these models, class label take the leaf's position and the intermediate branches represent conjunctions of features that lead to the class labels. After preprocessing, review versus common words matrix is generated for Amazon and TripAdvisor dataset. These matrices are used for training the classifiers. The classifiers are trained using one dataset and tested using the other in each dataset pair.

Table 1 Confusion matrix for SC

		Predicted sentiment	
		+ ve sentiment	– ve sentiment
Actual sentiment	+ ve sentiment	True positive (TP)	False negative (FN)
	– ve sentiment	False positive (FP)	True negative (TN)

3.1 Performance Analysis

Accuracy measure is used to evaluate the performance of SC. Accuracy is determined as the proportion of correct number of predictions given total number of text reviews. Let us assume 2×2 confusion matrix as mentioned in Table 1.

The accuracy of SC is,

$$Accuracy = \frac{TP + TN}{TP + TN + FP + FN} \tag{4}$$

4 Proposed Framework

The experimental evaluation of the cross-domain approach needs two entirely different training and testing datasets [3]. The first dataset reviews are from Amazon, [6] related to four product categories: books, DVDs, electronics and kitchen appliances which has 8000 reviews (1000 positive and 1000 negative reviews in each category). The second dataset is hotel reviews obtained from TripAdvisor, [8] which has millions of review. First N sampling technique is used to sample the data and consider only 4000 reviews in which 2000 positive and negative reviews each. This sampling method is opted because the database reviews are in random order. All the reviews are labelled.

We analyse the reviews in three stages—Preprocessing, Matrix Generation, Classification. Figure 2 shows the steps followed in the proposed method for the different feature models and classifiers considered.

4.1 Preprocessing

We preprocess the reviews using ML techniques such as Tokenization, removing stop words, Lemmatization, Stemming and Part-of-Speech (POS) Tagging.

Tokenization occurs at the word level, it is the process of breaking up the stream of text to words, phrases or symbols which are meaningful. These elements are termed as tokens [6, 11]. Stop words are words which don't express the sentiment of the sentences and so they are removed [6, 12]. In our work, we perform removal of stop

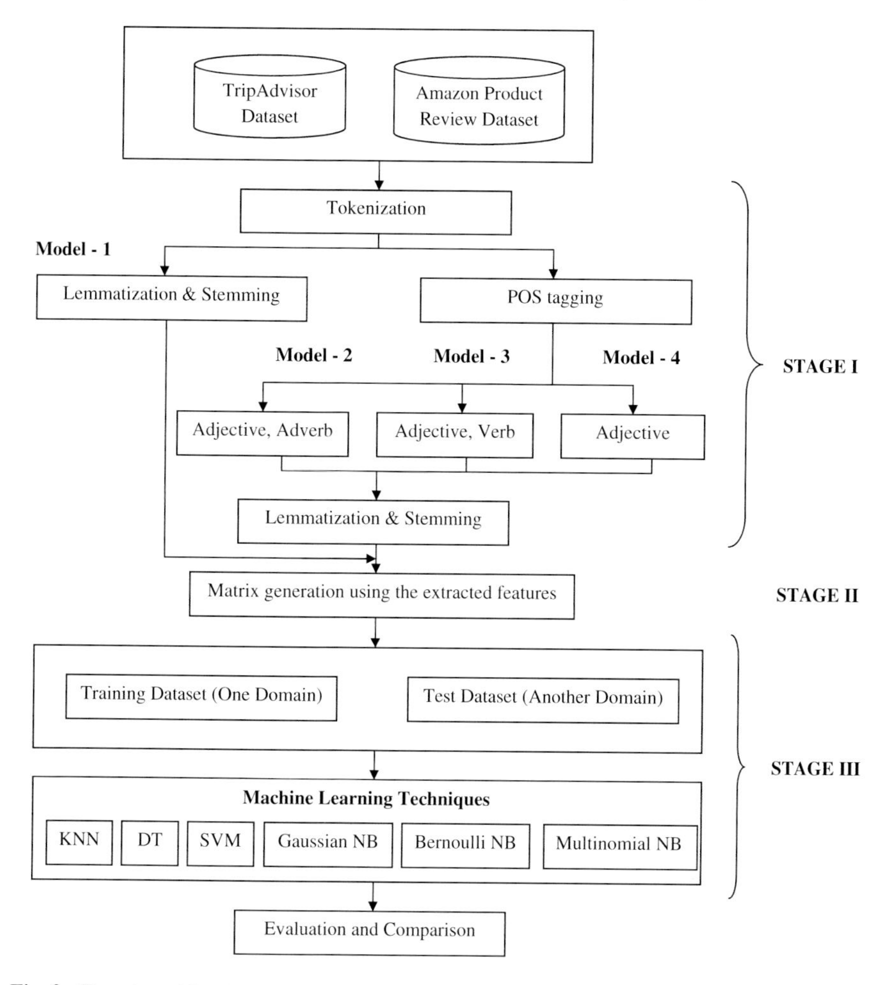

Fig. 2 Experimental setup

words after tokinizing the reviews. There is no single universal list of stop words. For a given purpose, any group of words can be chosen as the stop words. All words less than 3 characters, articles, prepositions and most common words can be part of the stop words [1, 12]. In Lemmatization, different inflected forms of a word are grouped together, so that they can be analyzed as a single item. In computational linguistics, it is the process of determining the lemma of a given word [6, 9]. Stemming is an informational retrieval process which reduces the words to their word stem or base form [6, 9]. Lemmatization and Stemming appear to be similar processes. The difference is that a stemmer operates on a single word without knowing the context and so it cannot discriminate between the words having different meanings depending upon the context unlike lemmatizer. The advantage of stemmer is that it can be easily implemented and is faster which overshadow the reduced accuracy in some applications. Part-of-Speech tagging which is also called grammatical tagging or word-category disambiguation, is the process of identifying the part of speech for each word in the review based on its definition and context. It considers the relationship with the adjacent and related words in a phrase, sentence or paragraph.

4.2 Model Generation

Following models are constructed for the analysis as mentioned in Fig. 2.

Model 1 Both the datasets are tokenized. Stop words are removed from the tokenized words and it is lemmatized. Lemmatized words are stemmed.

Model 2 Both the datasets are tokenized, pos tagging is applied and adjectives, adverbs are extracted and it is lemmatized and stemmed.

Model 3 Both the datasets are tokenized, pos tagging is applied and adjectives,verbs are extracted and it is lemmatized and stemmed.

Model 4 Both the datasets are tokenized, pos tagging is applied and adjectives are extracted and it is lemmatized and stemmed.

4.3 Matrix Generation

Common words between the two datasets are extracted from the preprocessing stage. In matrix generation, these common words are considered and term frequency matrix or binary matrix between reviews and the words is constructed for each of the datasets [3]. In our work, we consider frequency matrix for analysis.

In the classification stage, classifiers are trained and tested using this matrix.

5 Results and Analysis

The consistency of the result is checked by experimenting twice. First, the experiment is done using books and DVD reviews of Amazon (2000 pos and 2000 neg) and the hotel reviews from TripAdvisor (2000 pos and 2000 neg). Next, the experiment is done using electronics and kitchen appliances reviews from Amazon (2000 pos and 2000 neg) and the hotel reviews from TripAdvisor (2000 pos and 2000 neg). Each dataset pair is trained and tested. Amazon dataset is in xml format. Using regular expressions, reviews and labels (pos or neg) are extracted. Similarly, reviews are extracted from the TripAdvisor dataset, here the label is identified as positive if the rating (on a scale of 5) is 4 or 5 and negative if it is 1 or 2. The extracted reviews are preprocessed and common words are identified between the datasets for all the considered four models. Table 2 shows the number of common words after preprocessing in each model.

Table 2 Number of common words after preprocessing in each model

Dataset pair	Model 1	Model 2	Model 3	Model 4
Books, Dvd from Amazon, hotel from TripAdvisor	6889	3190	1818	1395
Electronics, kitchen appliances from Amazon, hotel from TripAdvisor	5254	2486	1444	1093

Table 3 Accuracy of classifiers—training Amazon and testing TripAdvisor

Classifiers	Model 1		Model 2		Model 3		Model 4	
	Dataset pair 1	Dataset pair 2	Dataset pair 1	Dataset pair 2	Dataset pair 1	Dataset pair 2	Dataset pair 1	Dataset pair 2
KNN (k = 15)	64.275	58.1	60.25	62.9	59.6	61.775	58.875	61.875
NB Gaussian	51.55	55.875	52.175	55.95	54.5	57.525	53.025	57.9
NB Multinomial	72.05	70.775	71.925	73.4	70.075	70.125	66.55	67.65
NB Bernoulli	69.35	73.25	68.45	74.275	69.8	68.575	64.25	65.125
DT	65.275	58.85	64.7	65.75	61.75	60.725	61.775	60.575
SVM rbf	66.85	66.275	65.8	65.375	62.225	60.725	59.275	58.325
SVM linear	70.65	68.15	68.675	70.7	66.15	70.125	61.65	65.85

Table 4 Accuracy of classifiers—training TripAdvisor and testing Amazon

Classifiers	Model 1		Model 2		Model 3		Model 4	
	Dataset pair 1	Dataset pair 2	Dataset pair 1	Dataset pair 2	Dataset pair 1	Dataset pair 2	Dataset pair 1	Dataset pair 2
KNN (k = 15)	60.625	63.225	59.875	61.85	57.25	59	56.875	56.625
NB Gaussian	53.55	54.15	53.675	55.35	51.975	53.225	51.75	53.275
NB Multinomial	58.925	66.8	65.725	72	64.65	69.075	63.175	67.3
NB Bernoulli	57.375	60.8	58.3	61.975	60.725	62.325	60.725	62.425
DT	64.275	68.95	61.75	64.325	58.65	59.45	57.45	56.9
SVM rbf	53.95	53.375	59.6	62.075	60.825	63.925	58.075	63.575
SVM linear	65.425	70.15	65.825	69.05	65.1	70.125	62.475	65.925

Review vs. Common word frequency matrix for both the datasets are constructed. Classification is done using the matrixes. One dataset is used for training and the other is used for testing. i.e. Classifier is trained with Amazon dataset and tested using TripAdvisor dataset and vice versa.

Table 3 shows the accuracy of classifiers when trained using Amazon dataset and tested with TripAdvisor dataset for each model. Table 4 shows the accuracy of classifiers when trained using TripAdvisor dataset and tested with Amazon dataset.

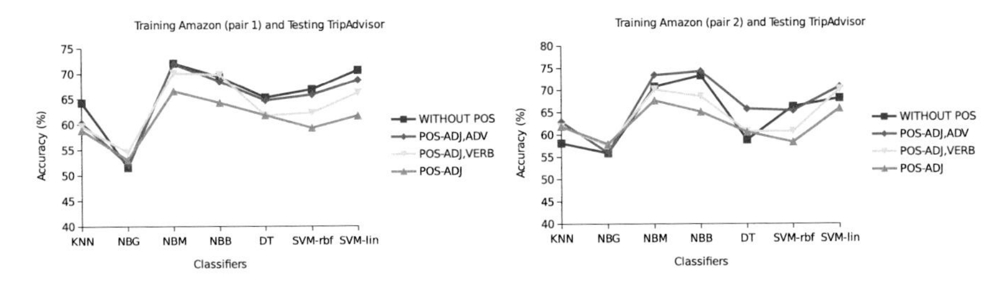

Fig. 3 Training with Amazon dataset and testing TripAdvisor dataset

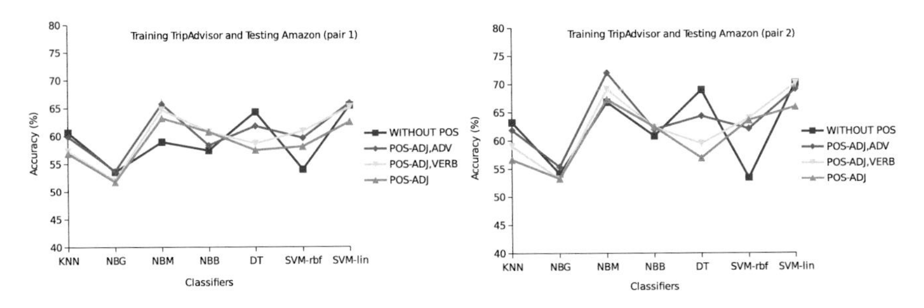

Fig. 4 Training with TripAdvisor dataset and testing Amazon dataset

From Figs. 3, 4 we observe that, for any model, NB Multinomial, SVM with linear kernel outperform the other classifiers with better accuracy. Performance of NB Bernoulli and DT are average, KNN, SVM with rbf kernel and NB Gaussian are poor with respect to accuracy.

Based on time,

1. With respect to models,
   Model 1 takes the highest time for training and testing, as it does not employ POS Tagging technique and so more common words between datasets, hence huge matrix and thus the impact in time. Model 2, 3, 4 take very less time compared to model 1.
2. With respect to classifiers,
   KNN takes the highest time to train and also to test the data irrespective of any model. All forms of NB, DT, SVM with linear kernel take almost the same time and they are the least compared to KNN and SVM with rbf kernel. SVM with rbf kernel takes more time compared to the above classifiers but considerably lesser time compared to KNN.

Considering both performance analysis and time analysis, we conclude that NB Multinomial, SVM with linear kernel are better than other classifiers for cross domain.

6 Conclusion

The overriding purpose of this analysis was to create an effective and efficient model for cross domain which can perform better with respect to computation and also space. To determine the best method, lot of models are considered and many pre-processing techniques are applied. Doing which determined, the model which incorporated POS Tagging performs better and Multinomial NB and SVM linear kernel classify the model very well compared to other classifiers. This work can be extended to big data with relatively large datasets. The results are promising with labelled dataset. The effectiveness of the same model on unstructured data can be worked upon.

References

1. Feldman, R.: Techniques and applications for sentiment analysis. Commun. ACM **56**(4), 82–89 (2013)
2. Dang, Y., Zhang, Y., Chen, H.: A lexicon-enhanced method for sentiment classification: an experiment on online product reviews. IEEE Intell. Syst. **25**(4), 46–53 (2010)
3. Bisio, F., Gastaldo, P., Peretti, C., Zunino, R., Cambria, E.: Data intensive review mining for sentiment classification across heterogeneous domains. In: IEEE/ACM International Conference on Advances in Social Networks Analysis and Mining (ASONAM), pp. 1061–1067, IEEE (2013)
4. Devroye, L., Wagner, T.J.: Nearest neighbor methods in discrimination. Handb. Stat. **2**, 193–197 (1982)
5. Li, T., Sindhwani, V., Ding, C., Zhang, Y.: Knowledge transformation for cross domain sentiment classification. In: Proceedings of the 32nd International ACM SIGIR Conference on Research and Development in Information Retrieval, pp. 716–717, ACM (2009)
6. Bollegala, D., Weir, D., Carroll, J.: Cross-domain sentiment classification using a sentiment sensitive thesaurus. IEEE Trans. Knowl. Data Eng. **25**(8), 1719–1731 (2013)
7. Jambhulkar, P., Nirkhi, S.: A survey paper on cross-domain sentiment analysis. Int. J. Adv. Res. Comput. Commun. Eng. **3**(1), 5241–5245 (2014)
8. Wang, H., Lu, Y., Zhai, C.: Latent aspect rating analysis on review text data: a rating regression approach. In: Proceedings of the 16th ACM SIGKDD International Conference on Knowledge Discovery and Data Mining, pp. 783–792, ACM (2010)
9. Cambria, E., Hussain, A.: Sentic computing: techniques, tools, and applications, vol. 2. Springer, Heidelberg (2012)
10. Cambria, E., Schuller, B., Xia, Y., Havasi, C.: New avenues in opinion mining and sentiment analysis. IEEE Intell. Syst. **2**, 15–21 (2013)
11. Pang, B., Lee, L.: Opinion mining and sentiment analysis. Found. Trends Inf. Retrieval **2**(1–2), 1–135 (2008)
12. Miller, G.A.: Wordnet: a lexical database for english. Commun. ACM **38**(11), 39–41 (1995)
13. Leoncini, A., Sangiacomo, F., Decherchi, S., Gastaldo, P., Zunino, R.: Semantic oriented clustering of documents. In: Advances in Neural Networks ISNN2011, pp. 523–529. Springer (2011)

Combining ELM with Random Projections for Low and High Dimensional Data Classification and Clustering

Abobakr Khalil Alshamiri, Alok Singh and Bapi Raju Surampudi

Abstract Extreme learning machine (ELM), as a new learning method for training feedforward neural networks, has shown its good generalization performance in regression and classification applications. Random projection (RP), as a simple and powerful technique for dimensionality reduction, is used for projecting high-dimensional data into low-dimensional subspaces while ensuring that the distances between data points are approximately preserved. This paper presents a systematic study on the application of RP in conjunction with ELM for both low- and high-dimensional data classification and clustering.

Keywords Extreme learning machine · Random projection · Classification · Clustering

1 Introduction

Extreme learning machine (ELM) is a new learning algorithm, proposed by Huang et al. [1, 2], inline with early proposals [3–5] for single-hidden layer feedforward neural networks (SLFNs). In ELM, the hidden layer parameters (input weights and biases) are randomly initialized and the output weights are analytically determined. Unlike the slow conventional gradient-based learning algorithms (such as back-propagation (BP)) for SLFNs, which require all parameters (weights and biases)

A.K. Alshamiri · A. Singh (✉) · B.R. Surampudi
School of Computer and Information Sciences, University of Hyderabad,
Hyderabad 500046, India
e-mail: alokcs@uohyd.ernet.in

A.K. Alshamiri
e-mail: abobakr2030@yahoo.com

B.R. Surampudi
Cognitive Science Lab, International Institute of Information Technology (IIIT),
Hyderabad 500032, India
e-mail: bapics@uohyd.ernet.in; raju.bapi@iiit.ac.in

V. Ravi et al. (eds.), *Proceedings of the Fifth International Conference on Fuzzy and Neuro Computing (FANCCO - 2015)*, Advances in Intelligent Systems and Computing 415, DOI 10.1007/978-3-319-27212-2_8

of the feedforward networks to be adjusted iteratively, ELM randomly assigns the hidden layer parameters and keeps them fixed during the learning process, and then determines the output weights analytically. The ELM hidden layer non-linearly transforms the input data into a high-dimensional space called ELM feature space. This transformation often renders the input data more separable in the ELM feature space, and hence simplifies the solution of the underlying or associated tasks. Due to its universal approximation capability (approximating any continuous target function) and classification capability (classifying any disjoint regions), extreme learning machine has been extensively used in regression and classification applications [2, 6]. Compared to traditional algorithms such as back-propagation (BP) and support vector machine (SVM), ELM not only provides good generalization performance, but also provides learning at extremely fast speed [2]. The good performance of ELM has encouraged researchers to integrate ELM method with SVM by replacing SVM kernel with normalized ELM kernel. Frénay and Verleysen [7] show that the integration of ELM with SVM has obtained good generalization performance. Recently, K-means algorithm has been combined with extreme learning approach to perform clustering in ELM feature space [8, 9]. This combination (ELM K-means) has been shown to obtain better clustering performance compared to classical K-means algorithm. Alshamiri et al. [10] incorporated the ELM method into an artificial bee colony (ABC) algorithm-based clustering approach to take advantage of the linear separability of the data in the ELM feature space and to overcome the shortcomings of ELM K-means algorithm.

Dimensionality reduction is the process of obtaining a low-dimensional representation of a given high-dimensional data. Generally, there are two types of dimension reduction techniques: feature selection and feature extraction. Feature selection techniques reduce the dimensions of data by finding a subset of relevant features from original dimensions. Feature extraction techniques reduce the dimensionality by constructing new dimensions either by using some transformations or by combining two or more original dimensions.

Random projection (RP) is a feature extraction-based dimensionality reduction technique which represents high-dimensional data in a low-dimensional space while approximately preserving the distances between the data points. The ability of random projection to approximately preserve the distances among the data points in the transformed space is an important property for machine learning applications in general. Fradkin and Madigan [11] have compared the performance of different machine learning methods such as nearest neighbor methods, decision trees and SVM in conjunction with the dimensionality reduction techniques RP and principle component analysis (PCA). In their work, PCA outperformed RP, but the computational overhead associated with PCA is high compared to RP. Deegalla and Bostrom [12] have first used RP and PCA for dimensionality reduction of data and then applied nearest neighbor classifier on the reduced data. Their experiments show that the performance of PCA depends heavily on the value we choose for the number of reduced dimensions. After reaching a peak, the accuracy for PCA decreases with the increase in number of dimensions. On the other hand, the accuracy for RP increases with the increase in number of dimensions. Fern and Brodley [13] investigated the application of RP integrated with ensemble methods for unsupervised learning of high-dimensional data. Cardoso and Wichert [14] proposed an iterative method for clustering high-dimensional data using RP and K-means algorithm. Recently, Gastaldo et al. [15] have proposed a new method which combines ELM with random projections. The performance of their method has been tested on two binary classification problems. The experiments demonstrated the ability of the proposed model to balance classification accuracy and computational complexity.

In this paper, we provide a systematic study on the application of RP in conjunction with ELM for both high- and low-dimensional data classification and clustering tasks. This study is based on the following intuition: the very fact that RP technique can do dimensionality reduction without distorting significantly the structure of data, can be used for different scenarios. First, for high-dimensional data, RP can be used to reduce the dimensionality by shrinking the number of neurons in the ELM hidden layer, as the solution may lie in a lower-dimensional subspace, and then we can use regression and K-means for classification and clustering, respectively. Second, for low-dimensional data, we use ELM to project the data into ELM feature space, where we expect the data to be linearly separable, and then we use RP to bring the data down to a lower space while preserving the linear separability of data. The second scenario tackles the problem of non-linear separability of data in the input space and reduces the computational cost of classification and clustering in the ELM feature space by projecting the data into a lower space while preserving the linear separability of the data.

The remainder of this paper is organized as follows: Sect. 2 describes the basics of the extreme learning machine (ELM). Section 3 introduces random projection (RP). In Sect. 4, the first scenario which combines ELM with RP for high-dimensional data classification and clustering is introduced. Section 5 presents the second scenario which combines ELM with RP for low-dimensional data classification and clustering. Experimental results and their analysis are presented in Sect. 6. Finally, Sect. 7 concludes this paper.

2 Extreme Learning Machine

ELM is recently proposed algorithm for training single-hidden layer feedforward neural networks (SLFNs). The algorithm randomly initializes the input weights of the networks and then determines the output weights analytically [1, 2, 16]. Research on SLFNs started long ago [3–5] and continues to attract attention. Unlike the slow gradient-based learning algorithms for SLFNs, which require all the parameters of the networks to be tuned iteratively, ELM requires no iteration when determining the SLFNs parameters as the hidden layer parameters are randomly initialized and remain fixed during the learning process and the output weights are analytically determined. ELM algorithm not only learns at extremely fast speed but also obtains good generalization performance [6]. For formally defining the ELM, we will follow the same notational convention as used in [2]. For instance, we are given a set of training examples $\aleph = \{(\mathbf{x}_i, \mathbf{t}_i) \mid \mathbf{x}_i \in R^d, \mathbf{t}_i \in R^m, i = 1, \ldots, N\}$, standard SLFNs with L hidden neurons and activation function $g(x)$, then the output of SLFNs can be represented as [2]:

$$\sum_{i=1}^{L} \beta_i g(\mathbf{w}_i \cdot \mathbf{x}_j + b_i) = \mathbf{y}_j, \ \ j = 1, \dots, N. \tag{1}$$

where $\mathbf{w}_i = [w_{i1}, \dots, w_{id}]^T$ and b_i are the input weight and the bias of the hidden neuron, respectively. $\beta_i = [\beta_{i1}, \dots, \beta_{im}]^T$ is the output weight. $\mathbf{w}_i \cdot \mathbf{x}_j$ is the inner product of $\mathbf{w}_i$ and $\mathbf{x}_j$. With that standard SLFNs, the parameters $\beta_i, i = 1, \dots, L$ can be estimated such that

$$\sum_{i=1}^{L} \beta_i g(\mathbf{w}_i \cdot \mathbf{x}_j + b_i) = \mathbf{t}_j, \ j = 1, \dots, N. \tag{2}$$

Equation (2) can be written as in [2]:

$$\mathbf{H}\boldsymbol{\beta} = \mathbf{T} \tag{3}$$

where

$$\mathbf{H} = \begin{bmatrix} g(\mathbf{w}_1 \cdot \mathbf{x}_1 + b_1) & \dots & g(\mathbf{w}_L \cdot \mathbf{x}_1 + b_L) \\ \vdots & \dots & \vdots \\ g(\mathbf{w}_1 \cdot \mathbf{x}_N + b_1) & \dots & g(\mathbf{w}_L \cdot \mathbf{x}_N + b_L) \end{bmatrix}_{N \times L} \tag{4}$$

$$\boldsymbol{\beta} = \begin{bmatrix} \beta_1^T \\ \vdots \\ \beta_L^T \end{bmatrix}_{L \times m} \quad \text{and} \quad \mathbf{T} = \begin{bmatrix} \mathbf{t}_1^T \\ \vdots \\ \mathbf{t}_N^T \end{bmatrix}_{N \times m} \tag{5}$$

$\mathbf{H}$ is the hidden layer output matrix of the network [17]. The output weights $\boldsymbol{\beta}$ can be computed from $\mathbf{H}$ and $\mathbf{T}$ by using a Moore-Penrose generalized inverse of $\mathbf{H}$, $\mathbf{H}^\dagger$ [18]. The illustration of the primal ELM architecture with L hidden neurons is shown in Fig. 1, which is adapted from [19]. The main steps of the ELM algorithm are given in Algorithm 1.

Algorithm 1: ELM Algorithm

input : A data set $\aleph = \{(\mathbf{x}_i, \mathbf{t}_i) \mid \mathbf{x}_i \in R^d, \mathbf{t}_i \in R^m, i = 1, \dots, N\}$, number of hidden neurons L, and activation function $g(x)$

1 Initialize hidden layer parameters $(\mathbf{w}_i, b_i)$, $i = 1, \dots, L$ randomly;
2 Compute the hidden layer output matrix $\mathbf{H}$;
3 Compute the output weight $\boldsymbol{\beta}$: $\boldsymbol{\beta} = \mathbf{H}^\dagger \mathbf{T}$;

In ELM theory, the number of neurons in the hidden layer of the ELM should be large enough to achieve good generalization performance. A detailed discussion on hidden neurons selection in particular and ELM in general can be found in [2, 20, 21].

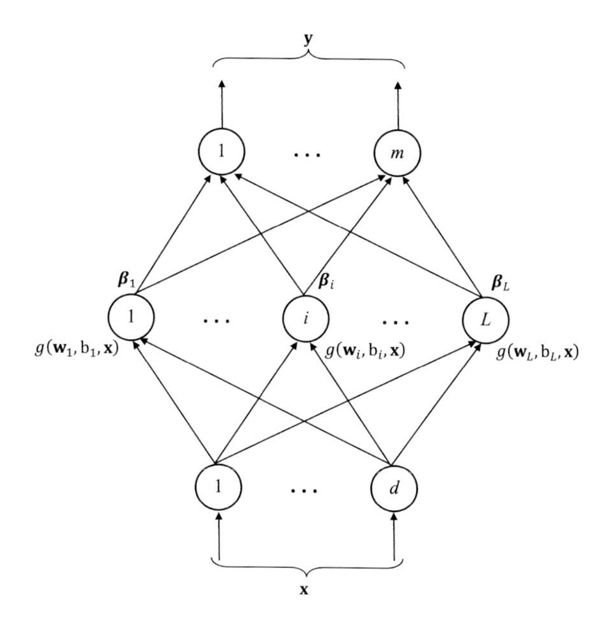

Fig. 1 ELM network

3 Random Projection

Random projection, as a simple and powerful technique for dimensionality reduction, is used to project high-dimensional data into low-dimensional subspace while approximately preserving the distances between the points.

Suppose we are given a data set $\mathbf{X} = \{(\mathbf{x}_i) \mid \mathbf{x}_i \in R^d, i = 1, \dots, N\}$, and a random $d \times r$ matrix $\mathbf{R}$, then the original d-dimensional data $\mathbf{X}$ is projected into r-dimensional ($r \ll d$) subspace using a matrix $\mathbf{R}$ as follows:

$$\mathbf{X}^{rp} = \mathbf{X}\mathbf{R} \tag{6}$$

where $\mathbf{X}^{rp}$ is the projection of the data into a lower r-dimensional subspace. The idea of random projection originates from the Johnson-Lindenstrauss (JL) lemma [22]. The JL lemma states that if we are given a set of N points in a d-dimensional space, then we can project these points down into r-dimensional subspace, with $r \geq O(\ln(N)/\epsilon^2)$, so that all pairwise distances are preserved up to a $1 \pm \epsilon$ factor, where $\epsilon \in (0, 1)$. Several algorithms have been proposed to construct a random matrix that satisfies the JL lemma [23]. In this paper, we use a matrix whose entries are independent realizations of ± 1 Bernoulli random variables [24]. Hence, The elements r_{ij} of the matrix $\mathbf{R}$ are assigned as follows:

$$r_{ij} = \begin{cases} +1/\sqrt{r} & \text{with probability } 0.5 \\ -1/\sqrt{r} & \text{with probability } 0.5 \end{cases} \tag{7}$$

In general, the elements of a matrix that maps the data from high-dimensional space to a low-dimensional subspace and satisfies the JL lemma can be assigned as follows:

$$\begin{cases} +1/\sqrt{\text{reduced dimension}} & \text{with probability } 0.5 \\ -1/\sqrt{\text{reduced dimension}} & \text{with probability } 0.5 \end{cases} \tag{8}$$

4 High-Dimensional Data Classification and Clustering Problems

4.1 *Classification of High-Dimensional Data*

In ELM network, the hidden layer maps the data from the input space R^d to the ELM feature space R^L. This ELM mapping performs either dimensionality reduction if $L < d$ or dimensionality expansion if $L > d$. The ELM algorithm (without combination with random projection) for the classification of high-dimensional data is shown in Algorithm 1, in which $L < d$. The integration of RP with ELM (ELM^{rp}) [15] for high-dimensional data classification is shown in Algorithm 2. In ELM^{rp} algorithm, the input weights and biases of the ELM hidden layer are initialized using Eq. (8) so that the mapping satisfies the JL lemma.

Algorithm 2: ELM^{rp} Algorithm

input : A data set $\aleph = \{(\mathbf{x}_i, \mathbf{t}_i) \mid \mathbf{x}_i \in R^d, \mathbf{t}_i \in R^m, i = 1, \ldots, N\}$, number of hidden neurons L, $L < d$, and activation function $g(x)$

1 Initialize hidden layer parameters $(\mathbf{w}_i, b_i)$, $i = 1, \ldots, L$ using Eq. (8);
2 Compute the hidden layer output matrix $\mathbf{H}$;
3 Compute the output weight $\boldsymbol{\beta} : \boldsymbol{\beta} = \mathbf{H}^{\dagger}\mathbf{T}$;

4.2 *Clustering of High-Dimensional Data*

Classical distance-based clustering algorithms, such as K-means algorithm, for low-dimensional spaces cannot produce clusters that are meaningful in high-dimensional data because several features are irrelevant and clusters are usually hidden in different low-dimensional subspaces [25, 26]. To enable conventional clustering algorithms to cluster high-dimensional data efficiently and effectively, the data is first projected into a low-dimensional space and then clustering is applied on the projected points. The ELM hidden layer perform this projection by setting the number of hidden neurons less than the number of input dimensions. In Algorithm 3, ELM is incorporated into K-means algorithm (ELM K-means) for data clustering. For clustering high-dimensional data, we set $L < d$ in Algorithm 3. In Algorithm 4, ELM combined with RP is first used to reduce the dimensionality of the data while approximately preserving the distances between the data points and then K-means algorithm performs clustering in the reduced space (ELM^{rp} K-means).

Algorithm 3: ELM K-means Algorithm

input : A data set $\aleph = \{(\mathbf{x}_i, \mathbf{t}_i) \mid \mathbf{x}_i \in R^d, \mathbf{t}_i \in R^m, i = 1, \ldots, N\}$, number of hidden neurons L, and activation function $g(x)$

1 Initialize hidden layer parameters $(\mathbf{w}_i, b_i)$, $i = 1, \ldots, L$ randomly;
2 Compute the hidden layer output matrix $\mathbf{H}$;
3 Apply K-means algorithm on the hidden layer output matrix $\mathbf{H}$;

Algorithm 4: ELM^{rp} K-means Algorithm

input : A data set $\aleph = \{(\mathbf{x}_i, \mathbf{t}_i) \mid \mathbf{x}_i \in R^d, \mathbf{t}_i \in R^m, i = 1, \ldots, N\}$, number of hidden neurons L, $L < d$, and activation function $g(x)$

1 Initialize hidden layer parameters $(\mathbf{w}_i, b_i)$, $i = 1, \ldots, L$ using Eq. (8);
2 Compute the hidden layer output matrix $\mathbf{H}$;
3 Apply K-means algorithm on the hidden layer output matrix $\mathbf{H}$;

5 Low-Dimensional Data Classification and Clustering Problems

5.1 *Classification of Low-Dimensional Data*

For low-dimensional data classification, ELM hidden layer maps the data from the d-dimensional input space to the L-dimensional ELM feature space, where $L > d$. By mapping the data to high-dimensional ELM feature space, the linear separability of the data often increases within the transformed space. The ELM algorithm (without combination with RP) for the classification of low-dimensional data is shown in Algorithm 1.

5.1.1 ELM Combined with RP (ELM-RP)

The ELM projection of the low-dimensional data into high-dimensional ELM feature space has advantage of increasing the linear separability of the data points in the ELM space but performing classification or clustering in this space is expensive due to the increase in dimensions. To tackle this problem, we propose to perform dimensionality reduction using RP on ELM feature space and then perform classification or clustering in the reduced space. The proposed approach has two advantages: first the linear separability of the data in ELM feature space is preserved in the reduced space, second the computational complexity of performing classification or clustering is reduced.

The network architecture of the proposed approach consists of two hidden layers instead of one as in the ELM network shown in Fig. 1. The network architecture of the proposed approach is similar to the ELM network shown in Fig. 1 with an extra hidden layer. The first hidden layer contains L neurons, whereas the second hidden layer contains r neurons with $r < L$. The second hidden layer output matrix $\mathbf{H}'$ is calculated as follows: from Eq. (4) $\mathbf{H} = \left[\mathbf{h}(\mathbf{x}_1), \ldots, \mathbf{h}(\mathbf{x}_N)\right]^T$ where $\mathbf{h}(\mathbf{x}) = \left[g(\mathbf{w}_1 \cdot \mathbf{x} + b_1), \ldots, g(\mathbf{w}_L \cdot \mathbf{x} + b_L)\right]$ then

$$\mathbf{H}' = \begin{bmatrix} g(\mathbf{w}'_1 \cdot \mathbf{h}(\mathbf{x}_1) + b'_1) & \ldots & g(\mathbf{w}'_r \cdot \mathbf{h}(\mathbf{x}_1) + b'_r) \\ \vdots & \ldots & \vdots \\ g(\mathbf{w}'_1 \cdot \mathbf{h}(\mathbf{x}_N) + b'_1) & \ldots & g(\mathbf{w}'_r \cdot \mathbf{h}(\mathbf{x}_N) + b'_r) \end{bmatrix}_{N \times r} \tag{9}$$

where $\mathbf{w}'_j = [w_{j1}, \ldots, w_{jL}]^T$ is the weight vector connecting the jth neuron in the second hidden layer and the neurons in the first hidden layer, and b'_j is the bias of the jth neuron in the second hidden layer.

The integration of RP with ELM (ELM-RP) for low-dimensional data classification is shown in Algorithm 5.

Algorithm 5: ELM-RP Algorithm

---

input : A data set $\aleph = \{(\mathbf{x}_i, \mathbf{t}_i) \mid \mathbf{x}_i \in R^d, \mathbf{t}_i \in R^m, i = 1, \ldots, N\}$, number of first hidden layer neurons L, $L > d$, number of second hidden layer neurons r, $r < L$, and activation function $g(x)$

1 Initialize first hidden layer parameters $(\mathbf{w}_i, b_i)$, $i = 1, \ldots, L$ randomly;
2 Calculate the first hidden layer output matrix $\mathbf{H}$;
3 Initialize second hidden layer parameters $(\mathbf{w}'_j, b'_j)$, $j = 1, \ldots, r$ using Eq. (8);
4 Compute the second hidden layer output matrix $\mathbf{H}'$;
5 Compute the output weight $\boldsymbol{\beta}' : \boldsymbol{\beta}' = \mathbf{H}'^{\dagger}\mathbf{T}$;

5.2 Clustering of Low-Dimensional Data

The data points which are not linearly separable in the input space often become linearly separable after mapping the data to high-dimensional ELM feature space, thereby improving the quality of clustering. The algorithm which integrates ELM method with K-means algorithm for clustering low-dimensional data is the same as Algorithm 3.

5.2.1 ELM Combined with RP (ELM-RP K-Means).

As it has been mention in Sect. 5.1.1, the data points which are non-linearly separable in the input space often become linearly separable after projecting them into high-dimensional ELM feature space. To avoid the computational cost of performing clustering in the high-dimensional ELM feature space, we use RP to reduce the dimensionality of data in ELM feature space and then we apply K-means algorithm in the reduced space.

In Algorithm 6, the low-dimensional data is first projected to the high-dimensional ELM feature space, then RP is used to reduce the dimensionality of ELM feature space while preserving the linear separability of the data and finally clustering using K-means is performed in the reduced space.

Algorithm 6: ELM-RP K-means Algorithm

input : A data set $\aleph = \{(\mathbf{x}_i, \mathbf{t}_i) \mid \mathbf{x}_i \in R^d, \mathbf{t}_i \in R^m, i = 1, \ldots, N\}$, number of first hidden layer neurons L, $L > d$, number of second hidden layer neurons r, $r < L$, and activation function $g(x)$

1. Initialize first hidden layer parameters $(\mathbf{w}_i, b_i)$, $i = 1, \ldots, L$ randomly;
2. Calculate the first hidden layer output matrix $\mathbf{H}$;
3. Initialize second hidden layer parameters $(\mathbf{w}'_j, b'_j)$, $j = 1, \ldots, r$ using Eq. (8);
4. Compute the second hidden layer output matrix $\mathbf{H}'$;
5. Apply K-means algorithm on the second hidden layer output matrix $\mathbf{H}'$;

6 Experimental Study

In this study, 8 high-dimensional data sets and 12 low-dimensional data sets have been used to evaluate the performance of the ELM combined with RP approach for classification and clustering. Here it is pertinent to mention that there is no consistency in the literature about the minimum number of dimensions that renders a data set high-dimensional. Data with as few as 10 dimensions are regarded as high-dimensional in some works. On the other hand, numerous other works, specially those pertaining to image processing or bio-informatics have hundreds or thousands of dimensions [26]. In this paper, we consider data sets with thousand or more dimensions as high-dimensional data.

6.1 Data Sets

The data sets used in this study are described as follows.

- *High-dimensional data sets:* The Columbia Object Image Library (COIL20) data set contains gray-scale images of 20 objects, where each object has 72 images.

Colon data set consists of 62 samples, where each sample has 2000 genes. The data set contains 22 normal and 40 tumor tissue samples. Global Cancer Map (GCM) data set consists of gene expression data for 190 tumor samples and 90 normal tissue samples. Leukemia data set contains 72 samples. Each sample has 7129 genes. The samples are either acute lymphoblastic leukemia (ALL) or acute myeloid leukemia (AML). Lung cancer data set consists of gene expression data for 181 samples. The samples are either malignant pleural mesothe-lioma (MPM) or adenocarcinoma (ADCA). ORL data set contains face images of 40 distinct subjects, where each subject has 10 different images. Prostate cancer data set contains 136 tissue samples among which 77 samples are tumor and 59 are normal. Each sample is described by 12600 genes. Yale is a face recognition data set which contains face images of 15 individuals. Each individual has 11 images.

- *Low-dimensional data sets:* Balance data set was generated to model psychological experimental results. There are three classes and the patterns are classified based on the position of the balance scale; tip to the left, tip to the right, or remain exactly in the middle. Cancer-Diagnostic data set is based on the "breast cancer Wisconsin - Diagnostic" data set and Cancer-Original data set is based on the "breast cancer Wisconsin - Original" data set. The patterns in both data sets belong to two different classes, benign and malignant tumors. Each pattern in both data sets is classified as either benign or malignant tumor. Cancer-Original data sat contains 699 patterns. After removing the 16 database patterns with missing values, the database consists of 683 patterns. Cardiotocography data set composes of measurements of fetal heart rate (FHR) and uterine contraction (UC) features on cardiotocograms. Classes of this data set were formed based on the opinions of expert obstetricians. These experts classify the data in two ways: a morphologic pattern (10 classes; 1–10) and a fetal state (three classes; normal, suspect and pathologic). Therefore the data set can be used either for 10-class (Cardiotocography-10) or 3-class (Cardiotocography-3) experiments. CNAE data set contains 1080 text documents. The documents describe the business of Brazilian companies grouped into 9 classes. Each text document is pre-processed to be represented as a vector. The values of the vector correspond to the frequency of the words in the document. The data set is very sparse as 99.22 % of the matrix values are zeros. Dermatology data set aims to determine the type of Eryhemato-Squamous Disease. The data set contains 366 patterns. After the removal of the 8 data set patterns with missing values, the data set consists of 358 34-dimensional patterns belonging to six different classes. Glass data set contains 6 glass types. The glass types are building windows-float processed, building windows-not float processed, containers, vehicle windows, head lamps and tableware . The Fisher Iris data [27] is one of the most popular data sets to test the performance of novel methods in pattern recognition and machine learning. There are three classes in this data set (Setosa, Versicolor and Virginica), each having 50 patterns with four features (sepal length, sepal width, petal length and petal width). One of the classes (viz. Setosa) is linearly separable from the other two, while the remaining two are not linearly separable. LIBRAS data set is based on the Libras Movement data set. The data set contains 15 classes of 24 patterns each. Each class references to a hand movement

Table 1 Specifications of high-dimensional data sets

Data sets	Patterns	Classification		Features	Classes
		Training	Testing		
COIL20	1440	1008	432	1024	20
Colon	62	30	32	2000	2
GCM	280	196	84	16063	2
Leukemia	72	38	34	7129	2
Lung	181	32	149	12533	2
ORL	400	280	120	1024	40
Prostate	136	102	34	12600	2
YALE	165	115	50	1024	15

Table 2 Specifications of low-dimensional data sets

Data sets	Patterns	Classification		Features	Classes
		Training	Testing		
Balance	625	475	150	4	3
Cancer-diagnostic	569	427	142	30	2
Cancer-original	683	512	171	9	2
Cardiotocography-3	2126	1595	531	21	3
Cardiotocography-10	2126	1595	531	21	10
CNAE	1080	810	270	856	9
Dermatology	358	270	88	34	6
Glass	214	142	72	9	6
Iris	150	100	50	4	3
LIBRAS	360	270	90	90	15
Spam	1534	1150	384	57	2
USPST	2007	1505	502	256	10

type in LIBRAS (Portuguese name 'LÍngua BRAsileira de Sinais', oficial brazilian sign language). The Spam data set consists of 1,534 patterns from two different classes, spam and not-spam. The USPST data set is a subset (the testing set) of the well-known handwritten digit recognition data set USPS.

The specifications of the high-dimensional data sets and the low-dimensional data sets have been provided in Tables 1 and 2, respectively. It is to be noted that in this paper, we consider data sets with thousand or more dimensions as high-dimensional data.

6.2 Results and Discussion

As mentioned already, the performance of ELM combined with RP approach for classification/clustering is evaluated on 8 High-dimensional and 12 Low-dimensional data sets. We have used the sigmoid function for non-linear mapping, and the input weights and biases were generated either from a uniform distribution over $[-1, 1]$ or using RP technique (i.e. Eq. (8)). All the simulations are carried out 20 times independently and the average performance is reported. For classification problems, the training and testing sets are randomly generated from the whole data set at each run according to Tables 1 and 2.

Figure 2 shows the classification performance of ELM and ELM^{rp} algorithms on different high-dimensional data sets with respect to different number of hidden neurons L. For each data set in Fig. 2, the highest value of L is equivalent to the compression ratio 1:20 in the feature-projection stage $L < d$. From the results shown in Fig. 2, it is obvious that ELM^{rp} algorithm outperformed the ELM algorithm consistently on all data sets. It can also be observed that the classification accuracy of ELM^{rp} in most cases increases gradually with the number of dimensions (i.e. neurons).

Figure 3 shows the clustering performance of ELM K-means and ELM^{rp} K-means on different high-dimensional data sets with respect to different number of hidden neurons L. From Fig. 3, ELM^{rp} K-means algorithm has obtained satisfying results on all data sets and outperformed ELM K-means on six out of the eight data sets. Compared to K-means clustering in the input space, ELM^{rp} K-means algorithm obtains similar or better results on four out of the eight data sets. The time required for ELM^{rp} K-means to perform clustering is significantly reduced. For example the time taken by K-means algorithm to perform clustering on COIL20 dataset in the input space is 5.84 s, whereas ELM^{rp} K-means algorithm with $L = 100$ takes 0.38 s. In addition, for RP-based clustering technique to obtain a high accuracy, a large number of dimensions is required. For example ELM^{rp} K-means algorithm obtains clustering accuracy of 62.9514 and 72.2222 on COIL20 and Leukemia data sets at $L = 200$ and $L = 700$, respectively. Thus, ELM^{rp} K-means algorithm provides a trade-off between the clustering accuracy and computational complexity.

We emphasize again that for high-dimensional data classification and clustering, the number of hidden neurons L in ELM and ELM K-means algorithms is less than the number of input dimensions d.

Figure 4 shows the classification performance of ELM and ELM-RP algorithms on different low-dimensional data sets with respect to different number of neurons. In Fig. 4, the ELM performs classification directly in the ELM feature space with different number of hidden neurons $L = [10, 20, \dots, 100]$. While for ELM-RP, the data sets are first projected to high-dimensional space $L = 1000$ and then reduced using RP to different number of neurons $r = [10, 20, \dots, 100]$. $\text{ELM}^{L=1000}$ refers to ELM algorithm performing classification in ELM feature space with number of hidden neurons $L = 1000$. From Fig. 4, we can observe that both ELM and ELM-RP algorithms have relatively similar performance on all data sets with different number of neurons $[10, 20, \dots, 100]$. Even though ELM-RP algorithm is perform-

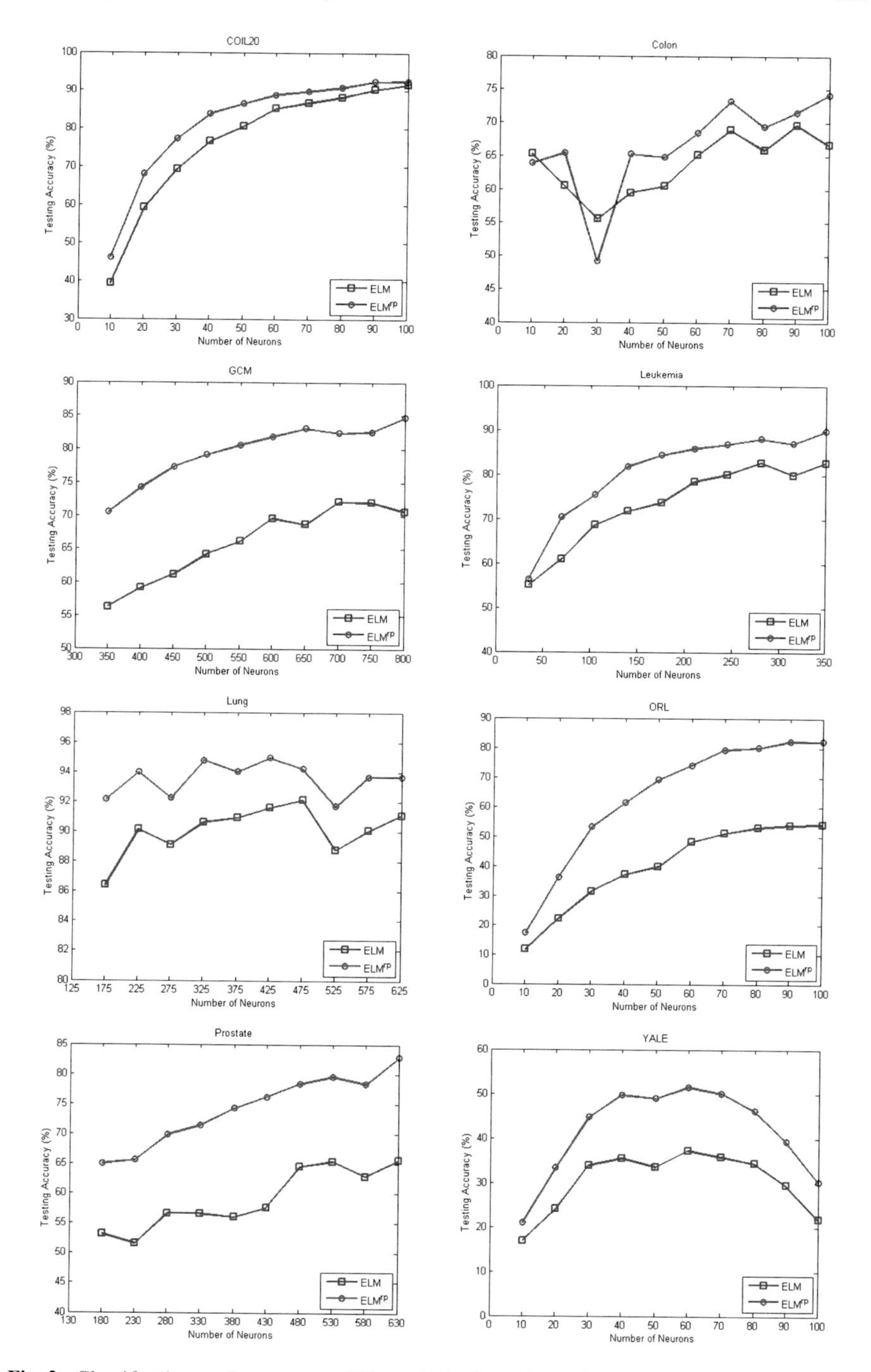

Fig. 2 Classification performance on different high-dimensional data sets

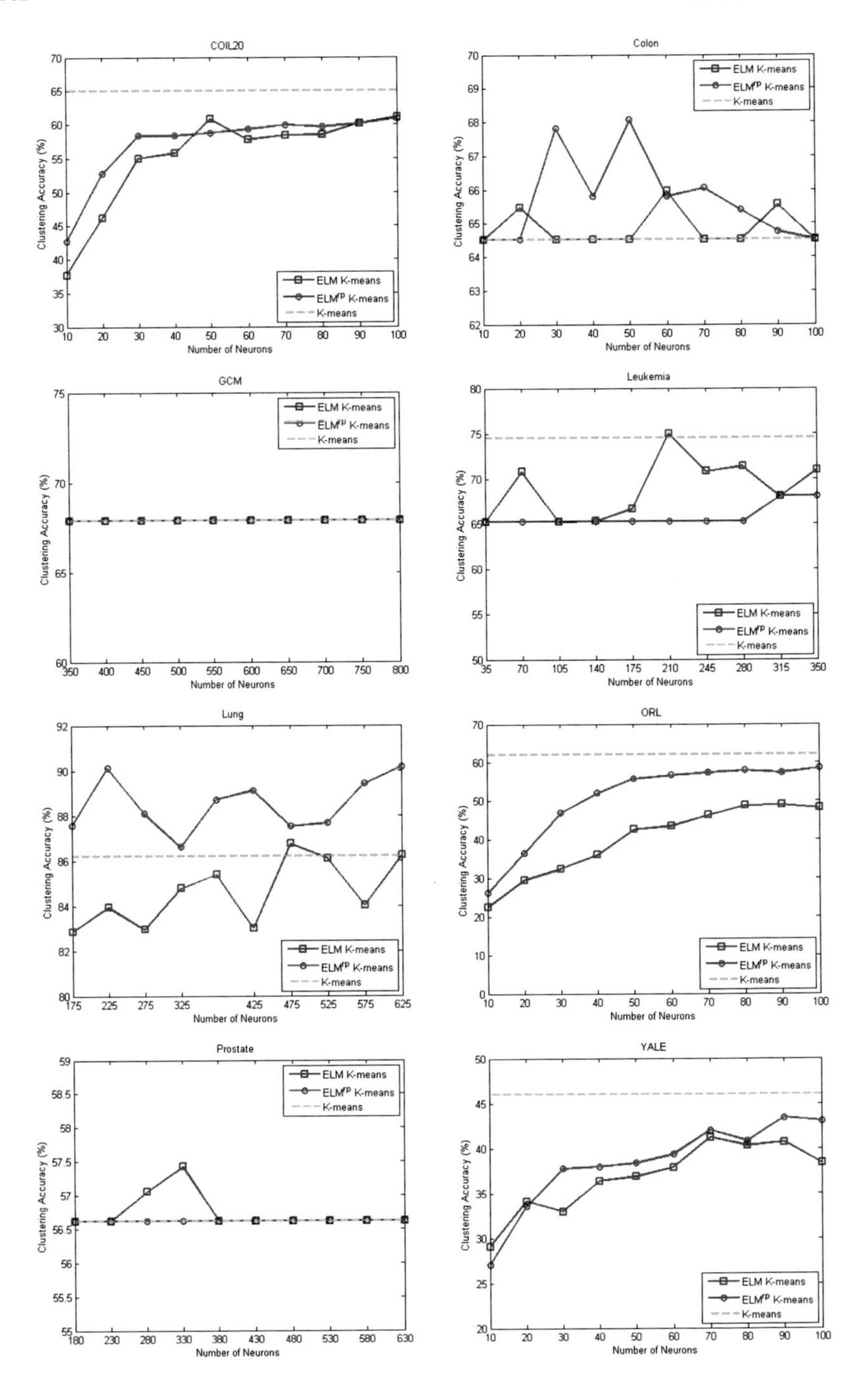

Fig. 3 Clustering performance on different high-dimensional data sets

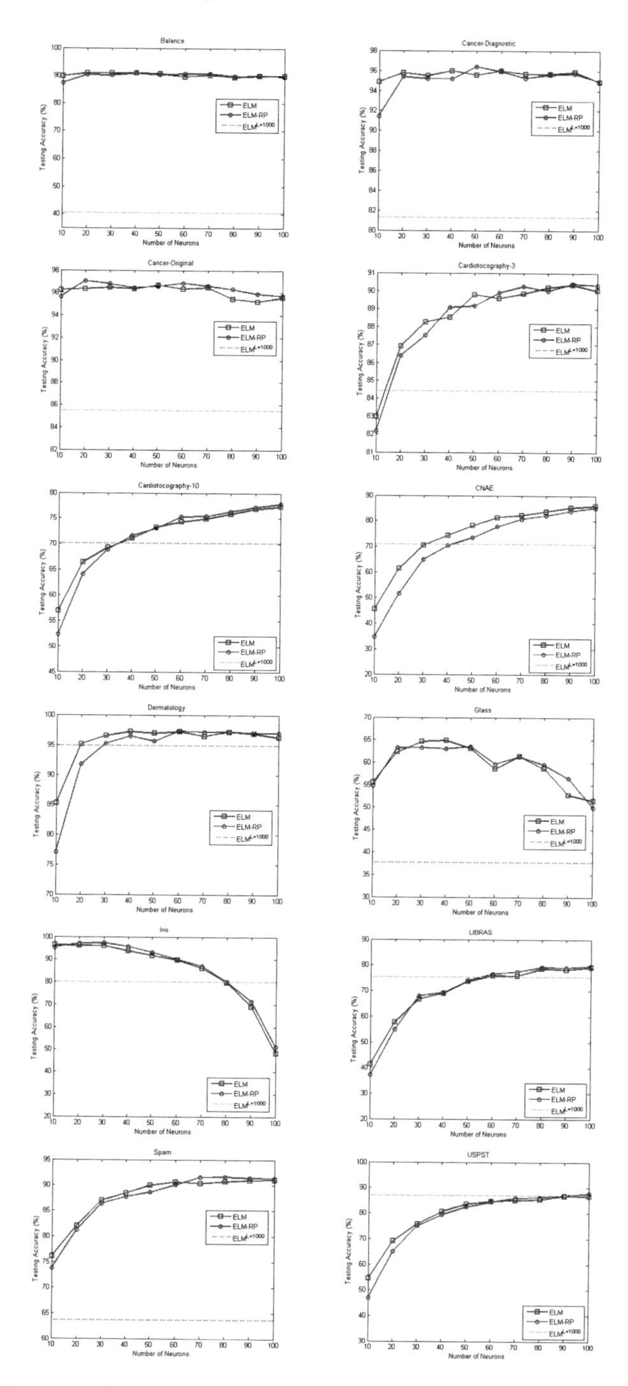

Fig. 4 Classification performance on different low-dimensional data sets

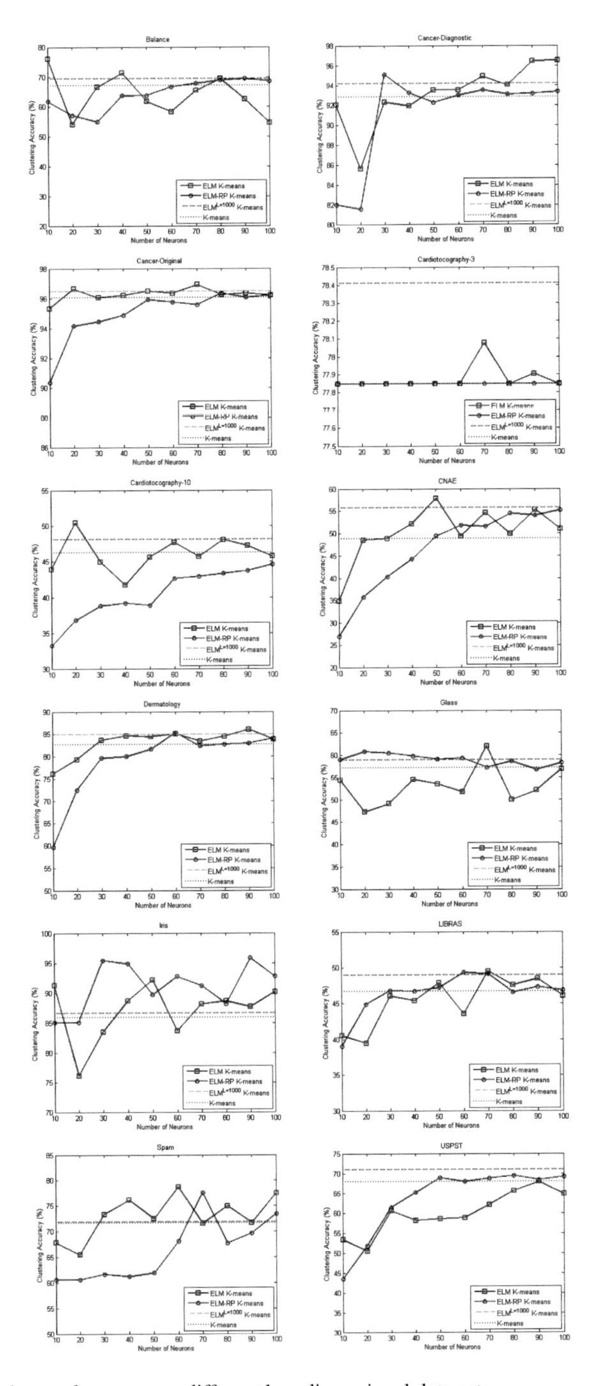

Fig. 5 Clustering performance on different low-dimensional data sets

ing classification in the reduced space of ELM feature space, it is capable of taking advantage of linear separability of data in the high-dimensional ELM space, where $L = 1000$, and avoid over-fitting. It has been observed that $\mathrm{ELM}^{L=1000}$ algorithm obtains excellent training accuracy, but fails to generalize to unseen data. The performance of $\mathrm{ELM}^{L=1000}$ can be much improved by using regularization technique [6].

Figure 5 shows the clustering performance of ELM K-means and ELM-RP K-means on different low-dimensional data sets with respect to different number of neurons. In Fig. 5, ELM-RP K-means algorithm performs clustering in the reduced space of the high-dimensional ELM space, where $L = 1000$ in ELM space. ELM K-means performs clustering directly in the ELM feature space with different number of hidden neurons $L = [10, 20, \ldots, 100]$. $\mathrm{ELM}^{L=1000}$ K-means refers to K-means algorithm performing clustering in ELM feature space with number of hidden neurons $L = 1000$. Figure 5 shows that ELM-RP K-means algorithm obtains comparable results with ELM K-means on most of the data sets and tends to obtain high accuracy with large number of neurons. It is also observed that ELM-RP K-means achieves satisfying results compared to $\mathrm{ELM}^{L=1000}$ K-means which indicates that ELM-RP K-means algorithm has the ability to balance clustering quality and computational complexity.

7 Conclusion

This paper delivers a systematic investigation on the application of RP in conjunction with ELM for both low and high-dimensional data classification and clustering. The capability of RP technique in performing dimensionality reduction without distorting significantly the structure of data has been investigated through different scenarios. The first scenario is for high-dimensional data where RP is used to reduce the dimensionality by shrinking the number of neurons in the ELM hidden layer and then classification or clustering is performed in the reduced space. The second scenario is for low-dimensional data in which the data is first projected to the high-dimensional ELM feature space, where the data is expected to be linearly separable and then RP is used to reduce the dimensions while preserving the linear separability of data. The experiments show that the integration of ELM with RP for data classification and clustering not only obtains satisfying results, but also provides a trade-off between generalization performance and computational complexity.

References

1. Huang, G.B., Zhu, Q.Y., Siew, C.K.: Extreme learning machine: a new learning scheme of feedforward neural networks. In: Proceedings of International Joint Conference on Neural Networks (IJCNN), Budapest, Hungary, vol. 2, pp. 985–990 (2004)

2. Huang, G.B., Zhu, Q.Y., Siew, C.K.: Extreme learning machine: theory and applications. Neurocomputing **70**, 489–501 (2006)
3. Broomhead, D.S., Lowe, D.: Multivariable functional interpolation and adaptive networks. Complex Syst. **2**(3), 321–355 (1988)
4. Schmidt, W.F., Kraaijveld, M.A., Duin, R.P.W.: Feedforward neural networks with random weights. In: Proceedings of 11th IAPR International Conference on Pattern Recognition Methodology and Systems, Hague, Netherlands, pp. 1–4 (1992)
5. Pao, Y.H., Park, G.H., Sobajic, D.J.: Learning and generalization characteristics of random vector functional-link net. Neurocomputing **6**, 163–180 (1994)
6. Huang, G.B., Zhou, H., Ding, X., Zhang, R.: Extreme learning machine for regression and multiclass classification. IEEE Trans. Syst. Man Cybern. B Cybern. **42**, 513–529 (2012)
7. Frénay, B., Verleysen, M.: Using SVMs with randomised feature spaces: an extreme learning approach. In: Proceedings of 18th European Symposium on Artificial Neural Networks (ESANN), Bruges, Belgium, pp. 315–320 (2010)
8. He, Q., Jin, X., Du, C., Zhuang, F., Shi, Z.: Clustering in extreme learning machine feature space. Neurocomputing **128**, 88–95 (2014)
9. Alshamiri, A.K., Singh, A., Surampudi, B.R.: A novel ELM K-means algorithm for clustering. In: Proceedings of 5th International Conference on Swarm, Evolutionary and Memetic Computing (SEMCCO), Bhubaneswar, India, pp. 212–222 (2014)
10. Alshamiri, A.K., Singh, A., Surampudi, B.R.: Artificial bee colony algorithm for clustering: an extreme learning approach. Soft Comput. (2015)
11. Fradkin, D., Madigan, D.: Experiments with random projections for machine learning. In: Proceedings of the 9th ACM SIGKDD International Conference on Knowledge Discovery and Data Mining, pp. 517–522 (2003)
12. Deegalla, S., Bostrom, H.: Reducing high-dimensional data by principal component analysis vs. random projection for nearest neighbor classification. In: Proceedings of the 5th International Conference on Machine Learning and Applications (ICMLA), Orlando, Fl, pp. 245–250 (2006)
13. Fern, X.Z., Brodley, C.E.: Random projection for high dimensional data clustering: a cluster ensemble approach. In: Proceedings of the 20th International Conference of Machine Learning (ICML), Washington DC, pp. 186–193 (2003)
14. Cardoso, A., Wichert, A.: Iterative random projections for high-dimensional data clustering. Pattern Recogn. Lett. **33**, 1749–1755 (2012)
15. Gastaldo, P., Zunino, R., Cambria, E., Decherchi, S.: Combining ELM with random projections. IEEE Intell. Syst. **28**(6), 18–20 (2013)
16. Huang, G.B., Chen, L., Siew, C.K.: Universal approximation using incremental constructive feedforward networks with random hidden nodes. IEEE Trans. Neural Networks **17**(4), 879–892 (2006)
17. Huang, G.B.: Learning capability and storage capacity of two-hidden-layer feedforward networks. IEEE Trans. Neural Networks **14**(2), 274–281 (2003)
18. Serre, D.: Matrices: Theory and Applications. Springer, NewYork (2002)
19. Huang, G.B., Ding, X., Zhou, H.: Optimization method based extreme learning machine for classification. Neurocomputing **74**, 155–163 (2010)
20. Lan, Y., Soh, Y.C., Huang, G.B.: Constructive hidden nodes selection of extreme learning machine for regression. Neurocomputing **73**, 3191–3199 (2010)
21. Zhu, Q.Y., Qin, A.K., Suganthan, P.N., Huang, G.B.: Evolutionary extreme learning machine. Pattern Recognit. **38**, 1759–1763 (2005)
22. Johnson, W.B., Lindenstrauss, J.: Extensions of Lipschitz mappings into a Hilbert space. In: Conference in Modern Analysis and Probability, pp. 189–206 (1984)
23. Achlioptas, D.: Database-friendly random projections. In: Proceedings of the 20th Annual Symposium on the Principles of Database Systems, pp. 274–281 (2001)
24. Baraniuk, R., Davenport, M., DeVore, R., Wakin, M.: A simple proof of the restricted isometry property for random matrices. Constr. Approximat. **28**, 253–263 (2008)

25. Bouveyron, C., Girard, S., Schmid, C.: High dimensional data clustering. Comput. Stat. Data Anal. **52**, 502–519 (2007)
26. Assent, I.: Clustering high dimensional data. Wiley Interdisc. Rev. Data Min. Knowl. Discovery **2**(4), 340–350 (2012)
27. Fisher, R.A.: The use of multiple measurements in Taxonomic problems. Ann. Eugenics **7**, 179–188 (1936)

Framework for Knowledge Driven Optimisation Based Data Encoding for Brain Data Modelling Using Spiking Neural Network Architecture

Neelava Sengupta, Nathan Scott and Nikola Kasabov

Abstract From it's initiation, the field of artificial intelligence has been inspired primarily by the human brain. Recent advances and collaboration of computational neuroscience and artificial intelligence has led to the development of spiking neural networks (SNN) that can very closely mimic behaviour of the human brain. These networks use spike codes or synaptic potentials as a source of information. On the contrary, most of the real world data sources including brain data are continuous and analogue in nature. For an SNN based pattern recognition tool, it is imperative to have a data encoding mechanism that transforms the streaming analogue spatiotemporal information to train of spikes. In this article, we propose a generalised background knowledge-driven optimisation framework for encoding brain data (fMRI, EEG and others). Further, we also formalise and implement a mixed-integer genetic algorithm based optimisation for background knowledge-driven data encoding for fMRI data and compare the performance with existing data encoding method like temporal contrast and Ben Spiker Algorithm.

Keywords Spike encoding · Spiking neural network · Mixed integer optimisation · Neucube

1 Introduction

Artificial neural network (ANN), from its beginning in [23], through to the present era of 'deep learning' has continuously been motivated by the principles of the human nervous system. Contrary to the relatively simplistic computational neuron models that were used in the earlier development (first and second generation), the present era of neural network uses network spiking neuron models. These spiking

N. Sengupta (✉) · N. Scott · N. Kasabov
Knowledge Engineering and Discovery Research Institute,
Auckland University of Technology, Auckland, New Zealand
e-mail: neelava.sengupta@aut.ac.nz
URL: http://www.kedri.aut.ac.nz

V. Ravi et al. (eds.), *Proceedings of the Fifth International Conference on Fuzzy and Neuro Computing (FANCCO - 2015)*, Advances in Intelligent Systems and Computing 415, DOI 10.1007/978-3-319-27212-2_9

models are biologically plausible and neuromorphic in nature. Although the major focal point of research in this area is on implementation of biologically realistic spiking neuron models and parameters, recently successful efforts have been made to apply these spiking neural networks (SNN) on pattern recognition problems that can be utilised in real-world engineering applications which deals with complex processes generating data ordered in time [15, 20, 22]. The main difference of spiking neuron model lies in the use of synaptic spike timing as a source of information as opposed to traditional perceptron models, which use static analogue information as the source. Therefore, to apply SNNs on real analogue data, a data encoding step is necessary (that transforms the analogue information to spikes) prior to learning the spike information in the SNN architecture.

The inspiration of a data encoding scheme derives from the ability of the brain cells to propagate signals rapidly over large distances. This is done through generating characteristic electrical pulses known as synaptic potentials. When presented with external stimuli such as light, sound, taste or smell, the input sensory neurons change their activities by firing a sequence of synaptic potentials in various temporal patterns. Such analogue to binary transformation of the signal in our brain is studied elaborately in [13, 14]. There exist several theories of how the continuous information is encoded in our brain. Two of the most prominent theories of the encoding scheme are 'rate' and 'temporal' coding scheme. The rate coding scheme purports the idea that information can be encoded by mean firing rate, that is, as the intensity of the stimulus increases, frequency of spike increases. The major drawback of this theory however, lies in the association of information with spike density, which in turn accounts for a level of inefficiency. On the other hand, temporal encoding is based on encoding information as precise spike timings. Hence, the information is carried by the structure of the spike pattern. That the temporal resolution of the neural code is on a millisecond time scale, highlights precise spike timing as a significant element in neural coding.

This article is further divided into four sections. Section 2 briefly describes the related work on data encoding. The proposed mixed integer optimisation problem for knowledge-driven data encoding is formalised and described in Sect. 3. In Sect. 4, we show, how the proposed data encoding technique can be applied to the fMRI data based on some assumptions, and discuss some of the experimental results. Section 5 includes a conclusion and discusses future work.

2 Related Work

The problem of encoding data is relatively less researched in the field of the spiking neural network as opposed to the dynamics and the learning aspect of the SNN. Data encoding plays a significant role in real world pattern recognition problem. Several research has investigated the implementation of different data encoding technique as part of the pattern recognition process using SNN [22]. ATR's artificial brain (CAM-Brain) project [6] used data encoding as part of its architecture for building large

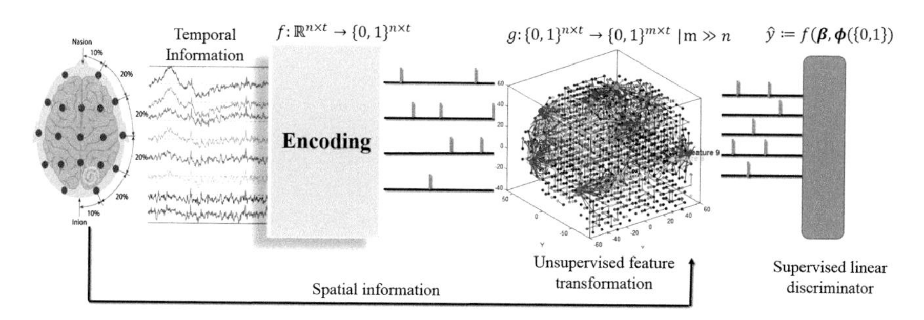

Fig. 1 The NeuCube neuromorphic architecture for spatiotemporal data is a pattern recognition and analysis tool for spatiotemporal data. The temporal information generated from a source (e.g. brain) is passed through a data encoder which transforms the continuous information stream to spike trains ($f : \mathbb{R}^{n \times t} \rightarrow \{0, 1\}^{n \times t}$). The spike train is then transformed to a higher dimensional space through unsupervised learning inside a SNN reservoir ($f : 0, 1^{n \times t} \rightarrow 0, 1^{m \times t} | m \gg n$). The higher dimensional spike train is then readout using a linear classifier like deSNN ($\hat{y} := h(\beta, \phi(o, 1))$) [21]

scale brain-like neural network system. Hafiz and Shafie [17] implemented a specific data encoding module for facial image recognition using SNN. Hardware accelerated implementation of spike encoding for image and video processing was put forward by Iakimchuk et al. [19]. Recently, Kasabov [20] proposed NeuCube Neuromorphic architecture to analyse and predict spatiotemporal pattern using SNN architecture. Figure 1 briefly describes the principles of pattern recognition for spatiotemporal data using the NeuCube architecture. It is quite evident from the description in Fig. 1, the linear classification/prediction model in NeuCube is dependent on the quality of both data encoding and unsupervised module to produce linearly discriminable spike trains. The literature on spike encoding technique is restricted to a few algorithms like Temporal contrast, Hough Spiker Algorithm (HSA) [18] and Ben Spiker Algorithm (BSA) [24]. The temporal contrast algorithm, which is inspired from the working principle of the visual cochlea, uses threshold based method to detect signal contrast (change). The spike events in temporal contrast are determined by a user-defined contrast threshold. The HSA and BSA algorithm however, determines a spike event using a deconvolution operation between observed signal and a predefined filter. The deconvolution in HSA is based on the assumption that the convolution function produces a biased converted signal which always stays below the original signal yielding an error [17]. BSA on the other hand, uses a finite reconstruction filter (FIR) for predicted signal generation. The BSA algorithm used a biologically plausible FIR described in [5]. Nevertheless, it must be noted that a single filter, as proposed in both [18, 24], may not fit every data source, as the reconstruction filters are data source dependent, and may have different filter parameters.

3 Formulation of Background Knowledge Driven Optimisation Problem

The problem of data encoding can be formally defined as a mapping $f : S \rightarrow B$ of real continuous signal $S \in \mathbb{R}^t$ to spikes $B \in \{0, 1\}^t$ over time t. In all of the literature discussed in Sect. 2, the events (spikes) are predefined by some rule (e.g. higher than a threshold change in signal, in case of temporal contrast). The major advantage of spatiotemporal data originating from the brain as opposed to other engineering data is the inherent existence of events, that generate the fluctuations in the signal produced from the neural activity. The process of transformation of neurologically defined events to signal can be defined as a reverse encoding or decoding method, formally described as a signal estimation function $\hat{S} := g(B)$. Several mathematical models $g(B)$ for signal estimation of fMRI or EEG data source exist in the research articles [11, 16]. However, the observed signal is an additive function of neural activity and the multi source noise generated from the device and the experimental setup, hence observed signal S can be formulated as:

$$S = f(B) + N \tag{1}$$

For encoding brain data, we propose a generalised optimisation-based framework based on minimisation of root mean squared error between observed signal S and predicted signal $\hat{S}$. It can be formalised as:

$$\begin{aligned} \min_{B,\theta} \quad & \sqrt{\frac{\sum_t (S-\hat{S}(B,\theta))^2}{t}} \\ \text{s.t.} \quad & B = \mathbb{I}^+ \\ & \sum_t B_t \leq a \\ & 0 \leq B \leq 1 \\ & b \leq \theta \leq c \end{aligned} \tag{2}$$

The above optimiser solves for the RMSE, subject to the following constraints:

1. Binary constraints for spikes: The binary constraint for the spikes are implemented by forcing B to be an integer and within a range of [0 1].
2. Constraint on the number of spikes: The $\sum_t B_t \leq a$ constraint enforces the maximum number of spikes to be limited to a. This constraint is of major importance from a biological plausibility perspective. Since the encoding scheme discussed here, aims to mimic the temporal coding behaviour of the human brain, it is always preferable to encode maximal information with the minimal number of spike.
3. Bounds can be set on the other parameters θ to be optimised as part of the prediction model $S(B, \theta)$

The aforementioned optimisation problem belongs to the paradigm of mixed-integer programming, where a subset of parameter or decision variables to be optimised, are integers. Numerous solvers have been developed over the years for dealing with this class of problems, such as for example, Simulated Annealing [10], Differential Evolution [2], Particle Swarm Optimisation [27] and Genetic algorithm [7, 8]. In our implementation, though, we have used the mixed-integer genetic algorithm solver [8] for solving the optimisation problem in Eq. 2. As opposed to traditional GA solvers that optimise a fitness function, mixed integer GA minimises a penalty function which include terms for feasibility. This penalty function is combined with binary tournament selection to select individuals for subsequent generations. The penalty function value of a member of a population is [8]:

- If the member is feasible, the penalty function is the fitness function.
- If the member is infeasible, the penalty function is the maximum fitness function among feasible members of the population, plus a sum of the constraint violations of the (infeasible) point.

4 Experiments and Evaluation

In this section, we will formalise a sample prediction model $f(B)$ for functional Magnetic Resonance Imaging (fMRI) data and will present experimental results and evaluation of data encoding by solving Eq. 2.

Functional Magnetic Resonance Imaging (fMRI) is most commonly acquired using Blood Oxygen Level Dependent (BOLD) response. The BOLD response $S(t)$, which is measured by the changes in deoxyhaemoglobin at time t, is caused by neural activation $u(t)$. The neural activations are caused by some sequence of events driven by experimental protocol [9]. Mathematically a BOLD response is modelled as a time invariant system. For example, a system whose output does not depend explicitly on time. Under the appropriate experimental protocol, however, BOLD response follows the superposition principle and thus acts as linear time invariant (LTI) [26]. According to [4], an LTI system can be completely characterised by the convolution integral function, which is given by,

$$\hat{S}(t) = \int_0^t B(\tau)h(t-\tau)d\tau \tag{3}$$

where $\hat{S}(t)$ represents the estimated BOLD response. $h(\tau)$ is the Haemodynamic response function (HRF) for BOLD response. $B(\tau)$ represents the neural events or the spikes in this case. It is clear from Eq. 3, under the LTI assumption, irrespective of the complexity, the BOLD response can be predicted by convolving the neural events (spikes) with a known HRF. Several flexible mathematical models of HRF are present in the literature [3, 12, 16], where the HRF models can be tuned by a number of free parameters. The free parameters can be adjusted to account for differences in the HRF across brain regions or subjects [1].

For encoding BOLD data, Eq. 3 can be substituted in Eq. 2, and the data encoding problem for fMRI data can be formally written as the following:

$$
\begin{aligned}
\min_{B,\theta} \quad & \sqrt{\frac{\sum_t (S-\hat{S}(B,\theta))^2}{t}} \\
\text{s.t.} \quad & B = \square^{+} \\
& \sum_t B_t \leq a \\
& 0 \leq B \leq 1 \\
& b \leq \theta \leq c \\
\text{where } & \hat{S}(B,\theta) = B \otimes H(\theta)
\end{aligned} \tag{4}
$$

where B and θ represent the spikes and the free parameters for the HRF model respectively. In the majority of the literature mentioned above, HRF is modelled by some form of gamma function with a set of free parameters controlling the shape and the span of the gamma curve. In our experiments, we have used the generalised gamma function described in Eq. 5 as HRF. Thus, the mixed integer solver jointly optimise for the spikes (B), shape (θ_1) and span (θ_2) parameters.

$$
H(t|\theta_1,\theta_2) = \frac{1}{\theta_1^{\theta_2}\Gamma(\theta_2)} t^{\theta_1-1} e^{\frac{-t}{\theta_2}} \tag{5}
$$

For all the experiments, preprocessed fMRI data of subject 04847 from the benchmark StarPlus fMRI dataset [25] is used. The implementation of the data encoding as a mixed integer optimisation problem is performed using matlab R2013b optimisation toolbox. The fMRI BOLD signals are normalised between 0 and 1 as part of preprocessing step.

Figure 2 demonstrates a comparison of reconstructed signal using the spikes generated by GA-gamma (proposed), temporal contrast and BSA [24]. It can be clearly observed, that the reconstructed signal generated by GA-gamma optimised spikes very closely follows the observed signal, while the signal generated by BSA encoded spikes, can capture the trend in the signal, but unable to capture the seasonality in the BOLD response. In Fig. 3, we have also compare the RMSE of signal reconstruction across 100 different voxels using spikes encoded by different methods, which shows that the proposed method consistently outperforms the baselines across voxels.

As part of the optimisation, the GA-gamma encoding method also optimise for the parameters of the response filter H. Figure 4 shows the gamma haemodynamic response filters learned by the model for a single voxel across 7 trials. It can be seen from the figure, that for a single voxel across trials, the shape of the HRF is nearly consistent, but varies in the scale. This result is consistent with the notion of the existence of minor variation of HRF across voxel and/or subject.

We have also analysed the effect of 'maximum number of spike' $\left(\sum_t B_t \leq a\right)$ constraint and 'HRF parameter bound' ($b \leq \theta \leq c$) on the optimal solution's spike count. Figure 5 shows a plot of 'maximum allowable spike count' versus the 'optimal number of spike' for two different constraints on θ. Choosing a low upper bound on θ in this case lowers the optimal spike count (spike density), contrary to a high upper bound. This experiment shows, how inclusion of knowledge (in choosing the bounds and constraints) in the optimisation effects the encoded optimal spike pattern. This

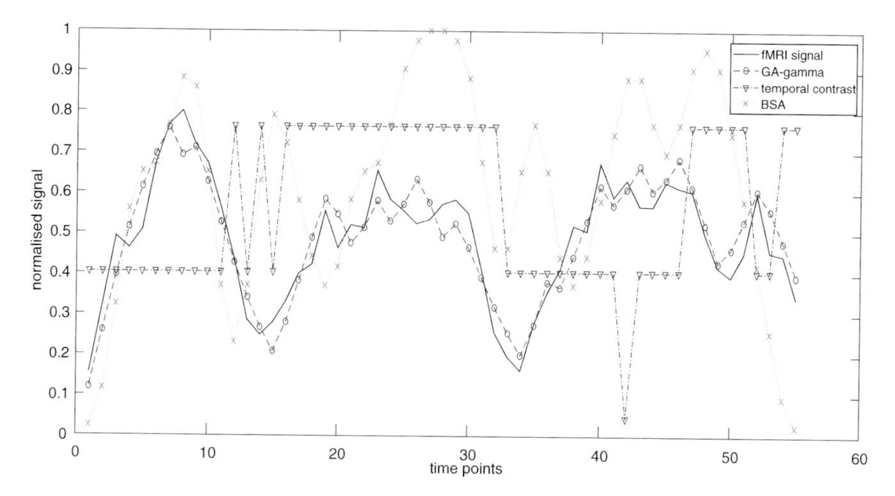

Fig. 2 Comparison of signal reconstruction by GA-gamma, temporal contrast and Ben Spiker Algorithm applied on 88th voxel of second trial in subject 04847

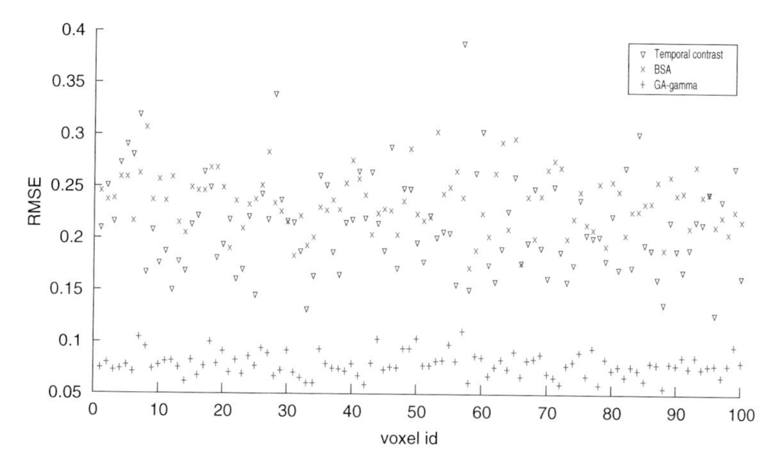

Fig. 3 Comparison of RMSE of reconstruction between GA-gamma, BSA and Temporal Contrast across 100 voxels of second trial in subject 04847

is also supportive of the fact that, this method encodes information in spike timings as opposed to rate coding where information is encoded in spike density.

As discussed in Sect. 3, noise contributes significantly in the observed signal. Hence preprocessing for noise reduction plays an important role for better encoding of signal. One of the preprocessing routine used for fMRI is data smoothing, where the data points of a signal are modified so individual points (presumably because of noise) are reduced, and points that are lower than the adjacent points are increased leading to a smoother signal.We have analysed the effect of noise reduction by linear smoothing on the quality of GA-gamma data encoding. Figure 6 compares the effect

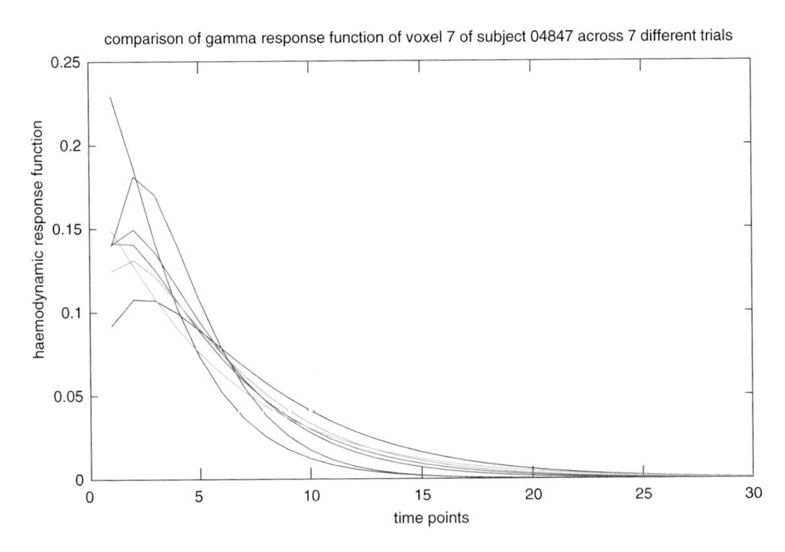

Fig. 4 Comparison of haemodynamic response function learned by GA-gamma encoding method for voxel 8 of subject 04847 across 7 different trials

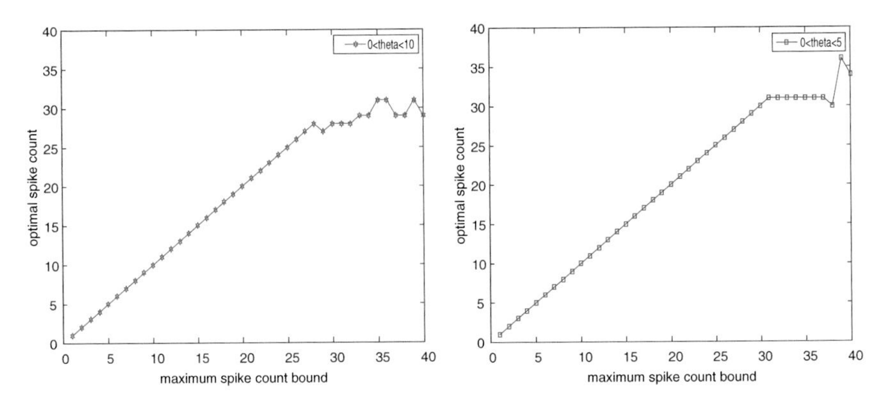

Fig. 5 Plot of the 'maximum allowable number of spike' (changing value of a in $\sum_t B_t \leq a$) versus 'actual number of spike' optimised by GA-gamma. The figure on the left depicts the result where θ is bound 0 and 5 and the figure on the right constrains θ within 0 and 10

of window size (in moving window based smoothing) on the reconstruction RMSE of GA-gamma optimised spikes. It is evident from the figure, that the increasing window size (increasing smoothness), decreases the RMSE, thus the quality of the data has a significant effect on signal reconstruction. However, it must be noted that, the proposed generalised framework can include specific noise models as part of the prediction model $\hat{S}$.

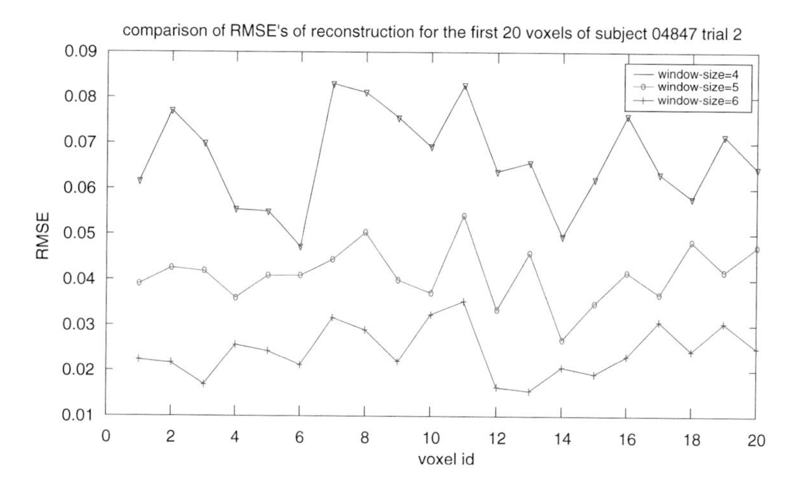

Fig. 6 Comparison of RMSE of reconstruction between GA-gamma, BSA and Temporal Contrast across 100 voxels of second trial in subject 04847

5 Conclusion and Future Work

In conclusion, in this research article, we have proposed a background knowledge driven generalised optimisation based framework for encoding brain data in spiking neural network architecture. In addition, we have also formalised the optimisation problem for fMRI data under the assumption of linear time invariance, and compared experimentally, the performance of the proposed method with the existing state of the art. In the future, we would like to further extend this framework for encoding EEG data. The next step towards encoding EEG data with this framework will include formalizing a prediction model for the EEG data. In another strand of research, the proposed framework, will be included as part of NeuCube framework and will be evaluated by the prediction performance of the NeuCube architecture.

References

1. Ashby, F.G.: Statistical Analysis of fMRI Data. MIT press, Cambridge (2011)
2. Babu, B., Jehan, M.: Differential evolution for multi-objective optimization. In: The 2003 Congress on Evolutionary Computation, 2003. CEC'03, vol. 4, pp. 2696–2703. IEEE (2003)
3. Boynton, G.M., Engel, S.A., Glover, G.H., Heeger, D.J.: Linear systems analysis of functional magnetic resonance imaging in human v1. J. Neurosci. **16**(13), 4207–4221 (1996)
4. Chen, C.T.: Linear System Theory and Design. Oxford University Press, Inc., Oxford (1995)
5. De Garis, H., Nawa, N.E., Hough, M., Korkin, M.: Evolving an optimal de/convolution function for the neural net modules of atr's artificial brain project. In: International Joint Conference on Neural Networks, 1999. IJCNN'99, vol. 1, pp. 438–443. IEEE (1999)
6. de Garis, H.: An artificial brain atr's cam-brain project aims to build/evolve an artificial brain with a million neural net modules inside a trillion cell cellular automata machine. New Gener. Comput. **12**(2), 215–221 (1994)

7. Deb, K.: An efficient constraint handling method for genetic algorithms. Comput. Methods Aappl. Mech. Eng. **186**(2), 311–338 (2000)
8. Deep, K., Singh, K.P., Kansal, M., Mohan, C.: A real coded genetic algorithm for solving integer and mixed integer optimization problems. Appl. Math. Comput. **212**(2), 505–518 (2009)
9. Frackowiak, Richard S.J., Friston, K.J., Frith, C.D., Dolan, R.J., Mazziotta, JC.: Human brain function. San Diego, CA (2004)
10. Dorigo, M., Maniezzo, V., Colorni, A.: Ant system: optimization by a colony of cooperating agents. IEEE Trans. Syst. Man Cybern. Part B Cybern. **26**(1), 29–41 (1996)
11. Fransson, P., Krüger, G., Merboldt, K.D., Frahm, J.: Mri of functional deactivation: temporal and spatial characteristics of oxygenation-sensitive responses in human visual cortex. Neuroimage **9**(6), 611–618 (1999)
12. Friston, K.J., Josephs, O., Rees, G., Turner, R.: Nonlinear event-related responses in fmri. Magn. Reson. Med. **39**(1), 41–52 (1998)
13. Gabbiani, F., Koch, C.: Coding of time-varying signals in spike trains of integrate-and-fire neurons with random threshold. Neural Comput. **8**(1), 44–66 (1996)
14. Gabbiani, F., Metzner, W.: Encoding and processing of sensory information in neuronal spike trains. J. Exp. Biol. **202**(10), 1267–1279 (1999)
15. Gerstner, W., Kistler, W.M.: Spiking neuron models: single neurons, populations, plasticity. Cambridge University Press, Cambridge (2002)
16. Glover, G.H.: Deconvolution of impulse response in event-related bold fmri 1. Neuroimage **9**(4), 416–429 (1999)
17. Hafiz, F., Shafie, A.A.: Encoding of facial images into illumination-invariant spike trains. In: 2012 International Conference on Computer and Communication Engineering (ICCCE), pp. 132–137. IEEE (2012)
18. Hough, M., De Garis, H., Korkin, M., Gers, F., Nawa, N.E.: Spiker: analog waveform to digital spiketrain conversion in atrs artificial brain (cam-brain) project. In: International Conference on Robotics and Artificial Life. Citeseer (1999)
19. Iakymchuk, T., Rosado-Munoz, A., Bataller-Mompean, M., Guerrero-Martinez, J., Frances-Villora, J., Wegrzyn, M., Adamski, M.: Hardware-accelerated spike train generation for neuromorphic image and video processing. In: 2014 IX Southern Conference on Programmable Logic (SPL), pp. 1–6. IEEE (2014)
20. Kasabov, N.: Neucube evospike architecture for spatio-temporal modelling and pattern recognition of brain signals. In: Artificial Neural Networks in Pattern Recognition, pp. 225–243. Springer (2012)
21. Kasabov, N., Dhoble, K., Nuntalid, N., Indiveri, G.: Dynamic evolving spiking neural networks for on-line spatio-and spectro-temporal pattern recognition. Neural Netw. **41**, 188–201 (2013)
22. Maass, W., Bishop, C.M.: Pulsed Neural Networks. MIT press, Cambridge (2001)
23. McCulloch, W.S., Pitts, W.: A logical calculus of the ideas immanent in nervous activity. Bull. Math. Biophys. **5**(4), 115–133 (1943)
24. Schrauwen, B., Van Campenhout, J.: Bsa, a fast and accurate spike train encoding scheme. In: Proceedings of the international joint conference on neural networks. vol. 4, pp. 2825–2830. IEEE Piscataway, NJ (2003)
25. Starplus dataset. http://www.cs.cmu.edu/afs/cs.cmu.edu/project/theo-81/www/. Accessed 10 Aug 2015
26. Vazquez, A.L., Noll, D.C.: Nonlinear aspects of the bold response in functional mri. Neuroimage **7**(2), 108–118 (1998)
27. Yiqing, L., Xigang, Y., Yongjian, L.: An improved pso algorithm for solving non-convex nlp/minlp problems with equality constraints. Comput. Chem. Eng. **31**(3), 153–162 (2007)

Short Circuit Evaluations in Gödel Type Logic

Raed Basbous, Benedek Nagy and Tibor Tajti

Abstract Short circuit evaluation techniques play important roles in hardware design, programming and other fields of computer science. In this paper one of the most known and used fuzzy logic system, the Gödel logic is considered. Various pruning algorithms are presented to quicken the evaluations of logical formulae in Gödel logic. Simulation results show the efficiency of the presented techniques.

Keywords Pruning techniques • Formula trees • Fast evaluation • Fuzzy logic • Many valued logic

1 Introduction

Logic is a base of computer science and information technology, electric engineering and other scientific and technical fields [3]. Apart from the most used classical Boolean logic there are several other branches developed for various purposes. In modal and temporal logics new operations (such as necessity or "sometime in the future") are introduced, while in many valued and fuzzy logic the

R. Basbous (✉)
Department of Administrative Affairs, Al-Quds Open University, Jerusalem, Palestine
e-mail: rbasbous@qou.edu

R. Basbous · B. Nagy
Department of Applied Mathematics and Computer Science, Eastern Mediterranean University, North Cyprus via Mersin 10, Famagusta, Turkey
e-mail: nbenedek.inf@gmail.com

B. Nagy
Faculty of Informatics, Department of Computer Science, University of Debrecen, Debrecen, Hungary

T. Tajti
Eszterhazy Karoly University of Applied Sciences, Eger, Hungary
e-mail: tajti@aries.ektf.hu

V. Ravi et al. (eds.), *Proceedings of the Fifth International Conference on Fuzzy and Neuro Computing (FANCCO - 2015)*, Advances in Intelligent Systems and Computing 415, DOI 10.1007/978-3-319-27212-2_10

set of truth values is extended from the classical two values. There are various paradoxes in Boolean logic, some of them can be avoided by using many valued and fuzzy logic [6]. We present only one such example, the famous liar paradox [1]. The sentence "This sentence is false." cannot have any of the classical two truth values. It cannot be true and cannot be false. However, a third, additional, truth-value can solve the paradox. From the 1930s various fuzzy logic were developed, e.g., Gödel type logic, Lukasiewicz logics and the product logic, just to mention the most known and important ones [5, 6]. Gödel introduced his logic to obtain an intuitionistic logic [4]. The most important feature of these type of logics that the classical law of double negation does not work in these systems as a logical law.

Evaluating logical expressions are important task when one is working with logical formulae. In programming languages usually short circuit evaluations are used. The logic used in conditions is very close to the Boolean logic even in some points (in some of the programming languages) it is not exactly the classical two-valued logic [9]. Fuzzy logic and fuzzy sets are also used in several applications [14].

In this paper, one of the most known fuzzy logic, the Gödel type logic is considered. After having some preliminaries (Sect. 2) including the semantics of Gödel type logic, various pruning techniques are presented (Sect. 3). By these algorithms one could cut some of the branches of the expression trees allowing a much faster method of evaluation. We also provide results for complex examples (Sect. 4) and several experiments (Sect. 5) showing the efficiency of our new techniques. Concluding remarks and further thoughts close the paper.

2 Preliminaries

In this section we briefly recall some well known areas, including expression trees, Boolean logic, and short circuit evaluations of Boolean formulae. Finally, in the last subsection the Gödel type logic is recalled.

2.1 Expression Trees

In several fields of mathematics and other sciences various formulae are used to describe the world. The used operators are mostly binary, and usually unary operators are also allowed. Mathematical, logical etc. expressions are usually displayed by tree graphs. The structure of the expression can easily been understood: The main operator is put to the root of the tree (it is the topmost node of the tree graph). Nodes with unary operators have exactly one child, while with binary operators have two children. In some cases, since associativity allows us to do so, some operators may have more than two children, e.g., one node with operator "+" could have four children, let us say a, b, c and d, representing the formula $a + b + c + d$. To evaluate an expression, in a way learned in school, one should start from the leaves of the tree. The values given there are used to evaluate every subformula and finally, the whole, original formula.

2.2 *Boolean Logic*

Classical two-valued logic was formalized by Boole in the 19th century. Because of the adequate formalism this logic is also called as mathematical logic. It works with algebraic technique: with only two values, 0 (false) and 1 (true) Boolean algebra gives the symbolic framework. It is used in electronic switching circuits, and thus, it gives the base of all our digital machines (e.g., electronic computers). The unit of the information is the bit, the answer to a yes-no question. It is true or false. Binary number systems are also based on Boolean logic.

Logical deductions are also expressed in algebraic way. In engineering textbooks the multiplication operator stands for conjunction (logical AND), and addition operator for disjunction (logical OR).

Readers not familiar to classical logic are referred to the textbook [3], or similar materials. Here we mention only some parts very briefly.

Usually Boolean logic is described by the operators conjunction, disjunction and negation (this latter operator is unary, all others are binary). Implication is also very frequently used. We note that, actually, one operator, e.g., the operator NAND, is enough in the Boolean logic to obtain a functionally complete system, i.e., to describe every possible formula by substituting other operators by special subformulae.

Applying special semantic properties of the system, to evaluate a formula, i.e., to know its truth value one may not need to do a full evaluation, i.e., some of the nodes of the formula tree may not need to be visited, their values may not have any effect on the final value. These types of evaluation techniques are highly used in programming languages helping the computation be (nearly) optimal. In the next part we show the two basic forms of these, so-called short circuit evaluations.

2.2.1 Short Circuit Evaluation

Short circuit evaluation or "shortcut evaluation" is widely used in programming languages to evaluate the logical expression in an optimal way. This technique is a type of time saving and, in some cases, it is also used for safety reasons [9]. In some programming languages the symbols && and || are used for the logical operations AND and OR, respectively. The idea of applying short circuit evaluation for these two operations is as follows: (see also Fig. 1, for the example.)

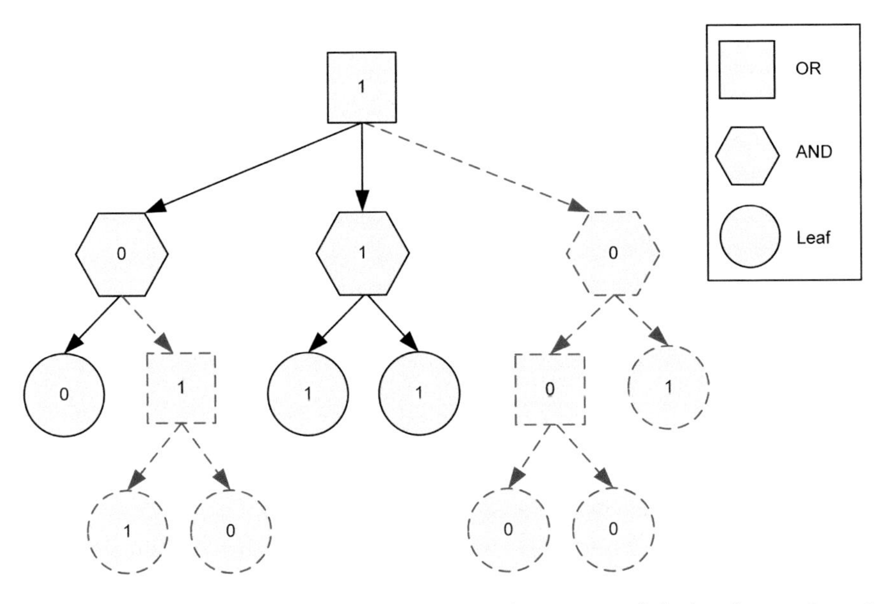

Fig. 1 An example of applying the short circuit evaluation on a tree. Only three inner nodes and three leaves (total 6 nodes) out of six inner nodes and eight leaves (total 14 nodes) are explored and evaluated to get the final result in the root

- In the case of AND, if it is known that all the conditions/arguments must be true in order to proceed, it is not necessary to check all the conditions if one of them is already known to be false.
- Similarly to the previously described AND operation, the OR operation can also be terminated early in some cases: If it is known that one condition is true, then there is no need to check the value of the other conditions.

2.3 *Evaluation of Game Trees*

Minimax theory is frequently used in game theory. In combinatorial (two-player, zero-sum, full information, finite and deterministic) games (with alternate moves of the players), the games are represented by trees containing all (theoretically) possible instances (matches) and outcomes. The minimax algorithm gives solution, i.e., it provides the best (i.e., optimal) strategies for both players. With these strategies players maximize their gains, while minimizing their losses. These optimal strategies and the optimal payoff, the value of the game, can be determined by the minimax algorithm. The evaluation is pretty similar to the evaluation of expressions, in this case the operators are MIN and MAX, representing the minimum and the maximum of the values of the children nodes. The algorithm uses simple

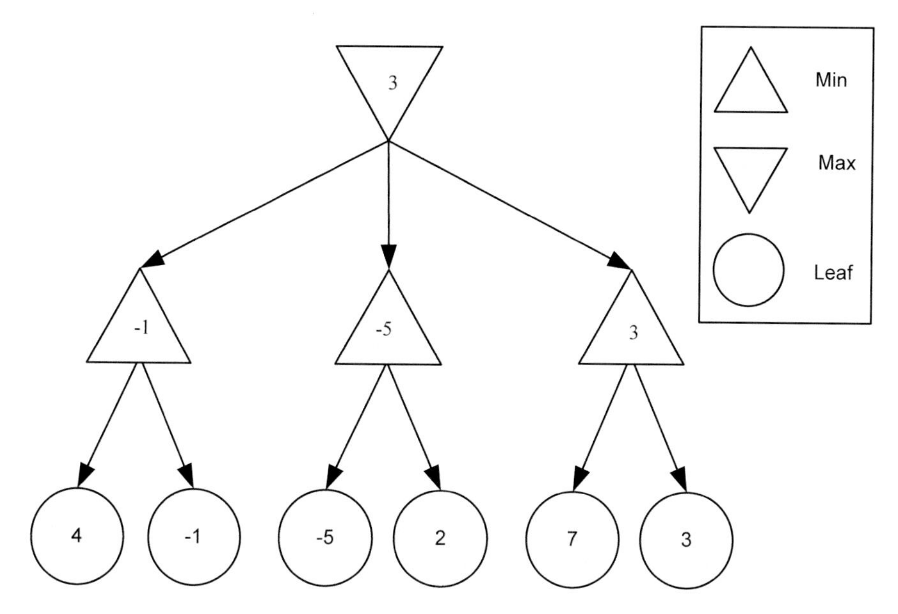

Fig. 2 A minimax tree

recursive functions to compute the minimax values of each successor state [12, 13]. Figure 2 shows an example.

Instead of detailing further the minimax algorithm we refer to [8, 13], and move to pruning techniques applied to quicken the evaluation of game trees. The alpha-beta pruning is somewhat similar to the short circuit evaluations, but they work with more general values (we do not need to restrict ourselves to 0 and 1).

2.3.1 Alpha Beta Pruning

The minimax algorithm evaluates every possible instance of the game, and thus to compute the value of the game, i.e., its optimal payoff, takes usually exponential time on the length of the instances of the game. To overcome on this issue special cut techniques can be used. The alpha-beta pruning (also called alpha-beta cutoff [12]) helps to find the optimal values without looking at every node of the game tree. While using minimax, some situations may arise when search of a particular branch can safely be terminated. So, while doing search, these techniques figure out those nodes that do not require to be expanded. The algorithm is described below based on [2, 12, 13].

Applying alpha-pruning (beta-pruning) means the search of a branch is stopped because a better opportunity for Max-player (Min-player) is already known elsewhere. Applying both of them is called alpha-beta pruning technique. Figure 3

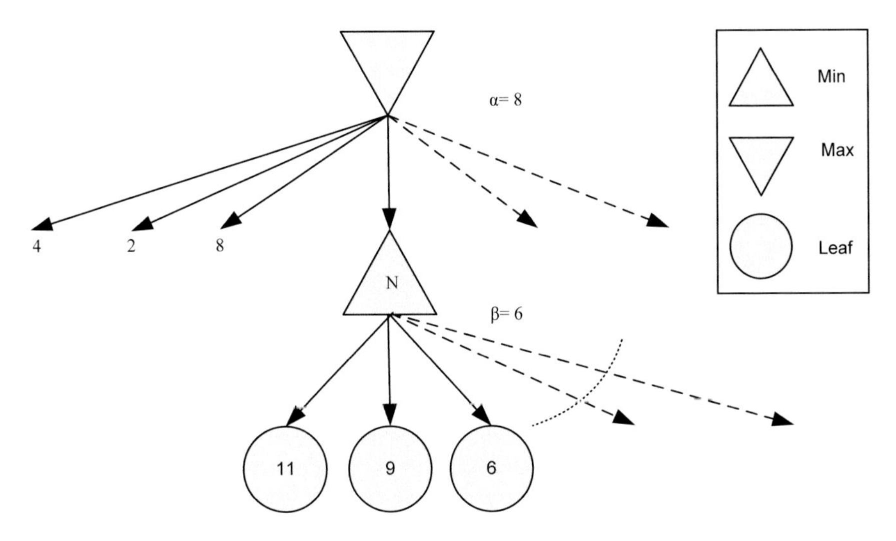

Fig. 3 A beta-pruning example

shows an example of beta-pruning, when β becomes smaller than or equal to α, we can stop expanding the children of N. These algorithms are presented in details in [2, 8, 13].

2.4 Gödel Type Fuzzy Logic

This system has been introduced by Kurt Gödel in 1932 [4]. The possible truth values are real numbers from the closed unit interval [0, 1], i.e., numbers $0 \leq x \leq 1$. The designated value, that is the value that is counted as true, is usually only the value 1. Gödel defined four main connectives for this system (implication, negation, disjunction and conjunction denoted to by the symbols →, ¬, ∨, and ∧, respectively). Their syntax is the same as in Boolean logic, and their semantics are defined in the following way [5, 10]:

$$(\mathrm{A} \rightarrow \mathrm{B}) = 1, \text{ if } \mathrm{A} \leq \mathrm{B}; \tag{1a}$$

$$(\mathrm{A} \rightarrow \mathrm{B}) = \mathrm{B}, \text{ otherwise} \tag{1b}$$

$$(\neg \mathrm{A}) = 1, \text{ if } \mathrm{A} = 0; \tag{2a}$$

$$(\neg \mathrm{A}) = 0, \text{ otherwise} \tag{2b}$$

$$(A \vee B) = \text{Max}\{A, B\} \quad (3)$$

$$(A \wedge B) = \text{Min}\{A, B\} \quad (4)$$

Note that, for simplicity, we use letters A, B for the variables and, also, for their values, without causing any misunderstanding. The system is infinitely many valued, and it fulfills the axioms of intuitionistic logic with one additional law, namely, the law of chain: The formula $((A \rightarrow B) \vee (B \rightarrow A))$ has value 1 independently of the values of subformulae A and B.

Expressions and so, expression trees in Gödel logic are very similar to Boolean expressions. The difference is that here the values at leaves (i.e., the truth values of variables) is not restricted to the Boolean set $\{0, 1\}$, but any real number between 0 and 1 (inclusively) can be used.

3 Cut/Pruning Techniques for Evaluation in Gödel Logic

In this section, various algorithms are proposed to quicken the evaluation of expression trees in Gödel logic in which the previously defined operations are used. We are dealing with trees with bounded set of truth (or payoff) values: the real numbers of the closed interval [0, 1] can be used. The conjunction and disjunction operations can be seen as operators AND and OR by restricting the values to the Boolean set, i.e., to $\{0, 1\}$. In our expression trees, in a similar manner to the Boolean expressions (actually, the syntax of expressions of Gödel logic is the same as the syntax of Boolean expressions), since both AND and OR are associative, without loss of generality we allow multiple children of nodes of these types. Nodes with negation must have exactly one child, while nodes with implications must have exactly two children, called left child and right child, respectively.

The proposed algorithms to quicken the evaluation of this kind of trees are described below in details in the next subsections.

3.1 Alpha Beta Pruning

A form of the classical short circuit evaluation for Gödel logic is exactly the usual alpha beta pruning. When OR and AND operators are used alternately by levels of the expression tree, alpha beta pruning can be applied and, actually, by knowing the possible minimal and maximal values, these information can also be used (reaching these values some neighbor branches can be cut without effecting the computed value).

3.2 Implication Pruning

Evaluating the implication nodes can be quickening using various techniques. These techniques are based on the possible left and right children of the implication node, and they are described in the following subsections in details.

3.2.1 Min-Max Children Pruning

We start with the case when the left child node is a conjunction and the right child node is a disjunction. The first idea to quicken this evaluation is to evaluate the branches connected to its successors one by one in parallel, that is, evaluating the first child of the min node and then the first child of the max node, then the second child of the min node, etc. This technique is very useful in case if minimum and maximum nodes are connected to implication node. Figure 4 shows an example of such case.

As shown in Fig. 4, min (left child) and max (right child) successors evaluated in parallel (one by one, after each other). After evaluating the second successor of max and min nodes, we have that min will be less or equal to 0.3 and max greater or equal to 0.6 which means that we can cut off the other successors and return back the value 1 for implication node. More generally, if it is known that the value of the right child (disjunction in this case) is at least as many as the value of the left child (conjunction in this case), we can eliminate the evaluation of the other brothers and sisters, the implication node has a value 1.

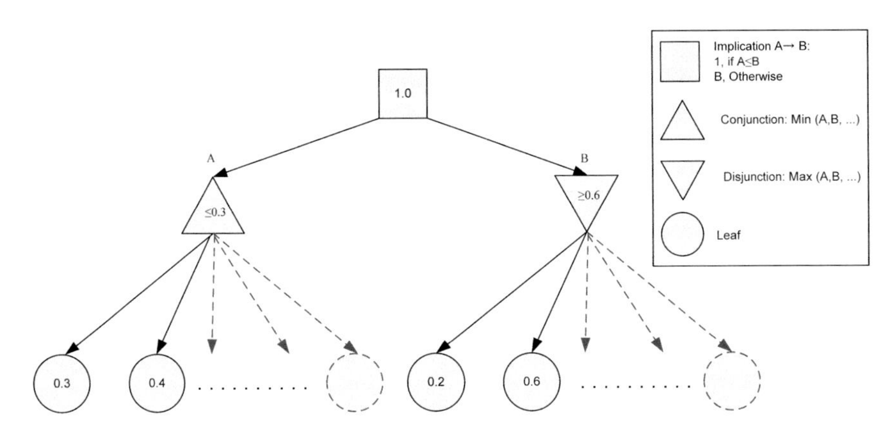

Fig. 4 An example of applying the pruning when evaluating an implication node. Two children, a min node to *left* side and a max node to *right* side are connected. The evaluation is done in parallel

3.2.2 Max-Min Children Pruning

The second pruning technique is applied when the successors of an implication node are max (disjunction at left child) and min (conjunction at right child). Figure 5 shows an example of this case.

As shown in Fig. 5, if we evaluating the max and min nodes branches in parallel (one by one in turns), we have that max node after evaluating the second successor will be always greater or equal to 0.5, while min node will be less or equal to 0.4 after evaluating the first branch. Since the value of min (left child) will be less than max (right child), we can cut all the subsequent children of max node and evaluate only the min node to return its value to implication node.

3.3 *Negation Pruning*

As mentioned in the subsections above, Eqs. (2a) and (2b) shows that negation node in Gödel expression has only two resulted values (0 and 1). The idea here is to check all the connected leaves in the lower layers of negation node before evaluating the connected nodes. If all the connected leaves has non-zero values and there is no more negation nodes connected, then cut-off can be applied, the subtree rooted at this negation node can be cut and 0 can be returned back the value to its parent node. Observe that with non-zero values using only conjunction, disjunction and implication the value cannot be zero. Thus, by a simple pattern matching on the subexpression it can be checked whether it contains other negation or a 0 value in a leaf.

This idea is described in Algorithm 1, see also, e.g., Figure 6 for examples. The Check_Leaves function exploring all the leaves of the subtree rooted at the negation

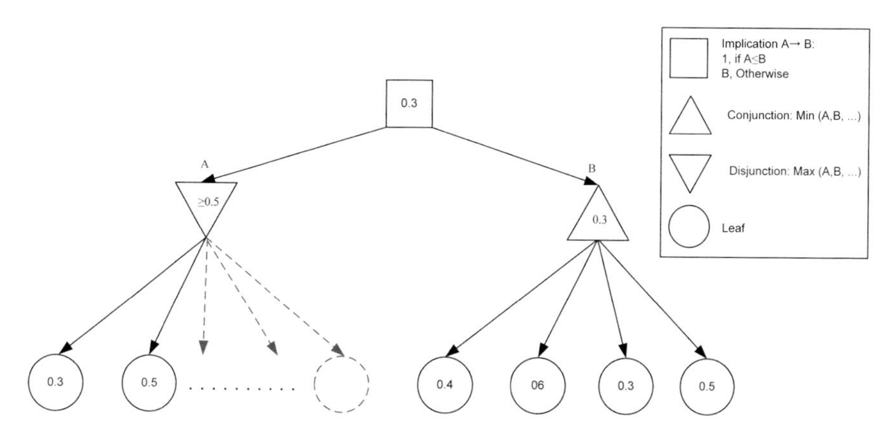

Fig. 5 An example of applying the pruning when evaluating an implication node. Two children are connected, a max node to *left* side and min node to *right* side. The evaluation is done in parallel

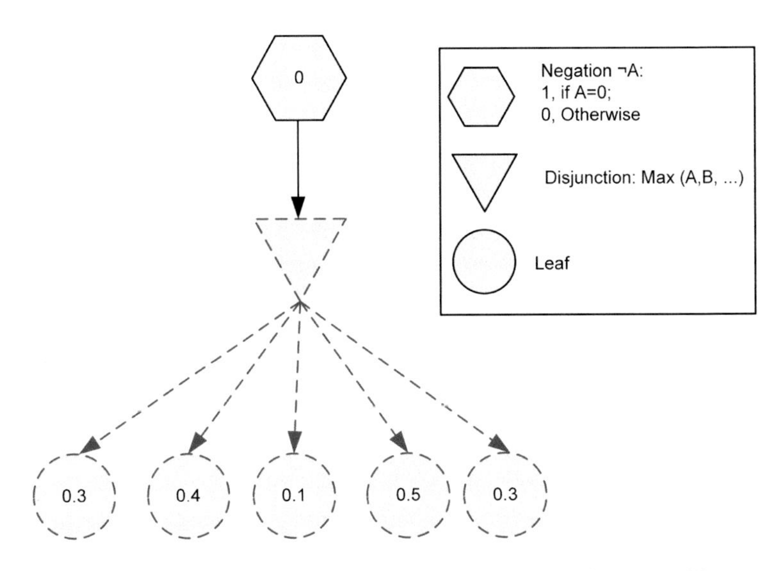

Fig. 6 An example of applying the pruning for a negation node when all connected leaves are non zeros

node. If all the connected leaves are non zeros and there is no other negation node in the subtree, then it returns the initiated flag value 1 meaning that the pruning can be done and the negated subexpression has a 0 value.

```
Algorithm 1 (Negation Prune)
function Check_Leaves(N1)
begin
initiate Flag=1
  if N1 is leaf then
    if N1 = 0 then
      Flag=0    %"If at least one leaf equals to zero or"
    return Flag
  else
  if N1 type is NEG then  %"if a NEG node is found, "
    Flag=0                %"then cut the search,    "
    return Flag           %"we cannot prune.        "
  else
     for every successor N1_i of N1 do
       if Check_Leaves(N1_i) = 0 then
         Flag=0
         return Flag
return Flag
end Check_Leaves
```

3.3.1 Implication with Negation Child Pruning

In addition to the above mentioned pruning technique for negation, more pruning techniques can be done for the cases when an implication node has a child with negation and max or min at the other child. The strategy here is to try first a pruning at the negation node.

Figure 7 shows an example where we have an implication node which has the successors max as left child and negation as right child. First, we can apply the negation prune since all the connected leaves are non zeros. Then, we evaluate the successors of the max node. After evaluating the second successor of max node we have that its value will be always greater that 0.6, while it is greater than the value of negation node (the right successor). A cut can be applied here, and there is no need to explore and evaluate the remaining successors of the max node for the exact evaluation of the implication node. A zero value will returned back as a value of the implication node.

Figure 8 shows an example where we have an implication node which has the successors min as left child and negation as right child. First, as in the previous example, we can apply the negation prune since all the connected leaves are non zeros. Then, we evaluate the min node successors. After evaluating the second successor of min node we have that its value is equal to 0 which is the minimum. The cut can be applied here, and no need to explore and evaluate the remaining successors of min node to evaluate the implication node. The value of the left and right successors are equal, so the value 1 will returned back as a value of the implication node.

Figure 9 shows an example in which the left child of the implication is a negation. In this case, independently of the type of the right child, if the pruning of the negation can be applied and gives a value 0 for the negation node, then the right

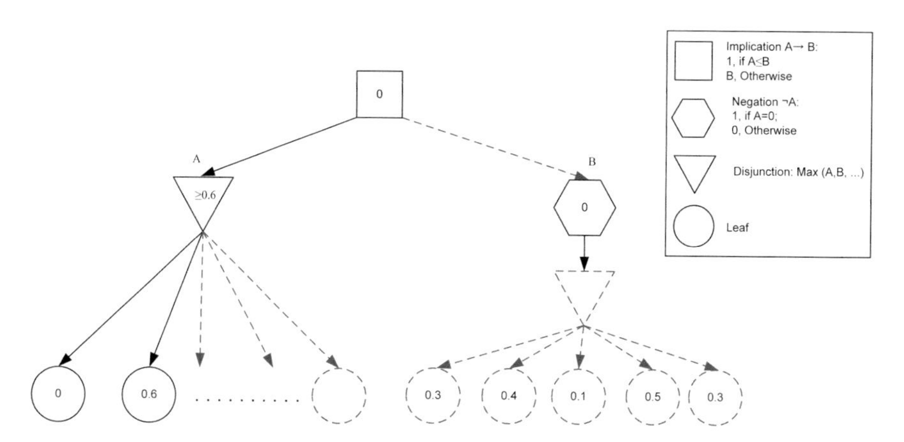

Fig. 7 An example of applying the pruning when evaluating an implication node. Two children are connected, a max node to *left* side and negation node to *right* side

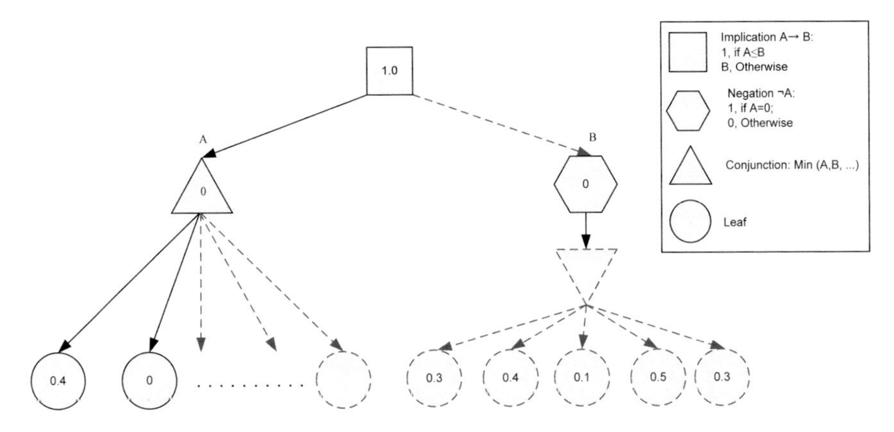

Fig. 8 An example of applying the pruning when evaluating an implication node. Two children are connected, a min node to *left* side and negation node to *right* side

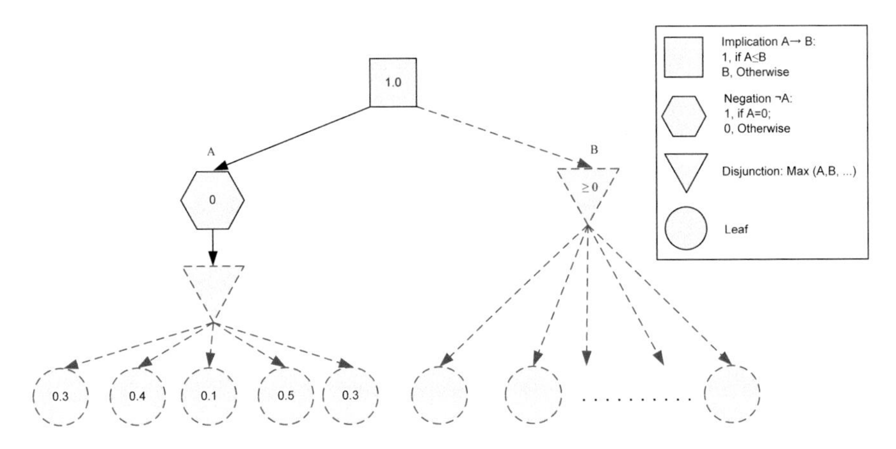

Fig. 9 An example of applying the pruning when evaluating an implication node with negation node connected to the *left* side

child of the implication can be cut off, independently of its value, the implication gets its value 1.

In the next section some complex examples are displayed.

4 Complex Examples

In this subsection we show two complex Gödel expression tree examples, and how the proposed pruning techniques can be used to evaluate these trees very efficiently. First, the expression trees are shown (see Figs. 10 and 12, respectively). These trees

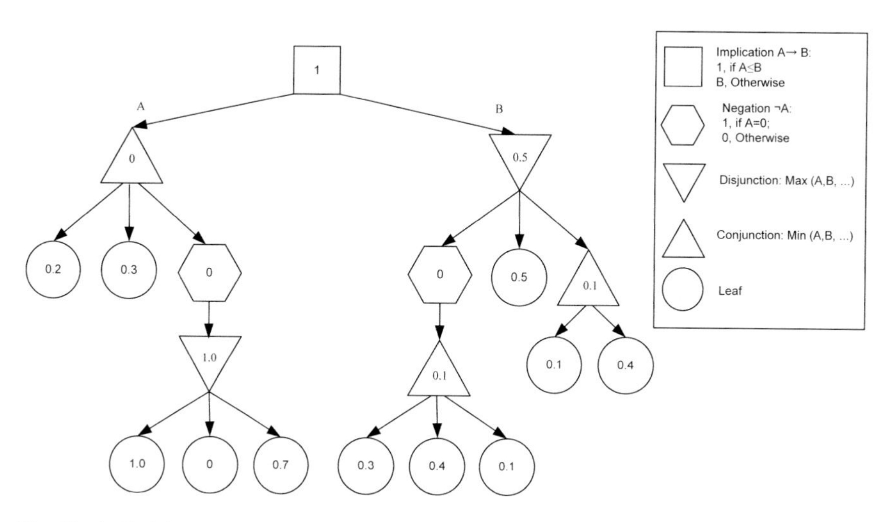

Fig. 10 A Gödel expression tree without pruning

include the four predefined operators (negation, implication, conjunction and disjunction). The order of the operators is various in these expressions, and also, leaves can be found in various levels (i.e., various depths) of the tree (Fig. 11).

To evaluate the tree of the example in Fig. 10, without applying our pruning algorithms, 8 inner nodes and 11 leaves must be explored and evaluated. Figure 11 shows the same tree evaluated after applying the proposed pruning algorithms. The same result of evaluation is provided at the root but by exploring and evaluating

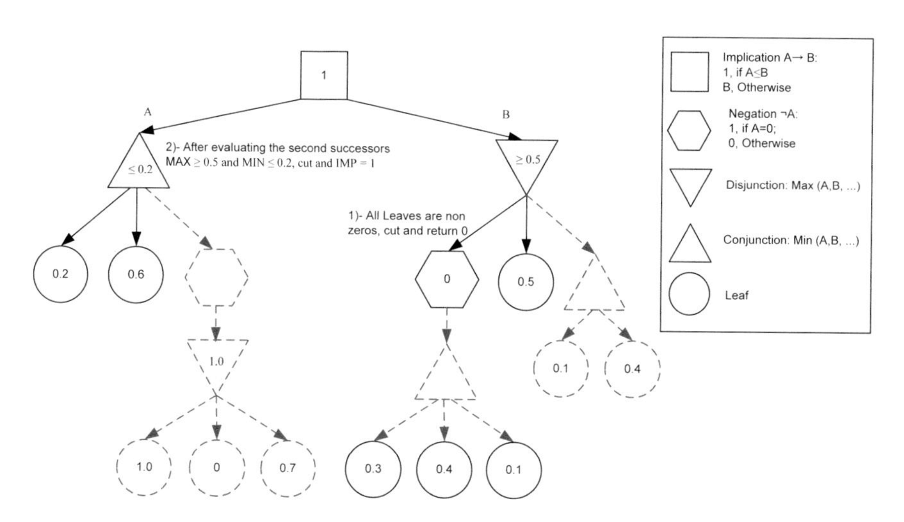

Fig. 11 A Gödel expression tree after applying the proposed pruning techniques (the part shown with *red* color is not evaluated due to cuts)

only 4 inner nodes and 6 leaves (total 10) out of 8 inner nodes and 11 leaves (total 19).

As shown in Fig. 11, the evaluation by applying the proposed pruning techniques has been done as follows:

- The evaluating process started in parallel to evaluate the Min and Max nodes (the two children of the implication at the root).
- While evaluating the first successor (negation node) of Max node, a negation prune has been applied for the reason that all the connected leaves are non zeros.
- After evaluating the second successor of Max and Min, the value of Max node (right branch) will be always greater than or equal to 0.5 and the value of Min node (left branch) is equal to or less than 0.2. A pruning can be applied here by cutting exploring and evaluating the remaining leaves and nodes (the third branch of Max and Min nodes) and return back the value 1 to the root (implication node).

In our other example, shown in Fig. 12, to evaluate the tree without applying the pruning algorithms mentioned above, 16 inner nodes and 18 leaves must be explored and evaluated (total 34). By using various pruning strategies the number of inner nodes that must be evaluated is 9, while the number of leaves that must be explored is 11 (total 19 nodes, see Fig. 13).

Figure 13 shows how the proposed pruning techniques used to evaluate the expression tree in Fig. 12. These techniques have been applied in the following way:

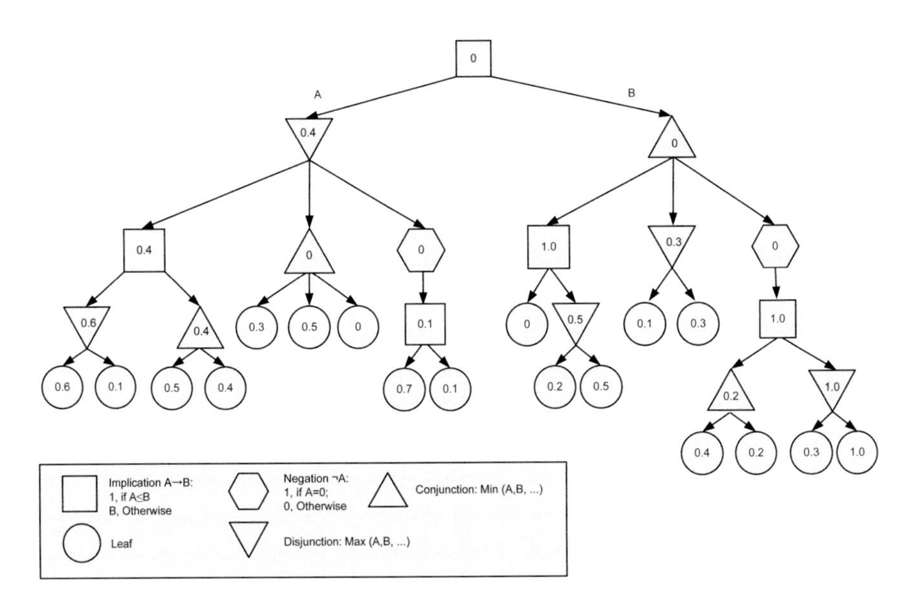

Fig. 12 A complex Gödel expression tree

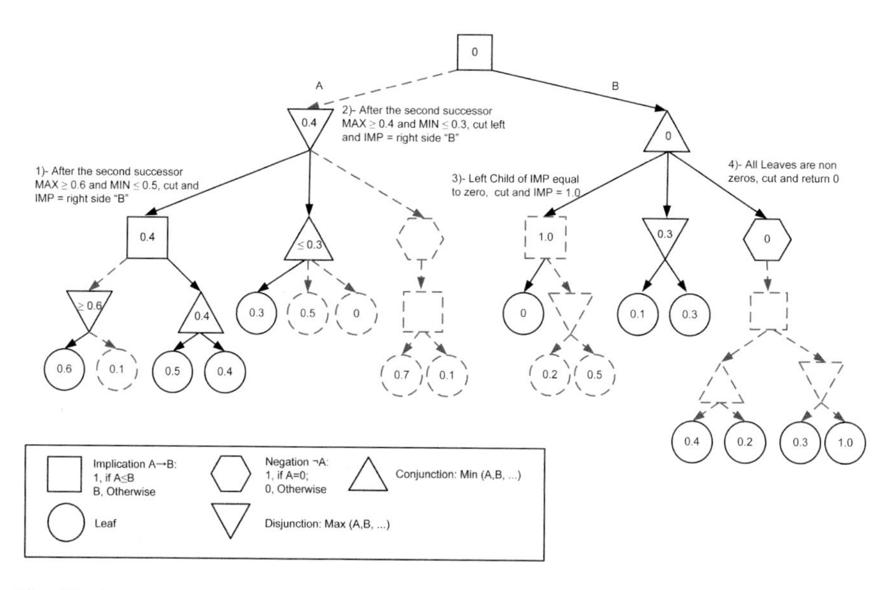

Fig. 13 A complex Gödel expression tree after applying the proposed pruning techniques

- The evaluating process started in parallel to evaluate the Max and Min nodes.
- While evaluating the first successor (implication node) of Max node, we have that it's right branch is greater than or equal to 0.6 while the left one is less than or equal to 0.5. The proposed algorithm cuts the left branch and returns back the value of the node at the right side.
- In the evaluation process of the first successor connected to the Min node (which is the right child of the root), a cut-off can be applied to evaluate the connected Implication node since its left branch is a zero leaf and the value 1 is returned back.
- After evaluating the second successor of both the main Max and Min nodes in the expression, we have that Max is greater than or equal to 0.4 and Min is less than or equal to 0.3. The proposed algorithm cut-off all the left branch of the root and continue to evaluate the right branch only. Furthermore, a cut-off has been applied while evaluating the second successor of the main Max node. The value of this node is at most 0.3 while the current value of the main Max node is at least 0.4; there is no need to explore the remaining leaves to evaluate it.
- A negation prune can be applied for the third successor of the main Min node. This is for the reason that all the connected leaves are non zeros.
- Finally, the value 0 returned back to the root as the final value of the expression.

These results show how fast the evaluation process can be after applying the proposed technique to evaluate an expression trees.

5 Simulation Results

The execution time of the simple evaluation algorithm of the expression trees for Gödel logic is linear to the number of nodes of the expression tree, because each node must be counted exactly once. The pruning algorithm we proposed is more complex, therefore in case of small expressions it requires more time to evaluate. However, when it is applied on large expression trees, its performance will be (much) better, because as the tree grows bigger, bigger part of the tree can be pruned in average. We have programmed the algorithm on Python language and there were 500 000 tests on formulae with various sizes. Since the tests were run on normal computer with non real-time operating system, the measurements for different node counts were run many times in mixed order to eliminate the effect of random other software running on the same computer.

Figure 14 shows the ratio of pruned nodes. The ratio of pruned leaf nodes to the number of leaf nodes has a pretty similar measure (Fig. 15 shows some details). For very small expressions it can happen that nothing or very minor part of the tree can be pruned. As the number of nodes increases the pruned ratio increases as well. For graphs with thousands of nodes very large portion of nodes of the graph can be pruned in average and even in the worst cases more than 80 or 90 % of nodes will be pruned. That means that the evaluation time in the pruned algorithm is not linear, but much better. Figure 16 shows the execution time of the two algorithms on the same examples. There we can see the difference we sentenced before. It looks as the execution time of the pruning algorithm is more like constant instead of linear to the number of nodes. This is because the number of evaluated nodes that are left after

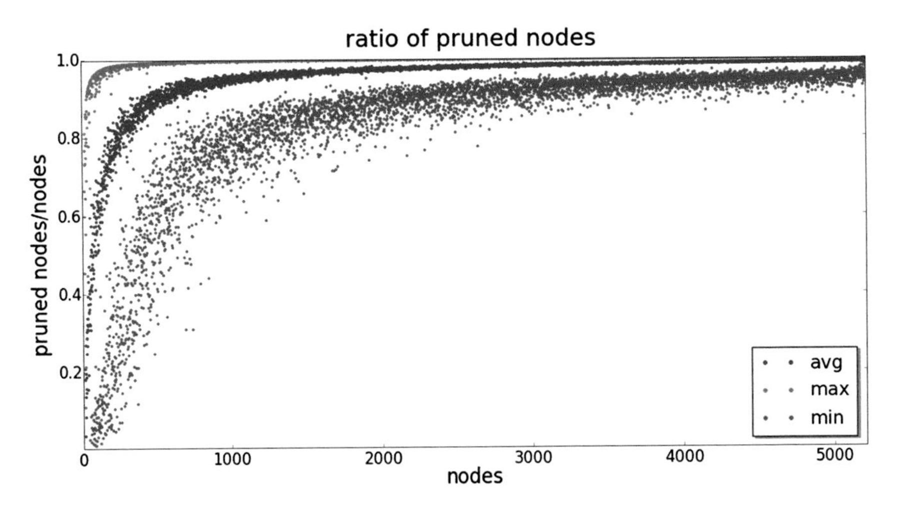

Fig. 14 The ratio of the number of pruned nodes with respect to the size of the expression (total number of nodes)

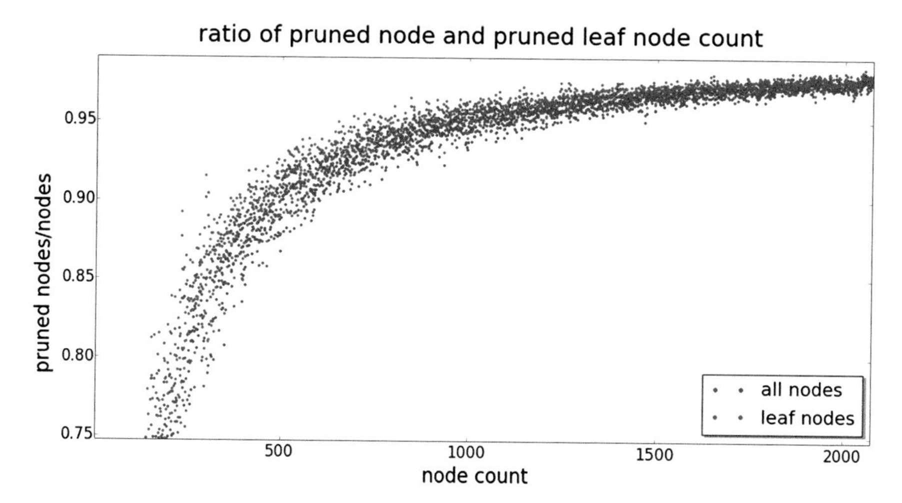

Fig. 15 The ratio of pruned nodes and pruned leaves with respect to the total number of nodes and total number of leaves, respectively

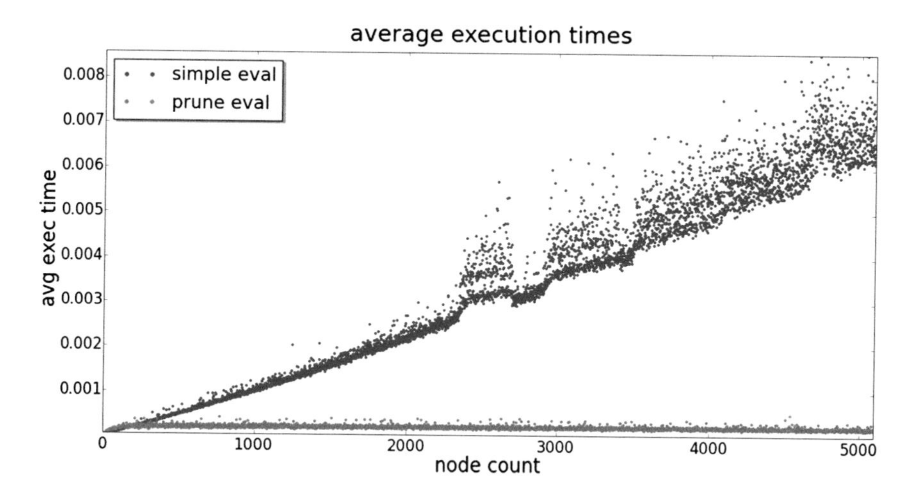

Fig. 16 The running time with respect to the size of the expression (total number of nodes)

the pruning is not linear to the number of all nodes. In fact in average the number of these nodes tends to be constant. This is shown in Fig. 17.

Figure 18 shows what is the ratio of the execution times of the pruning and the simple evaluation algorithms (the time of simple evaluation is used as unit). With smaller trees the execution time of the pruning algorithm is still above the execution time of the simple evaluation. Above about 200 nodes the pruning algorithm is faster and it stays so, because the simple evaluation completes in linear time and the pruning algorithm finishes in nearly constant time (in average). The standard

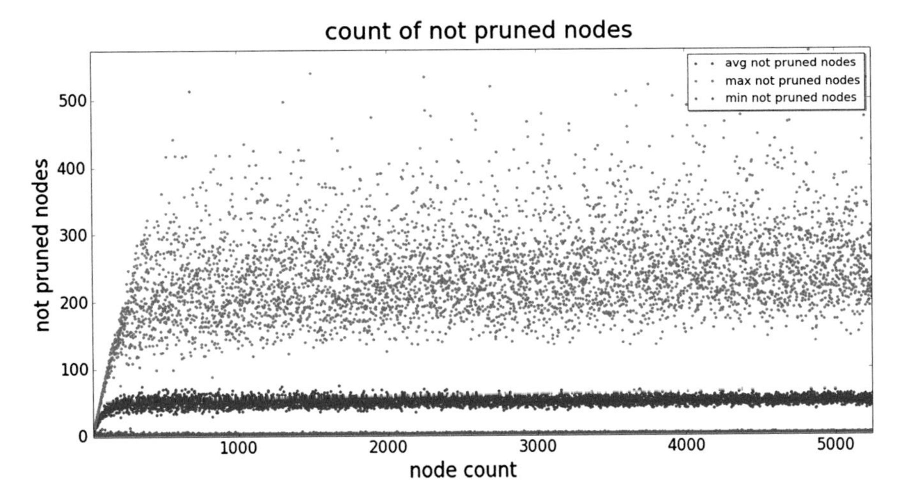

Fig. 17 The number of nodes left after pruning with respect to the size of the expression

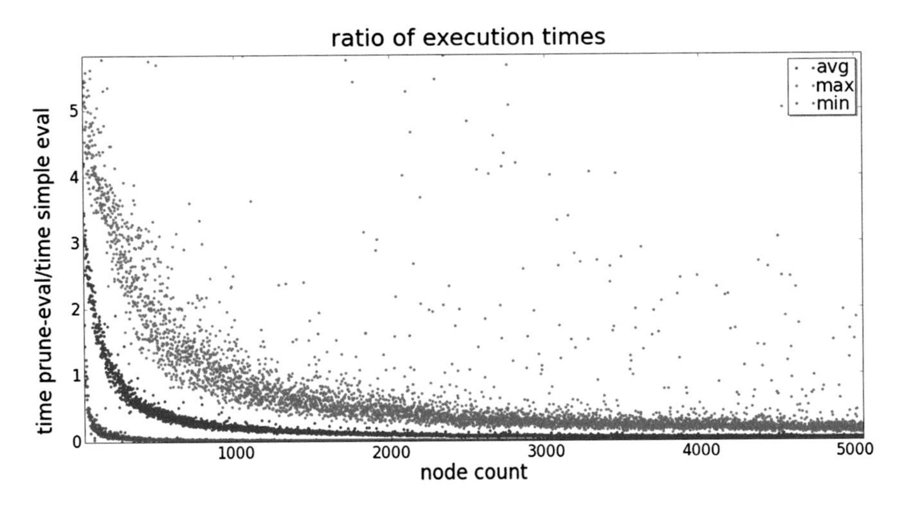

Fig. 18 The ratio of running time with respect to the size of the expression

deviation is relatively large on the running time since the amount of the pruned nodes could be very various. Also the difference between the measured minimum and maximum values are large. Fortunately the minimum measured pruned ratio values are also very good, with increasing the number of nodes also the minimum tends towards 100 %, in fact the measured minimum pruned ratio values are around 90 % if the number of nodes is about 5000.

6 Conclusions

Various logic systems are applied in various fields, e.g., in programming, hardware design, fuzzy technology. In this paper, one of the most known fuzzy logic is considered and various pruning techniques are shown to be very efficient to quicken the evaluation of logical expressions. We believe that for wide spread applications of fuzzy logic and fuzzy systems our results are very useful. As one could see depending the length of the formula its evaluation can be done very efficiently, in average.

In [10, 11] fuzzy logic systems are generalized by using interval-values. It is one of the plans for future work to deal with evaluations in this more general interval-valued logic. It is also an interesting task to deal with some kinds of generalizations of games, e.g., [2, 7] and work on various evaluation techniques.

Acknowledgment This work was partly supported by the research project GOP-1.1.1-11-2012-0026.

References

1. Barwise, J., Etchemendy, J.: The Liar: An Essay on Truth and Circularity. Oxford University Press, New York (1987)
2. Basbous, R., Nagy, B.: Generalized game trees and their evaluation. In: Proceedings of CogInfoCom 2014: 5th IEEE International Conference on Cognitive Infocommunications, pp. 55–60. Vietri sul Mare, Italy (2014)
3. Bell, J., Machover, M.: A Course In Mathematical Logic. North-Holland, New York (1977)
4. Gödel, K.: Zum intuitioonischen Aussagenkalkül. Anzeigner Akademie der Wissenschaften im Wien, Mathematish-Naturwissenschaftliche Klasse, 69, 65–66 (1932); "On the intuitionistic propositional calculus", reprinted in Kurt Gödel, Collected Works, vol. 1. Oxford University Press, New York (1986)
5. Gottwald, S.: Many-valued logic. In: Zalta, E.N. (ed.) The Stanford Encyclopedia of Philosophy, Spring 2015 edition. http://plato.stanford.edu/entries/logic-manyvalued/ (First published Tue Apr 25, 2000; substantive revision Thu Mar 5, 2015)
6. Hájek, P.: Metamathematics of Fuzzy Logic, Trends in Logic, vol. 4. Kluwer Academic Publishers, Dordrecht (1998)
7. Lakatos, G., Nagy, B.: Games with few players. In: Proceedings of ICAI'2004: 6th International Conference on Applied Informatics, Eger, Hungary, pp. II-187-196 (2004)
8. Melkó, E., Nagy, B.: Optimal strategy in games with chance nodes. Acta Cybern. **18**, 171–192 (2007)
9. Nagy, B.: Many-valued logics and the logic of the C programming language. In: Proceedings of ITI 2005: 27th International Conference on Information Technology Interfaces (IEEE), Cavtat, Croatia, pp. 657–662 (2005)
10. Nagy, B.: A general fuzzy logic using intervals. In: Proceedings of 6th International Symposium of Hungarian Researchers on Computational Intelligence, Budapest, Hungary, pp. 613–624 (2005)
11. Nagy, B.: Reasoning by intervals. In Proceedings of Diagrams 2006: Fourth International Conference on the Theory and Application of Diagrams, Stanford, CA, USA, LNCS-LNAI, vol. 4045, pp. 145–147 (2006)

12. Rich, E., Knight, K.: Artificial Intelligence. McGraw-Hill Inc., New York (1991)
13. Russell, S., Norvig, P.: Artificial Intelligence, a Modern Approach. Prentice-Hall, New Jersey (2003)
14. Zadeh, L.A.: Fuzzy Sets, Fuzzy Logic, and Fuzzy Systems: Selected Papers by Lotfi A. Zadeh. In: Klir, G.J., Yuan, B (eds.) Advances in Fuzzy Systems—Applications and Theory, vol. 6. World Scientific, River Edge (1996)

Coefficient of Variation Based Decision Tree for Fuzzy Classification

K. Hima Bindu and C. Raghavendra Rao

Abstract This paper considers a decision system with a fuzzy decision attribute (with finite set of values) to account for uncertainty. A novel fuzzy classification approach using a class of decision trees is developed. A decision tree is constructed for each decision category using Coefficient of Variation Gain as the attribute selection measure. A metric based on Residual Sum of Squares (RSS) to compare the fuzzy classifier is presented. The methodology of constructing the classifier and its performance aspects are presented.

Keywords Fuzzy decision tree · CvGain · CvDT · Residual sum of squares

1 Introduction

Fuzzy logic allows graded membership that helps to resolve conflicts arising due to imprecise and vague data of real world. Classification, a supervised learning approach tries to identify the class of a data object. In many applications, the decision attribute may be fuzzy: doctors may not be able to identify a disease with certainity based on patients case history, different doctors can predict different diseases for a patient, insurance company may not precisely classify an accident case, yield of a crop cannot be classified in crisp, Customer Relation Mangement may not be able to classify the customer in specific.

Decision tree is a supervised learning method used for classification and regression. It creates a model to predict the value of a target variable (class) by learning decision rules inferred from the data features.

When there is vagueness in decision-making (vagueness in identifying the target class), conventional decision tree confines to a class, based on some heuristics, which

K. Hima Bindu (✉)
Vishnu Institute of Technology, Bhimavaram 534202, Andhra Pradesh, India
e-mail: himabindu.k@vishnu.edu.in

C. Raghavendra Rao
University of Hyderabad, Central University, P.O., Prof. C.R.Rao Road, Gachibowli, Hyderabad 500046, Telangana, India

V. Ravi et al. (eds.), *Proceedings of the Fifth International Conference on Fuzzy and Neuro Computing (FANCCO - 2015)*, Advances in Intelligent Systems and Computing 415, DOI 10.1007/978-3-319-27212-2_11

may lead to wrong decisions. Instead, the decision can be fuzzy, with membership to all or some of the classes. Therefore, in spite of fitting the object into a wrong class, its membership to the classes helps to make better decisions. This approach is akin to human thinking and can lead to better decision making by human intervention if need arises to break the tie in favor of a single class. Fuzzy decision-making can improve the robustness and generalization capabilities of a classifier due to the elasticity of fuzzy sets formalism [12]. It is applicable for noisy and imprecise/vague data.

The paper is organized as follows. Section 2 recalls the literature on fuzzy decision trees. Section 3 describes the fuzzy classification method. Illustration of the fuzzy classification method is given in Sect. 4. Experimental results are presented in Sect. 5. The paper concludes with Sect. 6.

2 Literature Survey

Decision trees (e.g. ID3) [14] work with discrete attributes. Given the features (conditional attributes), the decision tree is traversed to reach a leaf node representing a class. Several revised algorithms are available for numerical data [2]. These algorithms follow various discretization or quantization techniques to convert numerical attributes to discrete attributes. These techniques have quantization error that affects the decision tree performance.

Few algorithms were proposed to use fuzzy intervals to overcome the quantization error [15, 19]. Neuro fuzzy technique was proposed in [5] for the same purpose. However, these approaches generate many fuzzy rules and their understandability is low [21]. Fuzzy ID3 Algorithm [21] overcomes these limitations by constructing decision tree using fuzzy entropy. Fuzzy entropy calculation is based on probability of fuzzy membership of data rather than probability of ordinary data. Fuzzy entropy is defined in various ways [7, 13, 21].

A fuzzy classifier with feature selection is proposed in [8]. It uses fuzzy entropy measure to partition the feature space into non overlapping decision regions. It uses k-means clustering to identify the number of fuzzy intervals and interval generation. Various fuzzy classification methods are available in [1, 3, 6, 9–11, 16, 20], including neuro fuzzy classifiers. However, neuro fuzzy techniques are considered less interpretable and they take more time to train the network.

A comparative study of three heuristics (fuzzy entropy, classification ambiguity, degree of importance of an attribute contributing to the classification) for generating a fuzzy decision tree is available in [22].

Soft decision tree [12] can automatically generate the best fuzzy split for the most discriminating attribute (splitting attribute) by minimizing the squared error function. The tree construction uses two separate sets of objects called Growing set and Pruning set. It follows pruning techniques to simplify the tree and reduce error. It uses refitting and back fitting techniques to improve generalization techniques of the tree. Classification with crisp decision tree is performed by traversing the tree from root until reaching a leaf. This traversal is always unique for a given data object.

Nevertheless, fuzzy ID3 decision tree traversal leads to multiple branches due to usage of fuzzy conditional attributes. The fuzzy ID3 approach performs multiplication, addition and normalization operations [21] to arrive at a result of the form "C1 with 0.69 and C2 with 0.31".

As enough literature exists for quantizing the attributes and due to the importance of generating fuzzy memberships for decision attribute, the method presented in this paper focuses on using discrete conditional attributes and fuzzy memberships for decision attribute.

3 Methodology

Assume a decision table DT with attributes $\{A_1, A_2, ..., A_n\}$. Attributes A_1, A_2, ... A_{n-1} are the conditional attributes and A_n is the decision attribute with m classes $C_1, C_2, ..., C_m$, denoted as $A_n = \{C_1, C_2, ..., C_m\}$.

A fuzzy decision table is formed by the crisp conditional attributes and the given fuzzy membership values for each of $C_1, C_2, ..., C_m$. Let $F = \{F_1, F_2, ..., F_m\}$ be the fuzzy membership values corresponding to the classes $C_1, C_2, ..., C_m$. m decision tables are built using the fuzzy membership values. In the decision table for a given class, its associated fuzzy membership values are used in place of the decision attribute. The decision tables formed are DT_1 with $\{A_1, A_2, ..., A_{n-1}, F_1\}$, DT_2 with $\{A_1, A_2, ..., A_{n-1}, F_2\}$,..., DT_m with $\{A_1, A_2, ..., A_{n-1}, F_m\}$. m decision trees are built ($Tree_1, Tree_2, ..., Tree_m$ for DT_1, DT_2, ..., DT_m) using these m decision tables with CvGain [4] as splitting criteria (using FuzzyCvDT Algorithm). The leaves in the decision tree $Tree_i$ will correspond to the fuzzy membership of C_i (i.e. F_i).

To predict the class of a new data object, it is fed to the m decision trees. From the output of these m decision trees w.r.t $C_1, C_2, ..., C_m$, fuzzy membership vector for the object will be generated with normalization.

3.1 CvGain

The decision trees $Tree_1, Tree_2, ..., Tree_m$ are constructed using CvGain [4] as the attribute selection measure. CvGain computation is based on Cv (Coefficient of Variation). Cv is a normalized measure of dispersion of a probability distribution. The computation of Cv is less expensive (when compared to Entropy using a logarithmic function) as it uses simple arithmetic operations and square root function. Coefficient of Variation of fuzzy membership of ith class (F_i) is computed using Eq. 1 where σ designates standard deviation and μ designates mean.

$$Cv(F_i) = \frac{\sigma(F_i)}{\mu(F_i)} \tag{1}$$

CvGain of attribute A_k, $k = 1, ..., n-1$ is computed using conditional *CvGain* of A_k on F_i as given in Eq. 2.

$$CvGain(F_i|A_k) = Cv(F_i) - Cv(F_i|A_k) \tag{2}$$

where

$$Cv(F_i|A_k) = \sum_{j=1}^{v} P_j \, Cv(F_i|A_k = a_j) \tag{3}$$

and a_j is the jth possible value of A_k with probability P_j. Considering the data set of Table 1 with DT_1 with attributes {Outlook, Temp, Humidity, Windy, F_{Yes} }, *CvGain*(Outlook) is computation based on Outlook = {Overcast, Rain, Sunny} as illustrated below.

$$CvGain(Outlook|F_{yes}) = Cv(F_{yes}) - Cv(F_{yes}|Outlook) \tag{4}$$

$$\begin{aligned} Cv(F_{yes}|Outlook) = P_{Outlook=Overcast}\ & Cv(F_{yes}|Outlook = Overcast)+ \\ P_{Outlook=Rain}\ & Cv(F_{yes}|Outlook = Rain)+ \\ P_{Outlook=Sunny}\ & Cv(F_{yes}|Outlook = Sunny) \end{aligned} \tag{5}$$

$Cv(F_{yes}|Outlook) = \frac{4}{14}Cv(0.12, 0.1, 0.41, 0.26) + \frac{5}{14}Cv(0.28, 0.82, 0.15, 0.12, 0.64)$ $+ \frac{5}{14}Cv(0.84, 0.83, 0.58, 0.44, 0.25) = 54.5905$. From Eqs. 4 and 5, $CvGain(F_{yes}|Outlook) = Cv(F_{yes}) - 54.5905 = 64.3344 - 54.5905 = 9.7439$.

3.2 *FuzzyCvDT Algorithm*

Normally decision trees work with categorical decision attribute (to predict the class label). Also, the conditional attributes need to be discrete to apply any decision tree algorithm. Many methods exist to convert conditional attribute to discrete [18]. This paper assumes conditional attributes to be discrete and presents an approach to build *m* decision trees (see Algorithm 1). Each decision tree models each class fuzzy membership value. The CvDT algorithm given in [4] is modified to handle the decision attribute being numeric (to model fuzzy membership), it is given in Algorithm 2. When *Cv* of the decision attribute is below α (used 0.33 for illustration) then the objects at the node are considered to be homogeneous, hence a leaf is generated. Its prediction value is the mean of the fuzzy memberships at the node. Similarly, when *CvGain* is negative, the node becomes a leaf with the mean of the fuzzy memberships at the node. Algorithm 3 presents the classifier. It outputs the fuzzy memberships of all classes for a given data object.

input : Decision table DT and Fuzzy memberships F where $|F| = m$
output: m fuzzyCvDTs $\{Tree_1, Tree_2, ..., Tree_m\}$

for $i \leftarrow 1$ **to** m **do**
  $DT_i \leftarrow A \setminus A_n \cup F_i$;
  $Tree_i \leftarrow$ `fuzzyCvDT`(DT_i);
end

Algorithm 1: BuildFuzzyClassifier

input : i^{th} Decision table DT_i with attributes $\{A_1, A_2, ..., A_{n-1}, F_i\}$
output: fuzzy decision tree $Tree_i$

create a node N ;
Let $A' \leftarrow \{A_1, A_2, ..., A_{n-1}\}$;
if `Cv`$(F_i) < \alpha \vee A' = \phi$ **then**
  return N as a leaf node with $\mu(F_i)$;
else
  for *each attribute* $A_i \in A'$ **do**
    $g_i \leftarrow$ `CvGain`(A_i, F_i) ;
  end
  Let $A_g \in A'$ be the attribute with highest CvGain ;
  if `CvGain` $(A_g, F_i) < 0$ **then**
    return N as a leaf node with $\mu(F_i)$;
  else
    create a decision node DN on A_g ;
    Let $\{v_1, v_2, ..., v_k\}$ be the k possible values of A_g;
    partition DT_i into $DT_1, DT_2, ..., DT_k$ based on values of A_g ;
    for $j \leftarrow$ *1 to* k **do**
      if $DT_j \neq \phi$ **then**
        create a branch node N_j for v_j as child of DN ;
        $N_j \leftarrow$ `fuzzyCvDT`(DT_j) ;
      end
    end
  end
end

Algorithm 2: FuzzyCvDT

input : Conditional attribute values of the data object d for attributes $\{A_1, A_2, ..., A_{n-1}\}$
output: Fuzzy memberships $f_1, f_2, ..., f_m$

for $i \leftarrow 1$ **to** m **do**
  $f_i \leftarrow Tree_i$ (d) ;
end

Algorithm 3: FuzzyClassifier

4 Illustration

Consider the Weather data set with fuzzy memberships for class = {Yes, No}. The dataset with fuzzy memberships of class = Yes (F_{Yes}) and class = No (F_{No}) is shown in Table 1. When the fuzzy membership is above 0.5 then the object belongs to the class. Hence first data object (Overcast, Hot, high, False) belongs to No class. The decision tables formed are DT_1, DT_2. Attributes Outlook, Temp, Humidity, Windy, F_{Yes} forms DT_1. Attributes Outlook, Temp, Humidity, Windy, F_{No} forms DT_2.

With DT_1, CvGain(Outlook) = 9.7439, CvGain(Temp) =1.8127, CvGain(Humidity) = 7.4535, CvGain(Windy) = −0.0412. Hence, the attribute 'Outlook' becomes the decision node at root. Later along the branch 'Rain', CvGain (Temp) − 0.4138, CvGain(Humidity) = 2.6224, CvGain(Windy) = 41.7516. Hence, 'Windy' becomes the decision node along this branch. Continuing this way, the decision trees are built. The fuzzy CvDT for the class = Yes is shown in Fig. 1 and the fuzzy CvDT for class = No is shown in Fig. 2.

Fuzzy memberships for a data object are obtained by taking the outputs of these trees. For the object (Sunny,Cool,normal,False), fuzzy membership output from FuzzyCvDT for F_Yes is 0.40 and fuzzy membership output from FuzzyCvDT for F_NO is 0.62. As can be observed from the Table 1, the original fuzzy memberships are 0.44 and 0.56. The fuzzy rules followed for this object are

```
IF Outlook is Sunny AND Humidity is Normal AND Temp is Cool
THEN 'Play = Yes' with Fyes = 0.40
IF Outlook is Sunny AND Humidity is Normal
THEN 'Play = No' with Fno = 0.62
```

Table 1 Weather dataset with fuzzy memberships for decision attribute

Outlook	Temperature	Humidity	Windy	YesFM	NoFM
Overcast	Hot	High	False	0.12	0.88
Overcast	Mild	High	True	0.10	0.90
Overcast	Hot	Normal	False	0.41	0.59
Overcast	Cool	Normal	True	0.26	0.74
Rain	Mild	High	False	0.28	0.72
Rain	Mild	High	True	0.82	0.18
Rain	Cool	Normal	False	0.15	0.85
Rain	Mild	Normal	True	0.12	0.88
Rain	Cool	Normal	False	0.64	0.36
Sunny	Hot	High	True	0.84	0.16
Sunny	Mild	High	False	0.83	0.17
Sunny	Hot	High	True	0.58	0.42
Sunny	Cool	Normal	False	0.44	0.56
Sunny	Mild	Normal	True	0.25	0.75

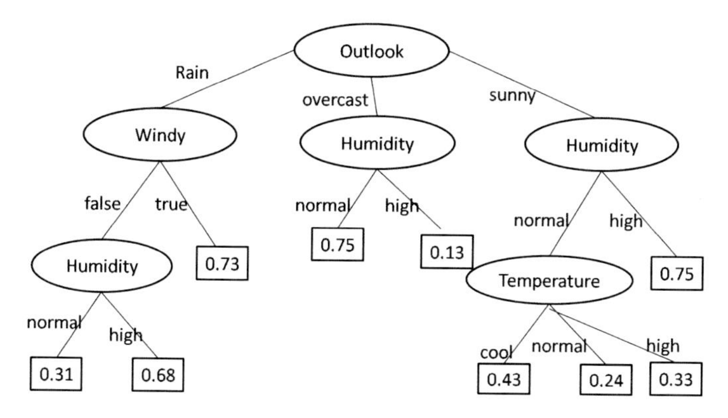

Fig. 1 Fuzzy CvDT for class = Yes

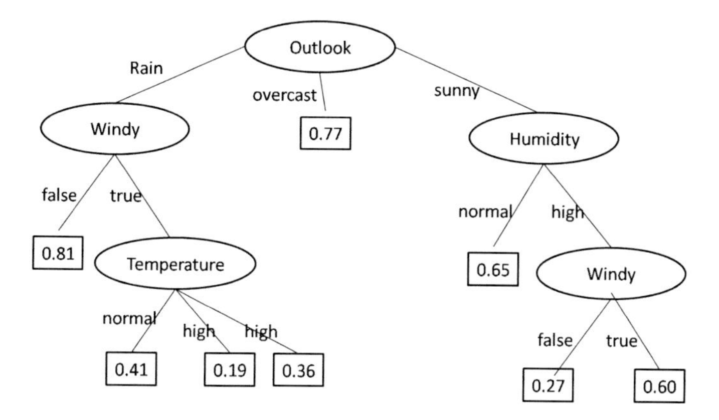

Fig. 2 Fuzzy CvDT for class = No

5 Experiment and Results

The proposed fuzzy decision classification system outputs the fuzzy membership of each class for the the data object. Hence, residual sum of squares (RSS) [17] is used to measure the departure in the fuzzy membership from actual to the predicted value. The experiment is carried with ten fold crossvalidation (10FCV) and the mean RSS value is reported for various datasets. The mean RSS for dataset of Table 1 is 0.0059. The mean RSS with 10FCV on Iris dataset is 0.015. The fuzzy memberships are assigned such that the fuzzy membership of the class to which the data belongs dominates the fuzzy memberships of other classes for demonstration.

The influence of fuzzy membership on the classification performance is observed by varying the fuzzy membership from 0.51 to 0.99. Figure 3 shows the average RSS behavior using the weather dataset of Table 1 with 10FCV. From Fig. 3, it is clear

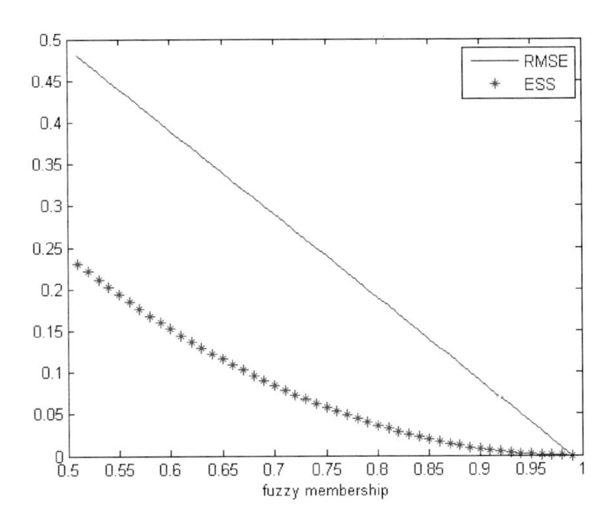

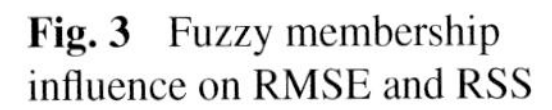
Fig. 3 Fuzzy membership influence on RMSE and RSS

that the RSS decreases as fuzzy membership for a class is nearing 1. Even though the uncertainty is high when fuzzy membership is near to 0.5, it can be observed that RSS value is low (below 0.25). As the uncertainty is reducing when fuzzy membership is increasing, RSS falls exponentially. Root Mean Squared Error (RMSE), the square root of RSS is falling linearly and is approaching 0 as fuzzy membership is approaching 1.

The performance of the fuzzy classification system using FuzzyCvDT with defuzzification is compared against ID3 and CvDT (with 10 fold cross validation). The fuzzy classification inference procedure is different from the crisp approaches. The experiment is carried out using ten datasets from UCI machine learning repository. The conditional attributes are discretized and fuzzy memberships are generated. The fuzzy memberships used are normalized and the dominating class has high fuzzy membership.The results are shown in Table 2. To find accuracy of the fuzzyCvDT, the fuzzy memberships are defuzzified and the dominating fuzzy membership decides the class of the data. This defuzzification without domain knowledge is used for the sake of experiment. It works well when the number of classes is low. As the number of classes increase, the entropy among the fuzzy memberships is high and the fuzzy memberships are close to each other. Hence, when the number of classes is high, defuzzification can force the data object to a wrong class. This can be observed from the accuracies of Ecoli, Yeast, Page-blocks and Wine red datasets. As reported in [4], the accuracies of ID3 and CvDT do not vary much. The accuracy of fuzzyCvDT is low when uncertainty of crisp classifier is low due to defuzzication error. But when the uncertainty is high, considering Blood Transfusion and Abalone datasets, the ac- curacy of fuzzyCvDT is high.

Defuzzification is needed to compare fuzzy classifier against convetional crisp classifiers. However, defuzzification has an error attached to it. Hence, Residual Sum of Squares(RSS) is adopted to measure the departure from the actual fuzzy memberships. The output of the crisp classifiers is considered as a bit vector (length of the

Table 2 Comparison with crisp decision trees

S.No	Dataset	#class	Accuracy			RSS(10E-2)		
			CvDT	ID3	FuzzyCvDT	CvDT	ID3	FuzzyCvDT
1	Iris	3	98	98	98.67	1.56	1.56	0.91
2	Wine	3	96.11	98.89	94.44	1.11	0.56	0.37
3	Breast cancer	2	98.86	99.43	95.85	0.71	0.86	0.93
4	Blood transfusion	2	82.07	82.07	88.65	17.37	17.37	4.48
5	Abalone	3	85.23	85.18	95.28	9.77	9.71	1.99
6	Ecoli	8	95.59	95	77.29	1.21	0.96	0.13
7	Yeast	10	93.11	92.57	69.93	1.16	1.2	0.09
8	Page-blocks	5	97.44	97.46	89.76	1.03	1	0.67
9	Wine red	6	95.44	95.5	88.43	1.22	1.26	0.25
10	Pima-Indians	2	94.68	94.94	96.36	3.7	4.03	1.49

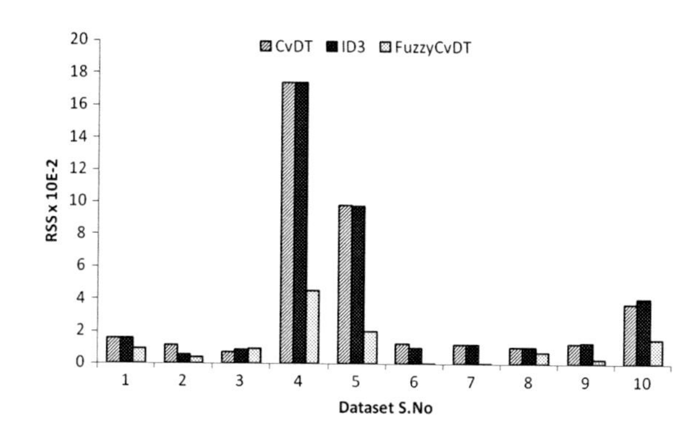

Fig. 4 RSS comparison

vector is equal to number of class categories). If the number of classes is 3, and the crisp classifier output is 2, then the vector is (0,1,0). RSS value is computed by considering the actual class vector and output vector of the classifier. In case of FuzzyCvDT the vector is made of output fuzzy memberships of each class. The average RSS values are reported in Table 2.

The fuzzy classification approach is able to handle uncertainty better as is evident from the results of Blood Transfusion and Abalone datasets (see Fig. 4). These are the datasets where uncertainty is high and the crisp classifiers are inferior. For these datasets the fuzzy classification method has better accuracy and average RSS is low. It is observed that fuzzy classifier RSS values are falling down with the number of class categories (see Fig. 5). Among all datasets, 'Blood Transfusion' performance is poor.

A simulation study with arbitrary fuzzy memberships is carried out to study the robustness of the approach w.r.t fuzzy membership values. The fuzzy membership values are chosen at random retaining the semantics of the class knowledge (i.e., the fuzzy membership values can vary but there is an agreement in semantics). The simulation (10 executions with different fuzzy memberships, each execution with 10FCV) results revealed that mean RSS is low with less standard deviation (see Fig. 6). Hence, the fuzzy classification approach developed can be considered as robust or consistent.

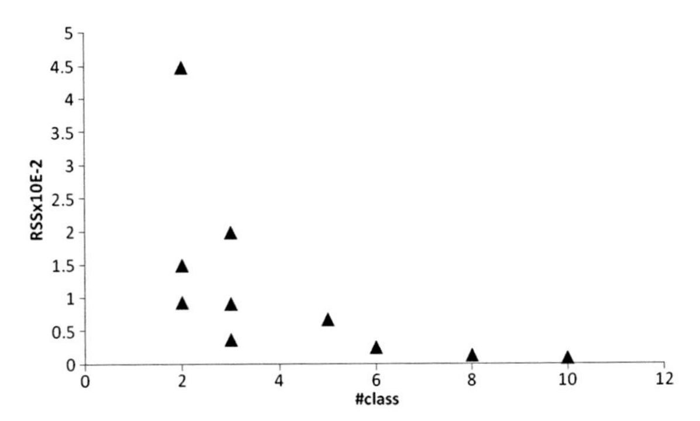

Fig. 5 RSS behaviour

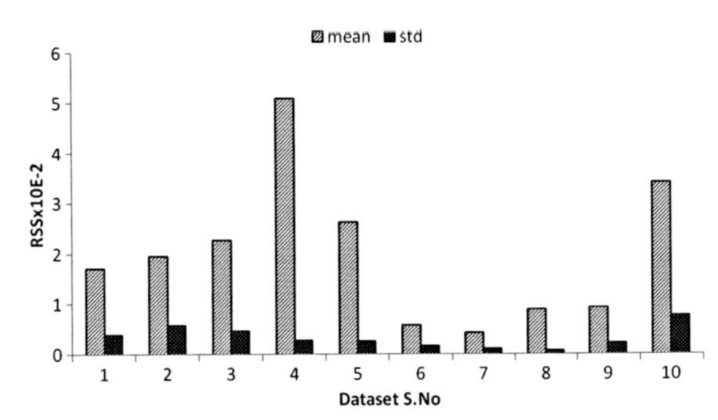

Fig. 6 RSS mean and std

6 Conclusion

A fuzzy classification system with a class of decision trees (FuzzyCvDTs) developed in this paper addresses uncertainty in decision attribute. The fuzzy CvDT method to build a decision tree based on fuzzy membership values is presented. RSS values demonstrated that when the dataset has high uncertainty, the fuzzy classifier is performing better.

Acknowledgments This work is supported by Grant No. SB/FTP/ETA-0194/2014 from Science and Engineering Research Board, India.

References

1. Abe, S., Thawonmas, R.: A fuzzy classifier with ellipsoidal regions. IEEE Trans. Fuzzy Syst. **5**(3), 358–368 (1997)
2. Berzal, F., Cubero, J.-C., Marín, N., Sánchez, D.: Numerical attributes in decision trees: a hierarchical approach. In: Berthold, M., Lenz, H.-J., Bradley, E., Kruse, R., Borgelt, C. (eds.) IDA 2003. LNCS, vol. 2810, pp. 198–207. Springer, Heidelberg (2003)
3. Genther, H., Glesner, M.: Advanced data preprocessing using fuzzy clustering techniques. Fuzzy Sets Syst. **85**(2), 155–164 (1997). http://www.sciencedirect.com/science/article/pii/0165011495003584 (methods for Data Analysis in Classification and Control)
4. Hima Bindu, K., Swarupa Rani, K., Raghavendra Rao, C.: Coefficient of variation based decision tree (cvdt). Int. J. Innovative Technol. Creative Eng. **1**(6), 1 (2011)

5. Ichihashi, H.: Tuning fuzzy rules by neuro-like approach. J. Jpn. Soc. Fuzzy Theor. Syst. **5**(2), 191–203 (1993)
6. Ishibuchi, H., Nozaki, K., Yamamoto, N., Tanaka, H.: Construction of fuzzy classification systems with rectangular fuzzy rules using genetic algorithms. Fuzzy Sets Syst. **65**(23), 237–253 (1994). http://www.sciencedirect.com/science/article/pii/0165011494900221 (fuzzy Methods for Computer Vision and Pattern Recognition)
7. Kosko, B.: Fuzzy entropy and conditioning. Inf. Sci. **40**(2), 165–174 (1986). http://www.sciencedirect.com/science/article/pii/002002558690006X
8. Lee, H.M., Chen, C.M., Chen, J.M., Jou, Y.L.: An efficient fuzzy classifier with feature selection based on fuzzy entropy. IEEE Trans. Syst. Man Cybern. Part B Cybern. **31**(3), 426–432 (2001)
9. Mandal, D.P.: Partitioning of feature space for pattern classification. Pattern Recogn. **30**(12), 1971–1990 (1997). http://www.sciencedirect.com/science/article/pii/S0031320397000125
10. Nauck, D., Kruse, R.: A neuro-fuzzy method to learn fuzzy classification rules from data. Fuzzy Sets Syst. **89**(3), 277–288 (1997), http://dx.doi.org/10.1016/S0165-0114(97)00009-2
11. Nozaki, K., Ishibuchi, H., Tanaka, H.: A simple but powerful heuristic method for generating fuzzy rules from numerical data. Fuzzy Sets Syst. **86**(3), 251–270 (1997). http://www.sciencedirect.com/science/article/pii/016501149500413O
12. Olaru, C., Wehenkel, L.: A complete fuzzy decision tree technique. Fuzzy Sets Syst. **138**(2), 221–254 (2003). http://dx.doi.org/10.1016/S0165-0114(03)00089-7
13. Pal, S., Chakraborty, B.: Fuzzy set theoretic measure for automatic feature evaluation. IEEE Trans. Syst. Man Cybern. **16**, 754–760 (1986)
14. Quinlan, J.R.: Induction of decision trees. Mach. Learn. **1**(1), 81–106 (1986). http://dx.doi.org/10.1023/A:1022643204877
15. Sakurai, S., Araki, D.: Application of fuzzy theory to knowledge acquisition. In: 15th Intelligent System Symposium (Society of Instrument and Control Engineers), pp. 169–174 (1992)
16. Simpson, P.K.: Fuzzy min-max neural networks-part 1: classification. IEEE Trans. Neural Netw. **3**, 776–786 (1992)
17. Snedecor, G.W., Cochran, W.G.: Statistical Methods, Seventh Edition. Iowa State University, Ames (1980)
18. Tan, P.N., Steinbach, M., Kumar, V.: Introduction to Data Mining, 1st edn. Addison-Wesley Longman Publishing Co. Inc., Boston (2005)
19. Tani, T., Sakoda, M.: Fuzzy oriented expert system to determine heater outlet temperature applying machine learning. In: 7th Fuzzy System Symposium (Japan Society for Fuzzy Theory and Systems), pp. 659–662 (1991)
20. Uebele, V., Abe, S., Lan, M.S.: A neural-network-based fuzzy classifier. IEEE Trans. Syst. Man Cybern. **25**(2), 353–361 (1995)
21. Umano, M., Okamoto, H., Hatono, I., Tamura, H., Kawachi, F., Umedzu, S., Kinoshita, J.: Fuzzy decision trees by fuzzy id3 algorithm and its application to diagnosis systems. In: Third IEEE Conference on Fuzzy Systems IEEE World Congress on Computational Intelligence, pp. 2113–2118. IEEE (1994)
22. Wang, X., Yeung, D., Tsang, E.: A comparative study on heuristic algorithms for generating fuzzy decision trees. IEEE Trans. Syst. Man Cybern. Part B Cybern. **31**(2), 215–226 (2001)

Predicting Protein Coding Regions by Six-Base Nucleotide Distribution

Praveen Kumar Vesapogu, Changchuan Yin and Bapi Raju Surampudi

Abstract Identification of protein coding regions in genomic sequences is a significant problem in bioinformatics. A significant number of techniques for identifying coding regions in genomic sequences are based on Fourier representation for the sequence of nucleotides and spectral analysis methods in order to detect protein coding regions. These methods usually consider 3-base periodicity signal. In this paper, we propose a method that analyzes the distribution of nucleotides at six positions (hexamer distribution). We present the six-base nucleotide distribution (SBND) algorithm and implementation results on genomic sequences of *Drosophila Melanogaster*, *Rat*, *Cow* and also from *H. sapiens*. The performance is compared with one of the best methods available in the literature and the results establish the viability of the proposed approach.

Keywords Exon · Intron · Hexamer

1 Introduction

In genome research predicting protein coding regions in DNA sequences is one of the primary tasks. A protein coding region is defined as any pattern in a DNA sequence under proper conditions which results in the generation of a protein product [8].

P.K. Vesapogu (✉)
School of Computer and Information Sciences, University of Hyderabad,
Hyderabad 500046, Telangana, India
e-mail: praveencs@uohyd.ac.in

C. Yin
Department of Mathematics, Statistics and Computer Science, The University of Illinois at Chicago, Chicago, IL 60607-7045, USA
e-mail: cyin1@uic.edu

B.R. Surampudi
Cognitive Science Lab, International Institute of Information Technology, Hyderbad 500032, Telangana, India
e-mail: raju.bapi@iiit.ac.in

V. Ravi et al. (eds.), *Proceedings of the Fifth International Conference on Fuzzy and Neuro Computing (FANCCO - 2015)*, Advances in Intelligent Systems and Computing 415, DOI 10.1007/978-3-319-27212-2_12

usually protein coding (exons) regions and non-coding regions (introns) are interrupted by each other. A variety of computational methods were developed, for identifying protein coding regions, but still it is difficult to identify short exons accurately.

During the past decades numerous methods were developed for discriminating protein coding regions (exons) and non-coding regions (introns) such as Fourier spectrum analysis [21], Markov models [11, 25], Spectral envelope [20], Neural networks [6, 19], EPND (Exon Prediction by Nucleotide Distribution) [21, 29]. Fourier analysis is based on extracting the distinct feature of protein coding regions i.e., 1/3 periodicity [21]. Tiwari et al. converted symbolic DNA sequences in to binary numerical sequences and applied Fourier Transform on the binary sequences to find the 1/3 periodicity. They employed sliding window approach for predicting the protein-coding regions. Stoffer et al. proposed spectral envelope which is frequency based principal components technique, for finding protein-coding regions in the DNA sequences [20]. Kotlar et al. extracted and applied spectral rotation measure for gene prediction [10]. Along with 3-base periodicity Gao et al. used fractal features of DNA sequences for gene identification [9]. Krogh et al. constructed Markov models for identifying protein coding regions [11]. Lapedes et al. implemented neural networks to predict the protein-coding regions [6]. Yin and Yau proposed EPND (Exon prediction by nucleotide distribution) algorithm [29]. They calculated the frequencies of occurrence, of the nucleotides A, T, G, and C in the three codon positions. They calculated 3-base periodicity for each DNA walk, based on the slope they decided the nucleotide as exon nucleotide or an intron nucleotide. They have improved the prediction accuracy of the EPND algorithm using different starting points in the DNA sequence. Saberkari et al. [15] used notch filter along with discrete Fourier transform (DFT) to ameliorate the quality of detecting protein coding sequences. Yin et al. [27] applied a nonlinear Tracking-Differentiator to denoise 3-base periodicity of protein coding regions.

Because of the rapid growth of raw genome sequence data, still there is a great need of novel methods for the accurate prediction of protein coding regions in DNA sequence. We propose a new algorithm which is based on six-base nucleotide distribution for finding protein coding regions.

2 Background

Discrete Fourier Transform (DFT) is used for spectrum analysis of DNA sequences. DNA sequence of length n is converted into four binary sequences each of length n [21]. Four binary sequences $U_A(n), U_T(n), U_G(n)$ and $U_C(n)$ for the four nucleotides, $A, T, G,$ and C are given as [21]:

$$U_\alpha(x_j) = \begin{cases} 1, \text{ if } x_j = \alpha \\ 0, \text{ otherwise;} \end{cases}$$

$$\text{for } j = 1, \ldots, n;$$

For each of the four binary sequences, power spectrum is calculated and the sum of the resulting power spectrum of the binary sequences is the Fourier power spectrum of the DNA sequence, is defined as [21]:

$$S(f) = \sum_{\alpha} S_{\alpha}(f) = \sum_{\alpha} 1/n^2 \left| \sum_{j=1}^{n} U_{\alpha}(x_j) \exp 2\pi i f j \right|^2$$

Here, frequency is denoted by $f = k/n$, with $k = 1, 2, ...n/2$. Based on the period-3 signal of the power spectrum Tiwari et al. [21] predicted the protein coding regions in DNA sequence. Yin and Yau [28] showed that period-3 signal is due to unequal distribution of A, T, G, and C nucleotides at first, second and third positions of the codon. They concluded that there is linear relationship at $n/3$ between nucleotide distribution on the three codon positions and power spectrum.

It is well known that sequence of six nucleotides called *hexamer* is biologically meaningful. Fickett et al. and Claverie et al. suggested that six nucleotide sequence (hexamer) would be effective for finding protein coding regions [4, 8]. Fickett et al. showed that simple counting of oligomers measure is more effective than sophisticated measures for finding protein coding regions and they concluded that in-phase hexanucleotides is the most efficient coding measure [8]. However in the literature currently the signal based methods are dominated by looking at periodicity at three bases. We can infer that if these methods are extended for six-base positions, i.e., looking at hexamer distribution it may lead to a novel approach for identifying protein coding regions. With this motivation, in this paper we propose a new algorithm that analyses distributions at hexamer positions in genomic sequence. We extend the EPND (Exon prediction by nucleotide distribution) in order to analyze nucleotide distribution at hexamer positions [29]. The results indicate that the proposed approach gives significantly better results compared to earlier methods such as Tiwari et al. [21], Anastassiou et al. [1] and EPND (Exon prediction by nucleotide distribution) [29]. The remaining paper is organized as: the description of the proposed algorithm, datasets utilized, results and discussion which includes the limitations of the proposed approach and finally the conclusions.

3 Proposed Method

The significance of oligomers, in particular, hexanucleotides was addressed by Fickett et al. [8] for finding protein coding regions in the DNA sequences. The proposed six-base nucleotide distribution (SBND) algorithm is based on the variance of the occurrence frequencies of the four nucleotides A, T, G and C at $1, 2, \ldots, 6$ hexamer positions. A sliding window approach is adopted. For each window of DNA sequence occurrence frequency of each nucleotide A, T, G and C at $1, \ldots, 6$ hexamer positions are calculated, then variance of these occurrence frequencies is calculated and then ratio of the window variance is calculated within the length of the window.

Let $F_{x1}, F_{x2}, \ldots, F_{x6}$ be the frequencies of occurrence of the nucleotide $x \epsilon \{A, T, G, C\}$ in the first, second, and up to six hexamer positions, respectively. For a given window ω of DNA sequence, the variance of the six-base periodicity can be calculated as:

$$S(w_i) = \sum_{x=A,T,G,C} \left[\frac{1}{n-1} \sum_{j=1}^{n} (F_{xj} - \mu_x)^2 \right]$$

Here, F_{xj} is the occurrence frequency of nucleotide x at jth position, μ_x is mean of occurrence frequency of nucleotide x in six hexamer positions and $n = 6$ the length of hexamer.

The average window variance of the DNA sequence is defined as follows:

$$S(a_i) = \frac{S(w_i)}{\omega}$$

for $i = 1, \ldots, N$; where N is the length of the DNA sequence, w_i is the ith window and ω is the length of the window.

Figures 1, 2, 3 and 4 show the average six base variance $S(a)$ on pure coding regions and pure non-coding regions of various lengths of different organisms. From Figs. 1, 2, 3 and 4, it is clear that the variance of nucleotides A, T, G, and C at six nucleotide positions has some signal in the protein coding regions and the signal is

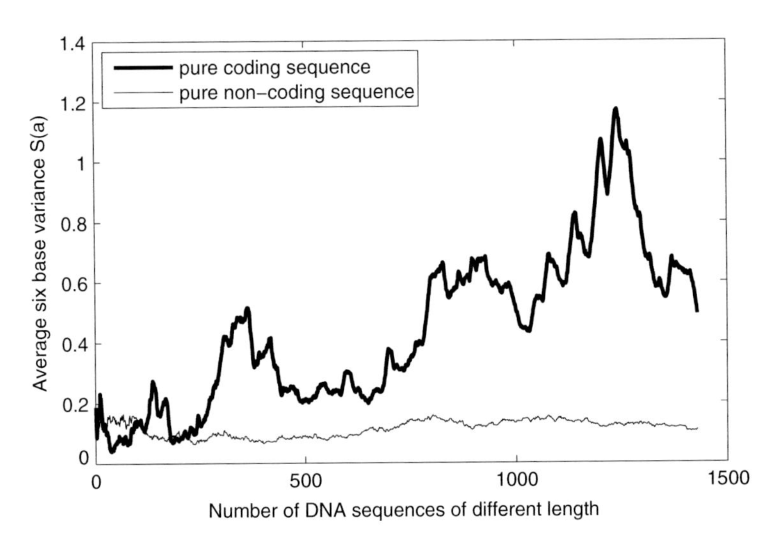

Fig. 1 Plot of average six base variance $S(a)$ on pure coding and pure non-coding sequences of *H. Sapiens*, of various lengths from 100 to 143000 bp. The *thick line* indicates the pure coding sequence and *thin line* indicates the pure non-coding sequence and can be separated at a threshold of 0.12

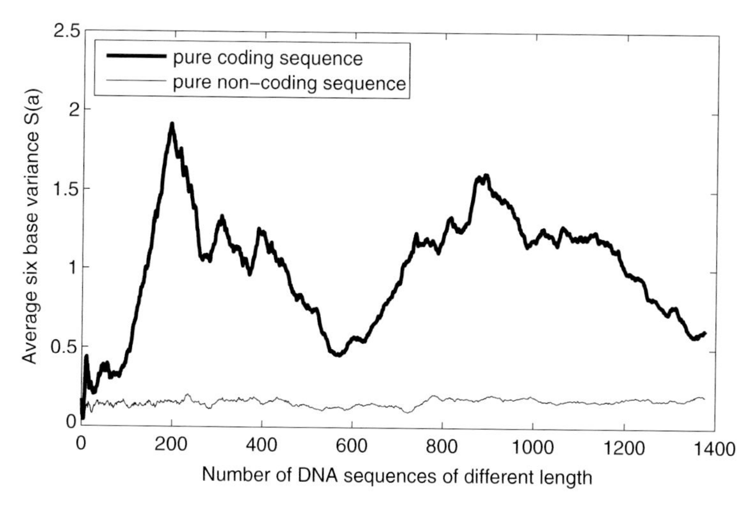

Fig. 2 Plot of average six base variance $S(a)$ on pure coding and pure non-coding sequences of *Cow*, of various lengths from 100 to 137000 bp. The *thick line* indicates the pure coding sequence and *thin line* indicates the pure non-coding sequence and can be separated at a threshold of 0.12

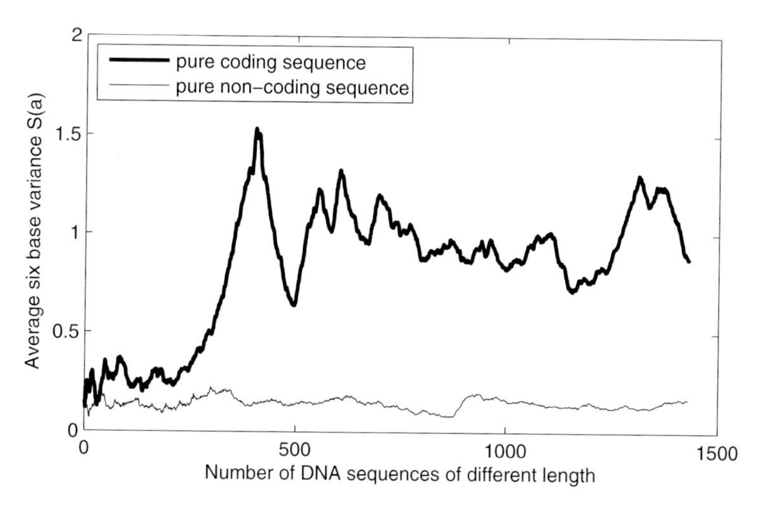

Fig. 3 Plot of average six base variance $S(a)$ on pure coding and pure non-coding sequences of *D. Melanogaster*, of various lengths from 100 to 143000 bp. The *thick line* indicates the pure coding sequence and *thin line* indicates the pure non-coding sequence and can be separated at a threshold of 0.12

absent in the non-coding regions. This gives motivation for the proposed Six-base nucleotide distribution (SBND) algorithm for finding protein coding regions in a DNA sequence.

It can be seen from the figures that coding and non-coding regions can be separated at a threshold value of $S(a) = 0.12$. The threshold value t is set to 0.12. Based

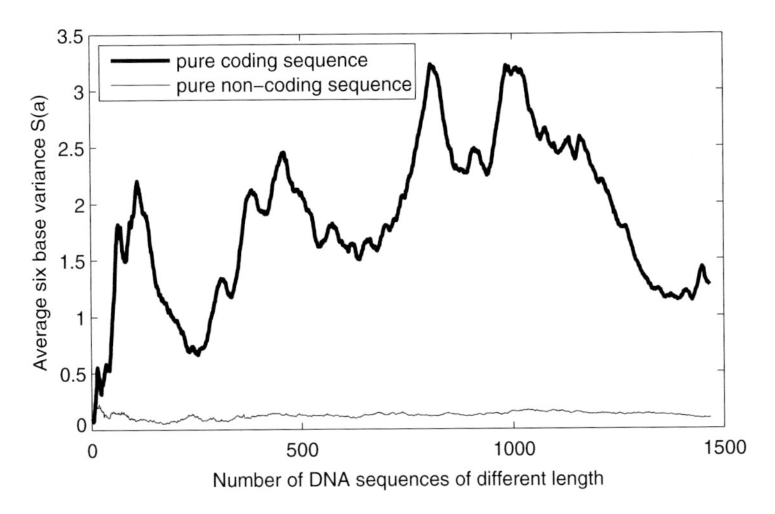

Fig. 4 Plot of average six base variance $S(a)$ on pure coding and pure non-coding sequences of *Rat*, of various lengths from 100 to 146000 bp. The *thick line* indicates the pure coding sequence and *thin line* indicates the pure non-coding sequence and can be separated at a threshold of 0.12

on these empirical observations, we formulated the following classification rule. If the ratio of six-base periodicity signal to the background noise of DNA sequence is less than 0.12 then the nucleotide which is at the middle of the window can be labeled as an intron, otherwise an exon.

4 Data Sets and Performance Measures

The data sets used for the performance evaluation of the proposed algorithm is Exon-Intron Database (EID) [17]. Data sets were downloaded from the website (www.meduohio.edu/bioinfo/eid/). Exon-Intron Database (EID) contains hEID file having fasta-formatted database of header information, exEID file having fasta-formatted database of exon sequences and intrEID file containing fasta-formatted database of intron sequences. To build full original gene structure, based on the position of exons and introns in the header sections, exons are combined with the relative introns. The full gene structure will end with a stop codon and precede with a start codon. Gene sequences in full-length are utilized to analyze the performance of the proposed algorithm.

The performance measures sensitivity, specificity and accuracy [3] are used to measure the performance of the proposed algorithm. The formulae for these measures are shown here: Sensitivity, $Sn = \mathrm{TP}/(\mathrm{TP}+\mathrm{FN})$, specificity, $Sp = \mathrm{TN}/(\mathrm{TN}+\mathrm{FP})$, and accuracy $\mathrm{Ac} = (\mathrm{TP}+\mathrm{TN})/(\mathrm{TP}+\mathrm{TN}+\mathrm{FP}+\mathrm{FN})$, where TP stands for the true positives, which is the number of exon nucleotides that have been correctly predicted as

Algorithm 1 Six-base nucleotide distribution (SBND)

Input: Read the DNA sequence D.
Output: Predicted DNA sequence with exon and intron labels.

N : Length of the DNA sequence D
ω : Window size
n : Length of hexamer
t : Threshold

1. set $windowsize = \omega$.
2. For each window w_i. $i = 1$ to N.
   (a) Calculate the occurrence frequencies $F_{x1}, F_{x2}, ..., F_{x6}$ of nucleotides A, T, G and C in 1, 2,..., 6 hexamer positions.
   (b) Calculate sum of six-base variance

$$S(w_i) = \sum_{x=A,T,G,C} \left[\frac{1}{n-1} \sum_{j=1}^{n} (F_{xj} - \mu_x)^2 \right]$$

   where F_{xj} is the occurrence frequency of nucleotide x at j^{th} position.
   μ_x is mean of occurrence frequency of nucleotide x in six hexamer positions.
   (c) Calculate average of window variance

$$S(a_i) = \frac{S(w_i)}{\omega}$$

   (d) If $S(a_i) >= t$, then set the ith nucleotide as exon, otherwise intron.
3. Slide the window to the next nucleotide position and go to step 2. Repeat until end of the sequence.

---

exons. TN is the true negatives, i.e., the number of intron nucleotides that have been correctly predicted as introns. FN is the false negatives, i.e., the number of intron nucleotides that have been predicted as exon. FP represents the false positives, which is the number of exon nucleotides predicted as introns. The percentage of protein coding segments that are correctly identified as coding cites is S_n. The percentage of protein noncoding segments that are correctly identified as noncoding is S_p. The average of S_n and S_p is accuracy Ac.

5 Results and Analysis

Figure 5 shows the plot of the proposed six-base nucleotide distribution method (SBND) on *H. sapiens* DNA sequence with exons and introns of length 2000 bp. If the $S(a)$ measure calculated for a nucleotide is greater than or equal to a threshold of 0.12, then that nucleotide is labelled as exon otherwise intron. It can be observed

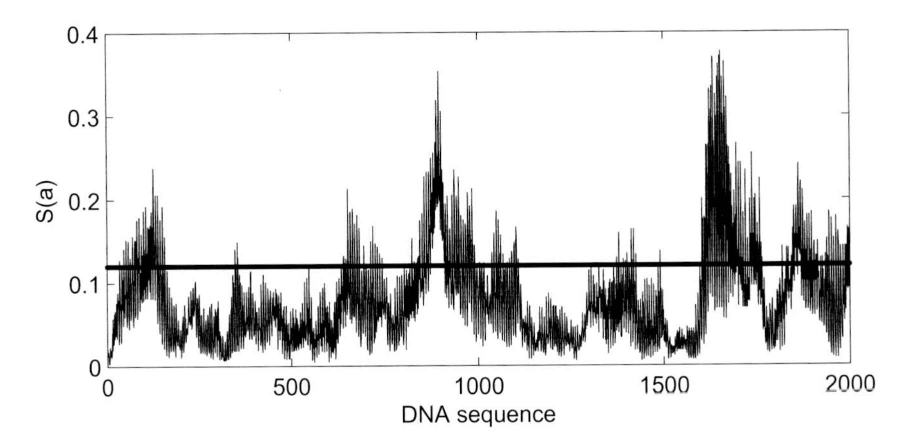

Fig. 5 Plot of the proposed method identifying coding and non-coding regions of *H. sapiens* DNA sequence. The horizontal line indicates the threshold set at 0.12 and all the nucleotide locations where the $S(a)$ value crosses the threshold are predicted as belonging to coding regions

Table 1 Performance evaluation of six-base nucleotide distribution (SBND) method for window size = 150

Organism	S_n	S_p	Ac
H. sapiens	0.33	0.89	0.72
Mouse	0.15	0.90	0.88
Rat	0.31	0.92	0.91

Table 2 Performance evaluation of six-base nucleotide distribution (SBND) method for window size = 350

Organism	S_n	S_p	Ac
H. sapiens	0.27	0.86	0.69
Mouse	0.02	0.89	0.87
Rat	0.44	0.93	0.92

from Fig. 5 that the proposed six-base nucleotide distribution (SBND) is good at predicting protein coding regions.

The results of the proposed algorithm with different window sizes are shown in Tables 1, 2 and 3. There is not much difference in the accuracy of the proposed algorithm with the window size between 150 and 500. The accuracy is decreasing when the window size is above 600 and when it is below 150. We have chosen window size equal to 150 for the rest of the experiments. Tables 1, 2 and 3 summarize the performance results of the algorithm with a window size of 150, 350, and 800, respectively across the three organisms of length 20,000 bp each.

Table 3 Performance evaluation of six-base nucleotide distribution (SBND) method for window size = 800

Organism	S_n	S_p	Ac
H. sapiens	0.24	0.83	0.66
Mouse	0.01	0.86	0.84
Rat	0.01	0.92	0.88

Table 4 Performance evaluation and comparison of EPND and Six-base nucleotide distribution

Oraganism	EPND			Six-base nucleotide distribution		
	S_n	S_p	Ac	S_n	S_p	Ac
H. sapiens	0.45	0.86	0.80	0.34	0.92	0.83
Mouse	0.44	0.85	0.84	0.28	0.90	0.89
Cow	0.18	0.84	0.81	0.10	0.91	0.89
D. melanogaster	0.47	0.84	0.75	0.34	0.89	0.76
Rat	0.18	0.86	0.84	0.11	0.89	0.87

We observed that the accuracy of the proposed Six-base nucleotide distribution (SBND) algorithm is better than that of Tiwari et al. [21], and that of Anastassiou [1]. These results are not shown here. We re-implemented one of the best available methods in the literature, namely, *EPND* (Exon prediction by nucleotide distribution) algorithm [29] and present only the comparative results with this method. Table 4 shows the performance of the EPND [29] and the proposed SBND method on five different organisms. Table 4 indicates that the proposed, SBND algorithm performed better than the EPND algorithm [29]. Although the SBND algorithm gives better performance results, a limitation is its computational complexity. Since six-base nucleotide distribution is estimated in SBND, the computational effort is more than when estimating 3-base periodicity distribution in other methods such as EPND. One of the future tasks is to improve the proposed method for finding protein coding regions in terms of reducing the computational complexity.

6 Conclusion

We proposed an ameliorated algorithm to identify exon and intron regions in the DNA sequence which is based on the variance of the nucleotides A, T, G and C in six hexamer positions. A sliding window approach is used to check whether the given subsequence is part of a coding or a non-coding region. It gives better performance than the earlier methods utilizing Fourier Transform approach [21] and also the EPND [29] method which is depended on the distribution of nucleotides in three codon positions. In terms of accuracy, the results indicate that the proposed method for predicting protein coding regions is an effective method. The limitations of the proposed method are its increased computational complexity. Future work would focus on reducing the computational complexity of SBND approach for the identification of protein coding regions.

References

1. Anastassiou, D.: Frequency-domain analysis of biomolecular sequences. Bioinformatics **16**, 1073–1081 (2000)
2. Bhavani, S.D., Rani, T.S., Bapi, R., S.: Feature selection using correlation fractal dimension: issues and applications in binary classification problems. Appl. Soft Compt. **8**(1), 555–563 (2008)
3. Burset, M., Guigo, R.: Evaluation of gene structure prediction programs. Genomics **34**, 353–367 (1996)
4. Claverie, J.-M., Tekaia, F., Sauvaget, I., Bougueleret, L.: Objective comparison of exon and intron sequences by the mean of 2 dimensional data analysis methods. Nucleic Acids Res. **16**, 1729–1738 (1988)
5. Claverie, J.-M., Bougueleret, L.: Heuristic informational analysis of sequences. Nucleic Acids Res. **14**, 179–196 (1986)
6. Farber, R., Lapedes, A.: Determination of Eukaryotic protein coding regions using neural networks and information theory. J. Mol. Biol. **226**, 471–479 (1992)
7. Fickett, J.W.: The gene identification problem: an overview for developers. Comput. Chem. **20**, 103–118 (1996)
8. Fickett, J.W., Tung, C.S.: Assessment of protein coding measures. Nucleic Acids Res. **20**, 6441–6450 (1992)
9. Gao, J., Qi, Y., Cao, Y., Tung, W.W.: Protein coding sequence identification by simultaneously characterizing the periodic and random features of DNA sequences. J. Biomed. Biotechnol. **2**, 139–146 (2005)
10. Kotlar, D., Lavner, Y.: Gene prediction by spectral rotation measure: a new method for identifying protein-coding regions. Genome Res. **13**, 1930–1937 (2003)
11. Krogh, A., Mian, I.S., Haussler, D.: A hidden Markov model that finds genes in E. coli DNA. Nucleic Acids Res. **22**, 4768–4778 (1994)
12. Lin, H., Li, Q.-Z.: Eukaryotic and prokaryotic prediction using hybrid approach. Theory Biosci. **130**, 91–100 (2011)
13. National Center for Biotechnology Information, National Institutes of Health, National Library of Medicine. http://www.ncbi.nlm.nih.gov/Genebank/index.html
14. Rani, T.S., Bapi, R.S.: Analysis of n-Gram based promoter recognition methods and application to whole genome promoter prediction. Silico Biol. **9**(1–2), s1–s16 (2009)
15. Saberkari, H., Shamsi, M., Sedaaghi, M.H., Golabi, F.: Prediction of protein coding regions in DNA sequences using signal processing methods. In: IEEE Symposium on Industrial Electronics and Applications, pp. 355–360 (2012)
16. Shah, K., Krishnamachari, A.: On the origin of three base periodicity in genomes. Biosystems **107**, 142–144 (2012)
17. Shepelev, V., Fedorov, A.: Advances in the Exon-Intron Database (EID). Briefings Bioinf. **7**, 178–185 (2006)
18. Silverman, B.D., Linsker, R.: A measure of DNA periodicity. J. Theor. Biol. **118**, 295–300 (1986)
19. Snyder, E.E., Stormo, G.D.: Identification of coding regions in genomic DNA sequences: an application of dynamic programming and neural networks. Nucleic Acids Res. **21**, 607–613 (1993)

20. Stoffer, D.S., Tyler, D.E., Wendt, D.A.: The spectral envelope and its applications. Stat. Sci. **15**, 224–253 (2000)
21. Tiwari, S., Ramachandran, S., Bhattacharya, A., Bhattacharya, S., Ramaswamy, R.: Prediction of probable genes by Fourier analysis of genomic sequences. CABIOS **113**, 263–270 (1997)
22. Tomlin, C., Axelrod, J.: Biology by numbers: mathematical modeling in developmenta biology. Nat. Rev. Genet. **8**, 331–340 (2007)
23. Vaidyanadhan, P.P., Yoon, B.J.: The role of signal-processing concepts in genomics and proteomics. J. Franklin Inst. **1**, 1–27 (2004)
24. Voss, R.: Evolution of long-range fractal correlations and 1/f noise in DNA base sequences. Phys. Rev. Lett. **68**, 3805–3808 (1992)
25. Yada, T., Hirosawa, M.: Detection of short protein coding regions within the cynobacterium genome: application of the Hiddden Markov Model. DNA Res. **3**, 355–361 (1996)
26. Yan, M., Lin, Z.S., Zhang, C.T.: A new Fourier transform approach for protein coding measure based on the format of the Z curve. Bioinformatics **14**, 685–690 (1998)
27. Yin, C., Yoo, D., Yau, S.S.-T.: Tracking the 3-Base periodicity of protein-coding regions by the nonlinear tracking-differentiator. In: 45th IEEE Conference on Decision and Control, pp. 2094–2097 (2006)
28. Yin, C., Yau, S.S.-T.: A fourier characteristic of coding sequences: origins and a non-fourier approximation. J. Comput. Biol. **9**, 1153–1165 (2005)
29. Yin, C., Yau, S.S.-T.: Prediction of protein coding regions by the 3-base periodicity analysis of a DNA sequence. J. Theor. Biol. **247**, 687–694 (2007)

Ontology for an Education System and Ontology Based Clustering

A. Brahmananda Reddy and A. Govardhan

Abstract Reference ontology is proficient to add considerably for reducing the issue of specificity of ontology applications. Predominantly taking superior education area into account, it is believed that a reference ontology committed to this expertise area can be considered as a significant tool for several stakeholders involved in examining the system of superior education as an entity, especially in the context of academic systems diversity all over the world. In this paper, the ontology construction procedure is explained from requirements elicitation procedure to the ontology assessment procedure. Reference ontology for learning field can be used as a tool for strategy development and University profiling apart from offering a non biased ranking instrument. The aim of this research work is to provide a qualitative progress over Vector Space Model (VSM)-based search by using ontology. The document clustering is done through Particle Swarm Optimization (PSO). A PSO-based ontology model of clustering knowledge documents is described and compared with the traditional vector space model. The proposed ontology-based framework provides enhanced performance and improved clustering compared to the traditional vector space model.

Keywords Ontology · Clustering · PSO · OWL · Concept tree · Education domain

A. Brahmananda Reddy (✉)
Department of Computer Science and Engineering, V N R Vignana Jyothi Institute of Engineering & Technology, Bachupally, Hyderabad, Telangana, India
e-mail: brahmanandareddy_a@vnrvjiet.in

A. Govardhan
School of Information Technology, Jawaharlal Nehru Technological University Hyderabad, Hyderabad, Telangana, India
e-mail: govardhan_cse@yahoo.co.in

V. Ravi et al. (eds.), *Proceedings of the Fifth International Conference on Fuzzy and Neuro Computing (FANCCO - 2015)*, Advances in Intelligent Systems and Computing 415, DOI 10.1007/978-3-319-27212-2_13

1 Introduction

Conceptual schemas known as ontologies are created for providing meaningful structure of data. Ontologies are the key technologies for enabling semantics-driven knowledge processing, and it is widely accepted that the next generation of knowledge management system will rely on conceptual models in the form of ontologies. Unfortunately, the development of real-world Enterprise wide Ontology-based Knowledge Management Systems is still in its infancy.

Domain ontology is used to characterize the knowledge for a meticulous type of application domain [1, 2]. On the other hand, concept maps are used to mine and represent the knowledge structure such as concepts and propositions as perceived by individuals [3, 4]. Concept maps are related to ontology such that both of these tools are used to denote concepts and the semantic associations amongst concepts. Nevertheless, ontology is an official knowledge depiction technique to make possible individual, computer communications and it can be articulated by using formal semantic markup languages like OWL and RDF, while concept map is an unceremonious tool for humans to state semantic knowledge composition [5, 6].

2 Literature Survey

The semantic web is known as intelligent web which offers the ability to work on the meaningful knowledge representation on the web. For the last few years the researchers have focused much on education domain due to its potential future of the semantic web technology and realized that it would lead to a revolution of education domain after a great detail and deep study on semantic web technology.

Farahat et al. [7] have described novel hybrid models that join unambiguous and dormant study to guess semantic resemblance amid documents. The proposed novel hybrid models unite unambiguous and dormant models of semantic resemblance. They are used to advance the performance of document clustering algorithms. The novel hybrid models for document illustration map the documents first to a semantic space in which resemblance amid the documents reflects how their provisions are statistically linked. Then dimension reduction methodologies are applied to attain a brief depiction that conserves semantic resemblance amid documents. The semantic similarities amid the terms are captured by applying dimension reduction methodologies. The hybrid model preserves semantic resemblance depending on term to term correlations. The experimental results have shown the noteworthy enhancement in clustering performance [7].

Jing et al. [8] have presented a clustering approach on the basis of ontology distance measure. The WordNet is used to compute the term mutual information matrix. The model combines the traditional term mutual information and term frequency information. The terms are treated as correlated in the ontology mechanism but uncorrelated in VSM-based clustering approach. The method integrates the term mutual information with the conceptual background knowledge given by ontologies [8].

Shehata et al. [9] have presented a new concept-based mining model. The model captures the semantic structure of every term within a document and sentence rather than measuring the frequency of the term within a document only. This approach uses the concept-based similarity parameter. The concept-based similarity measure combines the concepts weigh in the sentence, corpus level and document. The character of text clustering outcome is better accomplished by this model than by the customary single term-based methods [9].

3 Ontology for the Education System

In this section, the Ontology construction procedure is represented (Fig. 1):

The objective of building the Ontology is to provide a meaningful knowledge representation that can be processed by machines and humans equally and offer a consensual knowledge model of education field. The proposed educational domain

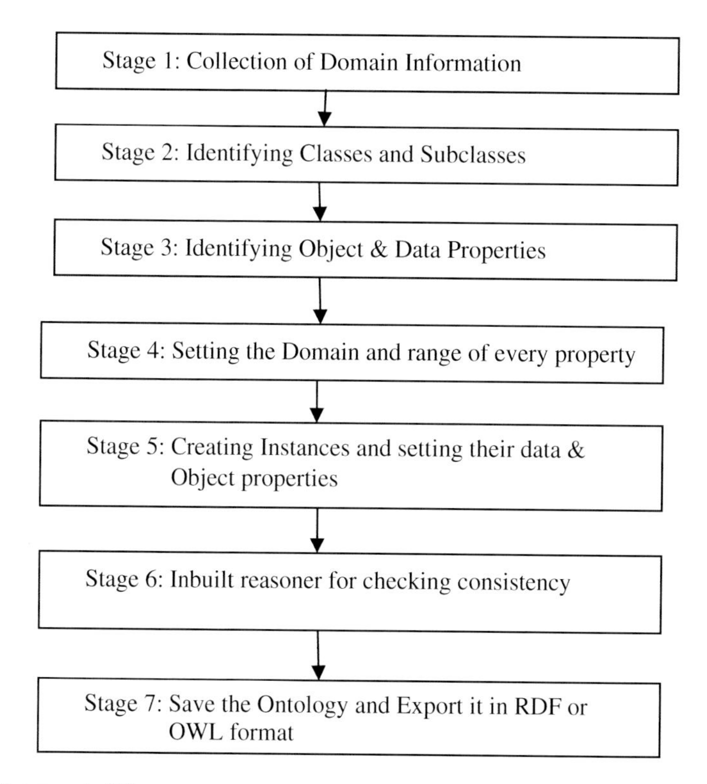

Fig. 1 Ontology building process

ontology explains numerous aspects and facilitates the organizational makeup, management, staff, students and other stake holders with their specific roles representing class and its several data properties to hold the corresponding information [10, 11] as shown in the Fig. 2.

RDF/RDFS and OWL 2.0 are used for development and implementation in order to facilitate the intent of human user and provide results that fulfill the information requirement accordingly by developing ontology (Fig. 1).

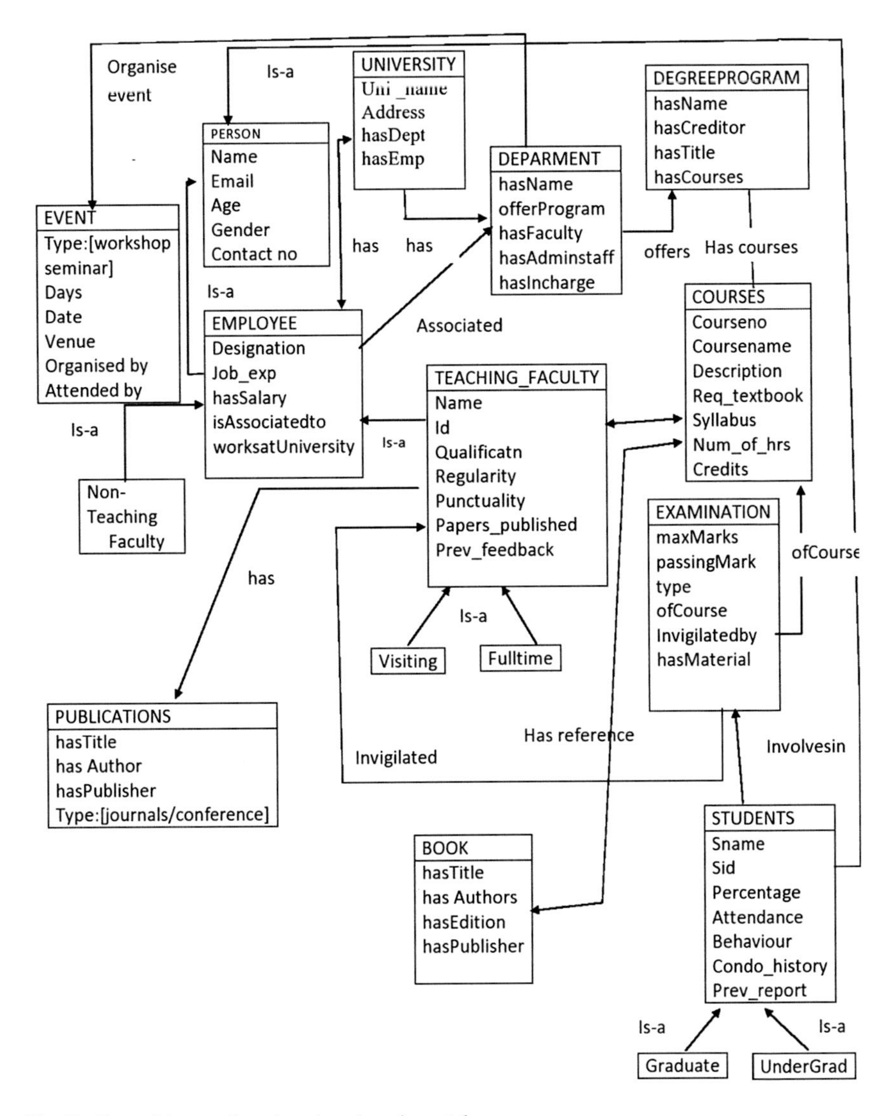

Fig. 2 General layout for education domain ontology

Table 1 Data dictionary

Attributes of concept: course
1. CourseCategory
2. CourseClassHours
3. CourseCode
4. CourseCreditsNumber
5. CourseGradingSystem
6. CourseLevel
7. CourseMaterial
8. CoursePrerequisites
9. CourseRoom
10. CourseSessionCode
10.1 SessionTiming
10.2 SessionType
11. CourseSyllabus
11.1. CourseDescription
11.2. CourseObjectives
12. CourseTitle
13. Lecture
13.1. LectureRoom
13.2. LectureSchedule

3.1 Conceptualization Phase

Conceptualization phase comprises of the concepts to subsist in the world and their associations. This step integrates the subsequent midway illustration methods: Data Dictionary (DD) (Table 1), Concepts tree (Fig. 3), Attributes Classification Trees (Fig. 4) and Object properties table (Table 2).

4 Ontology Based Clustering

Traditionally each document is represented by Vector Space Model (VSM) by representing weight of a particular vector/term. The similarity between the documents is found using the Euclidian distance measure. Document clustering is widely used in information retrieval. Among the Several document clustering techniques, K-Means clustering technique and PSO techniques are the well known document clustering techniques. The aim of PSO clustering is to find the centroid of cluster. A swarm represents the number of solutions for the candidate clustering.

When the user poses a query and if the query term is in the ontology then the inbuilt reasoner (stage 6 in Fig. 1) will list the equivalent classes and web documents are extracted using web crawler for the concepts in the ontology. The

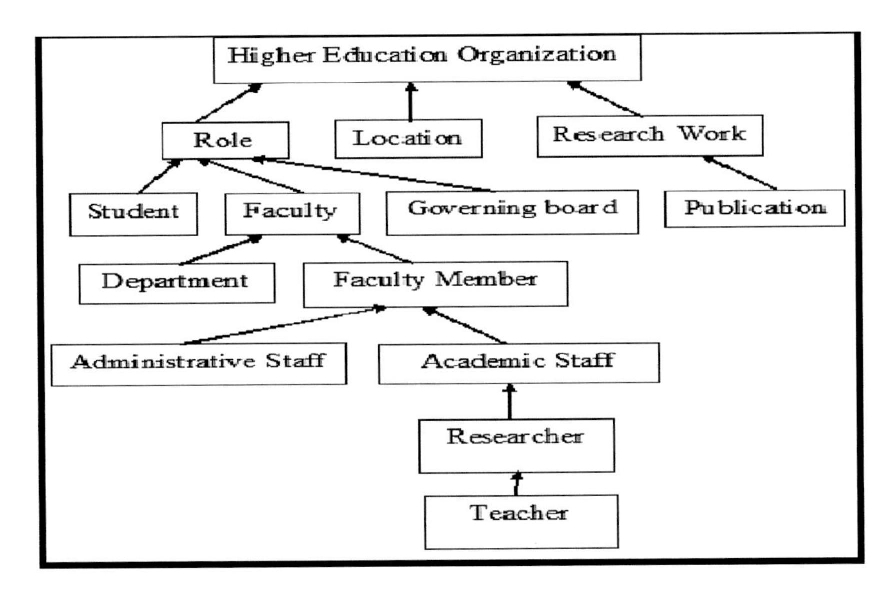

Fig. 3 Concepts tree of education domain ontology

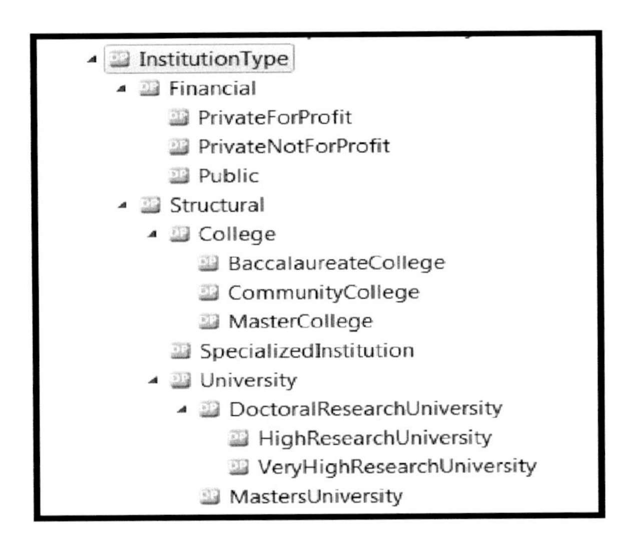

Fig. 4 Attribute classification tree (college type)

Table 2 Object properties

Object property	Area	Choice
AppointedTo	Teacher	Department
BelongsTo	Researcher	Research group
EnrolledBy	Student	Higher education
Writes	Researcher	Publication
SupervisedBy	Student	Teacher
Organizes	Laboratory	Seminar

semantic terms are extracted by applying PSO clustering algorithm to the classes found in fetched document and ontology.

4.1 Clustering Using PSO

The PSO algorithm is presented as follows:

Step 1: Randomly initialize the particle velocity and particle position. The K cluster centroids are randomly chosen for each particle.

Step 2: Estimate the fitness for each particle for the initial values

Step 3: Preserve the document cluster structure optimally. The cluster quality can be measured within cluster, within-cluster and mixture scatter matrix and it is given in Eqs. (1)–(3) respectively. The cluster quality is elevated if the cluster is firmly grouped. Equation (1) defines the within cluster and the Eq. (2) defines the similarity measure between clusters. The function conserves the cluster structure and maximizes the Eq. (3).

$$S_w = \sum_{i=1}^{k} \sum_{j \in N_i} (d_j - cc_i)(d_j - cc_i)^{\mathrm{T}} \quad (1)$$

$$S_b = \sum_{i=1}^{k} \sum_{j \in N_i} (c_j - \mathrm{gc})(c_j - \mathrm{gc})^T \quad (2)$$

$$S_m = \sum_{j=1}^{n} (d_j - \mathrm{gc})(d_j - \mathrm{gc})^T \quad (3)$$

where dj is the document that belongs to the cluster cc_i. cc_i is the ith cluster centroid, gc is the global cluster centroid, K is the number of clusters and Ni is the number of documents in the cluster c_i. S_w is the within cluster, S_b is amid clustering and S_m is the mixture scatter matrices which is sum of S_w and S_b. The fitness value of the particle is evaluated using S_b. The objective of clustering is to accomplish elevated intra-cluster resemblance and little inter cluster resemblance.

Step 4: Update the personal best position using the following equation.

$$P_pbest_g(t+1) = \begin{cases} P_pbest_g(t) & if \quad f(X_g(t+1) \geq f(P_pbest_g(t)) \\ X_g(t+1) & if \quad f(X_g(t+1) < f(P_pbest_g(t)) \end{cases} \tag{4}$$

Step 5: Revise the global finest position among all the individuals using Equation

$$\begin{aligned} P_pbest_g(t) &\in \{P_pbest_0,\ P_pbest_1,\ \ldots P_pbest_k\} \\ &= \min\{f(P_pbest_0(t) \ldots P_pbest_k(t))\} \end{aligned} \tag{5}$$

Step 6: Apply velocity update for each dimension of the particle using Equation

$$\begin{aligned} V_{gh} = \omega \times V_{gh} + C_1 \times rand_1 \times (P_pbest - X_{gh}) + C_2 \times rand_2 \\ \times (P_pbest - X_{gh}) \end{aligned} \tag{6}$$

Step 7: Update the position and generate new particle's location using Equation.

$$X_{gh} = X_{gh} + V_{gh} \tag{7}$$

Step 8: Reiterate steps 2–7 until a stop criterion, like adequately high-quality explanation is exposed or a highest number of generations is concluded. The entity particle that scores the finest fitness value in the population is considered as optimal solution.

Step 9: The relevance of documents is evaluated.

5 Results and Discussions

In the experiments, 20 random particles are generated. The fitness value is computed using the distance between the cluster centroid and the documents that are clustered. The fitness values are recorded for ten simulations. In the PSO module, the inertia weight w primarily set as 0.95 is chosen and the acceleration coefficient constants c1 and c2 are set to 2. If the fitness value is fixed, then it indicates the optimal solution. For each simulation, the initial centroid vectors of each particle are randomly selected. A set of 20 queries has been equipped manually for relative performance dimension. After 100 iterations, the same cluster solution is obtained. The F-measure values are the average of 20 runs. The parameters and the corresponding values are shown in Table 3.

Results show that the PSO clustering algorithm using ontology performs better than the PSO clustering algorithms without using ontology. To cluster the huge document data sets, PSO requires more steps to discover the best solution. As demonstrated in Fig. 5, the PSO method produces the clustering outcome with the

Table 3 PSO parameters and corresponding values

Parameter	Value
No. of clusters	10
No. of particles	20
Maximum no. of iterations	100
C1	2
C2	2
W	0.95

least fitness value for all four datasets using the ontology based similarity metric and Euclidian similarity metric.

The convergence of PSO depends on the particle's finest known location and the swarm's finest known global location. If the global position remains constant all through the optimization process, then the algorithm converges to the optimal. For a large data set, the PSO requires more iterations before optimal clustering. The PSO clustering algorithm can lead to more compact clustering outcome. The PSO approach in ontology-based VSM has enhancements compared to the outcome of the VSM-based PSO. Nevertheless, when the resemblance metric is distorted to the ontology-based metric, the PSO algorithm has a superior performance.

The convergence behavior of K-means and PSO algorithm are given in Table 4. Since 100 steps are not adequate for the PSO algorithm to come together to the most favorable solution, the outcome values in the Table 4 designate that the PSO method has enhancements compared to the consequences of the K-means method while using the Euclidian similarity metric. Nevertheless, when the similarity metric is altered to the ontology similarity metric, the PSO algorithm has a superior performance to the K-means algorithm.

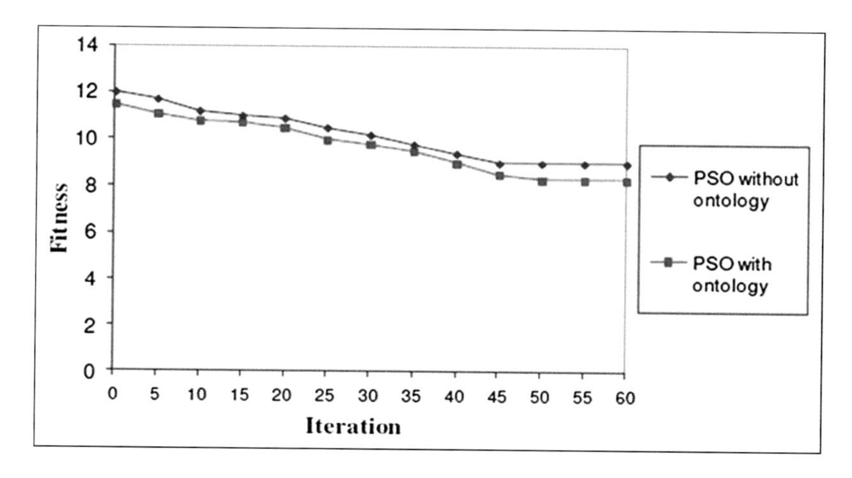

Fig. 5 Fitness curve comparison

Table 4 Performance comparison of K-Means and PSO

Data set		Fitness value	
		K-Means	PSO
Data Set1	Euclidian	8.11009	6.38039
	Ontology	10.41271	8.14551
Data Set2	Euclidian	6.25172	4.51753
	Ontology	9.57786	7.21153
Data Set3	Euclidian	4.14896	2.25961
	Ontology	5.71146	4.00555
Data Set4	Euclidian	8.62794	6.37872
	Ontology	12.8927	9.5379

6 Conclusion

A PSO-based ontology model of clustering knowledge documents is described and compared with the traditional vector space model. The results show that the proposed ontology-based framework provides enhanced performance and improved clustering compared to the conventional vector space model.

References

1. Dash, R., Mishra, D., Rath, A.K., Acharya, M.: A hybridized K-means clustering approach for high dimensional dataset. Int. J. Eng. Sci. Technol. **2**(2), 59–66 (2010)
2. Setchi, R., Tang, Q.: Concept indexing using ontology and supervised machine learning. Trans. Eng. Comput. Technol. **19**, 221–226 (2006)
3. Kang, J., Zhang, W.: Combination of fuzzy C-means and particle swarm optimization for text document clustering. Adv. Intell. Soft Comput. **139**, 247–252 (2012). (Springer, Berlin)
4. Zemmouchi-Ghomari, L., Ghomari, A.R.: Terminologies versus ontologies from the perspective of ontologists. Int. J. Web Sci. **1**(4), 315–331 (2013). (InderScience Publisher)
5. Laoufi, A., Mouhim, S., Megder, E.H., Cherkaoui, C.: An ontology based architecture to support the knowledge management in higher education. In: Proceedings of International Conference on Multimedia Computing and Systems, Ouarzazate, Morocco, pp. 1–6 (2011)
6. Burgun, A.: Desiderata for domain reference ontologies in biomedicine. J. Biomed. Inf. **39**(3), 307–313 (2006). (Elsevier Publisher)
7. Farahat, A.K., Kamel, M.S.: Enhancing document clustering using hybrid models for semantic similarity. In: Proceedings of the Eighth Workshop on Text Mining at the Tenth SIAM International Conference on Data Mining, Philadelphia, pp. 83–92 (2010)
8. Jing, J., Zhou, L., Ng, M.K., Huang, Z.: Ontology-based distance measure for text clustering. In: SIAM SDM Workshop on Text Mining, Bethesda, Maryland, USA (2006)
9. Shehata, S., Karray, F., Kamel, M.S.: An efficient concept-based mining model for enhancing text clustering. IEEE Trans. Knowl. Data Eng. **22**(10), 1360–1371 (2010)

10. Bawany, N.S., Nouman, N.: A step towards better understanding and development of university ontology in education domain. Res. J. Recent Sci. **2**(10), 57-60 (2013)
11. Sonakneware, P.S., Karale, S.J.: Ontology based approach for domain specific semantic information retrieval system. In: International Conference on Industrial Automation and Computing (April 2014)

Neural Networks for Fast Estimation of Social Network Centrality Measures

Ashok Kumar, Kishan G. Mehrotra and Chilukuri K. Mohan

Abstract Centrality measures are extremely important in the analysis of social networks, with applications such as identification of the most influential individuals for effective target marketing. Eigenvector centrality and PageRank are among the most useful centrality measures, but computing these measures can be prohibitively expensive for large social networks. This paper shows that neural networks can be effective in learning and estimating the ordering of vertices in a social network based on these measures, requiring far less computational effort, and proving to be faster than early termination of the power grid method that can be used for computing the centrality measures. Two features describing the size of the social network and two vertex-specific attributes sufficed as inputs to the neural networks, requiring very few hidden neurons.

Keywords Social network · Centrality · Eigenvector centrality · PageRank

1 Introduction

Social networks [1] have emerged as useful tools for understanding systems of high complexity and large size, involving people, machines, and organizations. One example of their use is by commercial organizations that perform target marketing by identifying the most *influential* people in a demographic sample, yielding the highest return for the resources spent in marketing efforts. Identification of influential vertices in a network is performed by computation of various *centrality* measures. While some centrality measures (e.g., degree) are easy to compute, other measures

A. Kumar (✉) · K.G. Mehrotra · C.K. Mohan
Department of EECS, Syracuse University, Syracuse13244, USA
e-mail: ashokjpr.kumar@gmail.com

K.G. Mehrotra
e-mail: mehrotra@syr.edu

C.K. Mohan
e-mail: mohan@syr.edu

V. Ravi et al. (eds.), *Proceedings of the Fifth International Conference on Fuzzy and Neuro Computing (FANCCO - 2015)*, Advances in Intelligent Systems and Computing 415, DOI 10.1007/978-3-319-27212-2_14

such as *Eigenvector centrality* [2] and *PageRank* [3] are considered more useful, but their computation requires considerable effort.

Influential vertex identification usually requires ordering (or sorting) vertices with respect to centrality values. The fact that vertex x has a centrality value of $c(x)$ is not very useful by itself; what is important is that $c(x) > c(y)$, the centrality value of another vertex y, enabling us to infer that x is a more influential vertex than y. This ordering can be computed efficiently, by training a neural network that attempts to estimate centrality values.

An example social network is shown in Fig. 1, with the desired neural network inputs and outputs (eigenvector centrality values) shown in Table 1, to be estimated using a neural network with the architecture shown in Fig. 2.

We also explored an alternative approach, estimating eigenvector centrality values by early termination of the iterative computation algorithm used for exact computation. However, the neural network approach proved to be faster, especially for large networks with thousands of vertices.

Section 2 of this paper introduces social networks and relevant centrality measures. In Sect. 3, we present the neural network methodology we used for centrality estimation. Section 4 presents implementation details and results based on well-known benchmark networks as well as large networks that were randomly generated. Concluding remarks are presented in Sect. 5.

Fig. 1 A small social network with 5 vertices

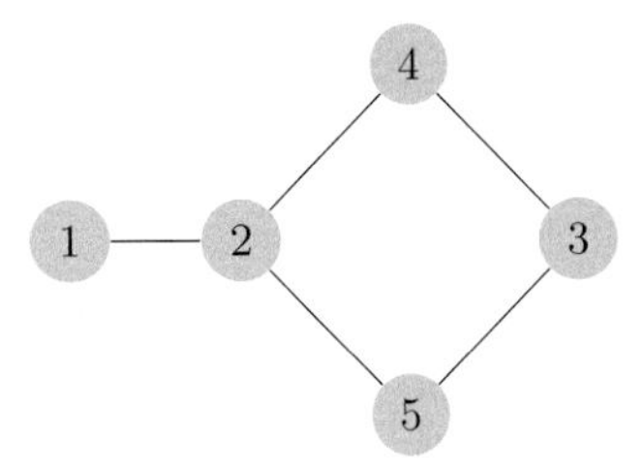

Table 1 Inputs and Outputs for the neural network used to predict eigenvector centrality for the social network in Fig. 1

Node ID I	Number of vertices: $\|V\|$	Number of edges: $\|E\|$	Degree of vertex	Number of vertices at distance ≤ 2	Eigenvector centrality
1	5	5	1	4	0.113
2	5	5	3	5	0.294
3	5	5	2	4	0.198
4	5	5	2	5	0.197
5	5	5	2	5	0.197

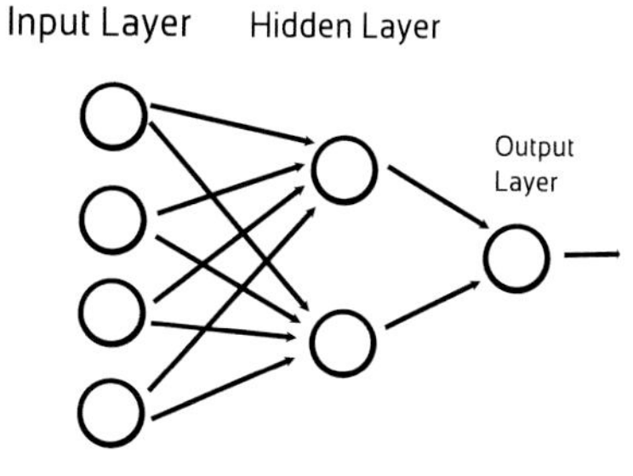

Fig. 2 Architecture of Neural network used for experimentation

2 Centrality Measures for Social Networks

A social network is a graph-theoretic abstraction in which the vertices often represent autonomous *actors* (such as people) and the edges represent relationships between them, e.g., who communicates with whom in a given period of time. Some of the earliest studies of social networks occurred over eighty years ago, with work initially performed by sociologists on relatively small networks. Recent years have seen an explosion of research results discussing properties of social networks. A key question of interest in multiple applications is the identification of important vertices in the network, as manifested in the following examples:

1. Who can influence other people to purchase products made by a company?
2. Who are the leaders of a terrorist or criminal organization?
3. Which researchers are held in greatest esteem by their peers?
4. Whose blogs or tweets are likely to be reproduced, re-broadcast, or cited?
5. Whose opinions sway the votes of most people participating in an election?
6. Through which vertices do most messages flow, in a communication network?

Many network centrality measures have been proposed in the literature, and capture the large variety of roles exhibited by a vertex in a network. The following are examples of measures frequently used:

- *Degree centrality*, which counts the number of vertices to which a given vertex is adjacent (directly connected).
- Expanded neighborhood size, which measures the number of vertices at a short distance (less than a specified number of hops) from a given vertex.
- *Betweenness centrality*, the number of shortest paths (between all pairs of vertices in the network) that pass through a given vertex.
- *Eigenvector centrality*, which incorporates the relative importance of vertices to which a given vertex is connected, defined recursively as follows:

$$f_e(x) = \sum_{v \in V} A_{x,v} f_e(v)/\theta,$$

where V is the set of vertices, $A_{x,v} = 1$ if an edge exists from $x \in V$ to $v \in V$, and θ is a constant. This is equivalent to computing the eigenvectors of the adjacency matrix, i.e., solutions to the matrix equation

$$Ax = \lambda x.$$

The eigenvector centrality of the kth vertex in the network is the kth component of the eigenvector corresponding to the largest eigenvalue.

- *PageRank* is a similar measure that emphasizes the number of vertices that point "towards" a given vertex, which indicates relative importance of a vertex in a directed network, and has the following recursive definition:

$$f_p(x) = \alpha \sum_{v \in V} A_{v,x} f_p(v) / \sum_{k \in V} A_{k,v} + \ (1 - \alpha)/|V|$$

PageRank differs from Eigenvector centrality, using a denominator term that accounts for the number of immediate neighbors of each vertex.

In many applications, the degree centrality measure is not adequate in assessing the importance of a vertex; a vertex of low degree may be critical in that it connects different components of a network, or because all of its immediate neighbors are of high centrality. Hence the more complex centrality measures such as Eigenvector centrality and PageRank are preferred in many applications. Computation of eigenvalues is expensive, and a frequently used approach to compute eigenvector centrality is the *Power iteration* or *Von Mises* iteration method [4]. In this approach, we start with a non-zero vector b and repeatedly pre-multiply b by the adjacency matrix A, normalizing components by the magnitude of the resulting vector:

$$b_{new} = Ab_{old}/|Ab_{old}|.$$

The algorithm does not always converge, but can be restarted with a different initial value of b. The number of iterations required for satisfactory approximation (to the eigenvector) cannot be determined *a priori.*

3 Neural Network Approach

Feedforward neural networks, often trained by the *error back-propagation* approach, have been applied to numerous supervised learning tasks requiring function approximation or classification. We now discuss the applicability of this approach to estimating relative eigenvector and PageRank centrality measures.

3.1 Neural Network Architecture

The previously mentioned questions 1–6 (posed at the beginning of Sect. 2) require only that we assess relative centrality values, i.e., which vertex is more important than which other vertex; the exact centrality values are not important. Therefore, an estimator model (such as a feedforward neural network) can be evaluated using correlation measures such as the well-known Pearson correlation measure, instead of the mean squared error (MSE) that is typically used in evaluating neural networks. Training may still be performed using MSE, but the testing results can be assessed using correlation values.

We have experimented with multiple sets of features that could be used as inputs to the neural networks for assessing (relative) centrality of vertices. Minimally, we need features that describe the size of the social network (number of vertices, and number of edges), as well as easily computable features that describe the local behavior of any vertex, such as the number of vertices in its immediate neighborhood and the number of vertices that can be reached within two hops from a vertex.

The first two features ($|V|$ and $|E|$) cannot be omitted since vertices with very similar local behavior may have substantially different centrality values depending on the network: a vertex with ten neighbors may be considered highly central in a network with 20 vertices and 40 edges, but a similar vertex would be considered to be much less important in a network with 200 vertices and 4000 edges. The third feature, a vertex's degree, is clearly important, but is not sufficient to describe a vertex's centrality, e.g., when a low-degree vertex connects high-degree vertices with no other path between them. The fourth feature, number of nodes at distance ≤ 2 from a vertex was also considered. However. the time complexity of exact evaluation of the number of vertices, that can be reached within two hops from a vertex is high. Therefore, throughout the rest of the paper, the fourth feature is summation of degree of all immediate neighbors.

This minimal set of four features was found to result in satisfactory performance of the neural networks used in simulations, with no additional improvement obtained when other features were added as inputs to the neural network.

We used a feedforward neural network with a single hidden layer and one network output, keeping the model as simple as possible to avoid "overfitting" problems. We experimented with different choices for the number of neurons in the hidden layer. Since sigmoid neuron functions (such as the hyperbolic tangent) are monotonic in their inputs, at least two hidden neurons are necessary. Using more neurons did not result in any significant improvements in neural network performance.

Motivated by these considerations, a 4-2-1 feedforward neural network architecture was used to generate simulation results reported in the next section. The output of the neural network estimates the network centrality of a particular vertex in the social network. For example, for the network in Fig. 1, we use a neural network such as the one shown in Fig. 2.

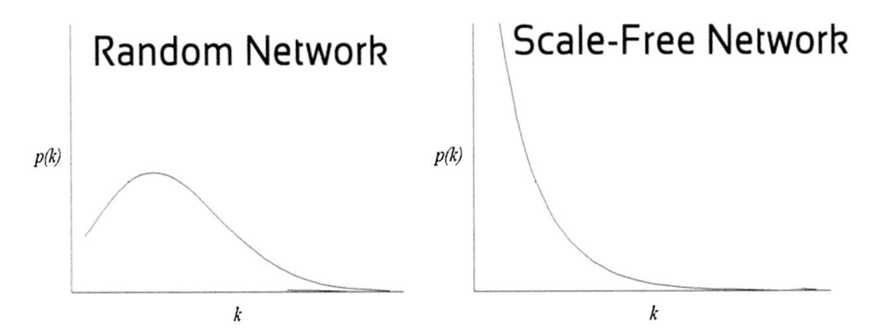

Fig. 3 Degree distributions of scale free and random networks [9]. $p(k)$ denotes the fraction of nodes with degree k

3.2 Degree Distribution

Researchers have explored social networks generated using different mechanisms, with substantially different degree distributions, such as the following (cf. Fig. 3):

- **Random Networks** are generated by repeatedly adding edges between pairs of randomly selected vertices [5].
- **Scale Free Networks** are generated using a model proposed by Barabasi and Albert [6], and their degree distribution follows a *power law* [7]. Many real-life social networks display the characteristics of scale free networks; a well-known example is the internet [8].

Since these classes of social networks display substantially different properties, we separate the centrality measure estimation task into two steps. Given a new social network, we must first categorize it as belonging to Random, Scale-free, or another class, by examining statistics of its degree distribution. This categorization can also be performed using a feedforward neural network. For very large social networks, sampling a small portion of the network may be sufficient to categorize it.

3.3 Predicting Centrality

After classifying the input social network (as scale-free or random), we use a neural network trained on the dataset of networks of same class with feature set described in Table 1, to predict the relative centrality of different vertices. A neural network trained on random networks should be used for testing a social network categorized as random, and similarly another neural network trained only on scale free networks should be used for testing a social network categorized as scale free.[1]

[1] All the results reported in this paper were from simulations executed on an Intel Corei5@2.60 GHz machine with 8 GB RAM.

4 Results

This section presents simulation results obtained using neural networks to predict relative centrality values in multiple social networks.

We tested our approach on some popular networks, such as [10–14], whose characteristics are described in Table 2, including an indication of the correlation between easy-to-compute degree centrality and hard-to-compute Eigenvector centrality, which varies considerably between social networks. For these popular social networks, we estimated relative centrality values using a neural network trained on scale-free networks generated using algorithms described in [6, 15]. We obtained the average correlation of 0.881 between neural network outputs and actual centrality values. The last two columns in this table show correlations between actual and predicted centrality values, which are better than correlations with degree centrality alone, shown in the fourth column. In other words, the additional features used in the neural network do help to improve predictability of eigenvector centrality.

We also generated 100 social networks, selecting the number of vertices in each network to be in the range between 50 and 2000. Table 3 presents average correlation values between neural network outputs and exact centrality values on these social networks obtained using classical algorithms, where training and testing were conducted separately on scale-free networks and random networks.[2]

Table 2 Results on well-known benchmark social networks

Network name	Number of vertices	Number of edges	Correlation (EV, DC)	Correlation (PR, NN)	Correlation (EV, NN)
Zachary's Karate Club	34	78	0.917	0.999	0.988
Dolphin Social Network	62	159	0.720	0.987	0.826
Copperfield Word Adjacencies	112	425	0.957	0.992	0.998
American College Football	115	616	0.762	0.997	0.870
Books about US Politics	105	441	0.670	0.998	0.727

Notation EV EigenVector centrality values, *PR* PageRank centrality values, and *NN* centrality values generated by the Neural Network

Table 3 Average correlation value between predicted centrality using neural network and actual centrality of social networks

Network category	Eigenvector centrality	PageRank entrality
Random networks	0.926	0.997
Scale-free networks	0.965	0.996

These test results are obtained on 100 social networks with number of vertices in each network in the range between 50 and 2000

[2]Random network generation algorithms were obtained from [5, 16], and scale-free network generating algorithms were from [6, 15], using the implementation in the Python software package, NetworkX (http://networkx.github.io).

Table 4 Time (in seconds) required to compute actual PageRank vs. neural network predictions

Network size	Number of edges	Time required using exact algorithm	Time required using neural network
500	24984	1.6	0.047
1000	99782	13.1	0.132
1500	225043	49.5	0.289
2000	399621	105.9	0.506
2500	624753	226.5	0.827

Note These results are average over 10 trails

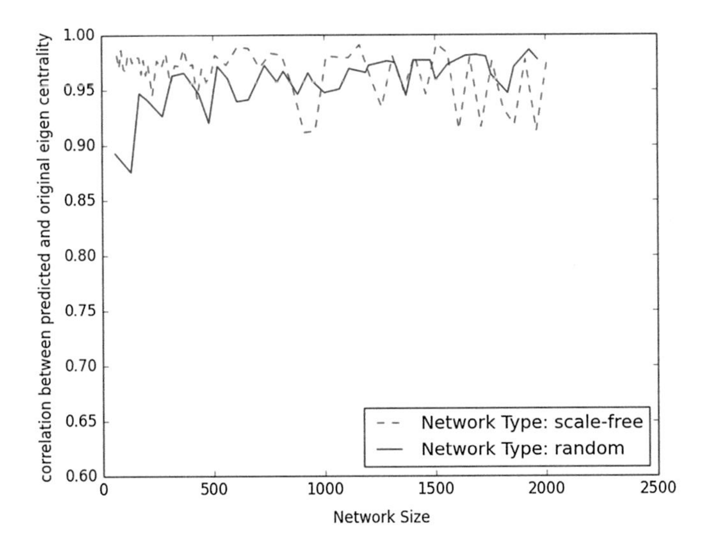

Fig. 4 Behavior of correlation with network size

The next set of simulations was conducted on larger scale free social networks with 500–2500 vertices and 25–625 k edges, generated using the algorithms mentioned earlier, to determine the amount of computational time costs incurred for PageRank computation. The last two columns in Table 4 demonstrate that the neural network approach required two orders of magnitude less time than the exact PageRank computation algorithm, i.e., a hundred-fold reduction in computational effort. The correlation values improved with social network size, as shown in Fig. 4.

We then conducted further simulations on larger social networks with 3000–7000 vertices and 1–5 million edges, computing eigenvector centrality using the exact algorithm as well as the neural network approach. In addition, we also evaluated the approach of terminating the exact algorithm early (after two iterations). Columns 3–5 of Table 5 show that the neural network outperforms the latter approach as well, with computation times reduced by a factor of approximately 3 on these example social networks.

Table 5 Time (in seconds) required to compute Eigenvector centrality using exact algorithm without early termination, exact algorithm terminated after two iterations, and neural network approach

Network size	Number of edges	Time required for exact computation	Time taken with early termination of exact algorithm	Time taken by neural network approach
3000	900003	5.5	4.0	1.2
4000	1599512	12.1	8.2	2.0
5000	2499445	24.3	9.9	2.7
6000	3600226	34.4	15.1	4.0
7000	4898875	72.2	20.2	6.0

The last column includes time for feature computation of each vertex and time taken to compute centrality of each vertex, but not network training time
Note These results are average over 10 trails

5 Concluding Remarks

Efficient computation of Eigenvector and PageRank centrality measures is an important concern in the analysis of large social networks. In this paper, we have shown that an ordering of vertices using approximations to these measures can be achieved efficiently using neural networks. Our approach is successful in obtaining results with high correlation values (>0.9) while requiring orders of magnitude less computational effort. This work can be extended to compute other computationally expensive network centrality measures using learning algorithms, using easily measurable attributes of a social network to estimate high level properties of vertices.

References

1. Newman, M.: Networks: An Introduction. Oxford University Press, Oxford (2010)
2. Bonacich, P.: Factoring and weighting approaches to status scores and clique identification. J. Math. Sociol. **2**(1), 113–120 (1972)
3. Page, L., Brin, S., Motwani, R., Winograd, T.: The pagerank citation ranking: bringing order to the web (1999)
4. Lanczos, C.: An Iteration Method for the Solution of the Eigenvalue Problem of Linear Differential and Integral Operators. Press Office, United States Governm (1950)
5. Erdös, P., Rényi, A.: On the evolution of random graphs. Publ. Math. Inst. Hung. Acad. Sci. **5**, 17–61 (1960)
6. Barabási, A.-L., Albert, R.: Emergence of scaling in random networks. Science **286**(5439), 509–512 (1999)
7. Clauset, A., Shalizi, C.R., Newman, M.E.J.: Power-law distributions in empirical data. SIAM Rev. **51**(4), 661–703 (2009)
8. Faloutsos, M., Faloutsos, P., Faloutsos, C.: On power-law relationships of the internet topology. In: ACM SIGCOMM Computer Communication Review, vol 29, pp. 251–262. ACM (1999)
9. Degree distribution of scale free and random network. https://en.wikipedia.org/wiki/Scale-free_network
10. Books about us politics. http://networkdata.ics.uci.edu/data.php?d=polbooks

11. Girvan, M., Newman, M.E.J.: Community structure in social and biological networks. Proc. Natl. Acad. Sci. **99**(12), 7821–7826 (2002)
12. Lusseau, D., Schneider, K., Boisseau, O.J., Haase, P., Slooten, E., Dawson, S.M.: The bottlenose dolphin community of doubtful sound features a large proportion of long-lasting associations. Behav. Ecol. Sociobiol. **54**(4), 396–405 (2003)
13. Newman, M.E.J.: Finding community structure in networks using the eigenvectors of matrices. Phys. Rev. E **74**(3), 036104 (2006)
14. Zachary, W.W.: An information flow model for conflict and fission in small groups. J. Anthropol. Res. **33**, 452–473 (1977)
15. Holme, P., Kim, B.J.: Growing scale-free networks with tunable clustering. Phys. Rev. E **65**(2), 026107 (2002)
16. Batagelj, V., Brandes, U.: Efficient generation of large random networks. Phys. Rev. E **71**(3), 036113 (2005)

Scale-Free Memory to Swiftly Generate Fuzzy Future Predictions

Karthik H. Shankar and Marc W. Howard

Abstract A flexible access to how the current state of memory would evolve into the future is extremely valuable in predicting events to occur in distant future. We propose a neural mechanism to non-destructively translate the current state of temporal memory into the future, so as to construct a timeline of future predictions. In a two-layer memory network that encodes the Laplace transform of the past, time-translation can be accomplished by modulating the weights between the layers. Computationally, such a network appears to be extremely resource-conserving, and could prove useful in AI systems. We hypothesize that such a mechanism is neurally realistic in the sense that the brain performs it. We propose that within each cycle of hippocampal theta oscillations, the memory state is swept through a range of time-translations to yield a future timeline of predictions. A physical constraint requiring coherence in time-translation across memory nodes results in a Weber-Fechner spacing for the representation of both past (memory) and future (prediction) timelines.

Keywords Time-translated memory states · Inverse laplace transformed memory

1 Introduction

A living system that can swiftly compute possible future states and organize them along a future timeline will have a huge advantage in decision making and goal planning. Computing predictions for the distant future can be accomplished if one has immediate access to a future state of memory. The critical problem, then, is to flexibly time-translate the current memory state into a temporally distant future state without destroying the current memory. Given a mechanism for time-translation, a complete future timeline could be built by sequentially translating to different moments in the future.

K.H. Shankar(✉) · M.W. Howard
Center for Memory and Brain, Boston University, Boston 02215, USA
e-mail: kshankar79@gmail.com

V. Ravi et al. (eds.), *Proceedings of the Fifth International Conference on Fuzzy and Neuro Computing (FANCCO - 2015)*, Advances in Intelligent Systems and Computing 415, DOI 10.1007/978-3-319-27212-2_15

This paper proposes a specific computational mechanism for swiftly time-translating the current memory state to a sequence of distant future states. We build on a formalism for representing temporal memory in a two layer network that encodes and inverts the Laplace transform of the past events [13]. This network naturally represents memory in fuzzy fashion wherein the temporal inaccuracies in memory for distant past grows in a scale-invariant way. Such a memory representation has been shown to be extremely resource conserving, requiring linearly growing resources to represent exponentially long timescales in memory [14]. In effect, the memory network trades-off accuracy for the capacity to encode extremely long timescales in memory, which seems to give a significant boost in predictive power in forecasting scale-invariant natural world signals [14]. Furthermore, any network representing temporal memory in a scale-invariant fashion can be constructed through linear combinations of such two layer memory networks [15], making this two layer memory architecture very generic.

Here we exploit the fact that the memory representation encoded in the Laplace domain can be instantly time-translated by an amount δ through a point-wise multiplication by the function $\exp(-s\delta)$, where s is the Laplace domain variable that would parametrize the nodes in the network. Since the point-wise multiplication operation can be parallelized, the time-translation operation on the memory state can be literally instantaneous. Moreover, the fuzziness inherited from the parent network will be infused into the translation operation so that the future prediction due to translation will also show scale-invariant fuzziness. We propose that this feature could be automatically lead to temporal generalizations with minimal computational requirements. Although our proposition can be built purely at a computational or neuromorphic engineering level, we believe that the brain can actually perform these operations.

We pursue the hypothesis that construction of a future timeline is accomplished neurophysiologically within each hippocampal theta cycle. *Theta oscillations* are voltage oscillations in the hippocampus with a characteristic 4–8 Hz frequency. A population of neurons in the rodent hippocampus—*time cells*—consistently fire during circumscribed intervals of a delay period [8, 11]. When rodents navigate through an environment, a population of neurons—*place cells*—consistently fire when the animal is in a circumscribed region of space. The set of locations where a particular neuron fires is referred to as its *place field*. Critically, there is a systematic relationship between the rat's position within a place field and the phase of the theta oscillation at which the neuron fires [10], known as *phase-precession*. The firing of a neuron at later phases in the theta cycle conveys information about the future position of the animal, as if the predicted-position is being translated along a future trajectory [1]. This hints that phase precession could underlie the mechanism for translating the present state of memory into a future state.

To connect the neurobiological phenomena to the computational hypothesis, we propose that in each theta cycle as the phase θ sweeps from 0 to 2π, the current memory state is translated from the present time τ to a future time $\tau + \delta$, where δ sweeps from $\delta \simeq 0$ to δ_{max}, generating a future timeline within the theta cycle (Fig. 1a). At the end of each cycle, δ is reset back to 0 so as to rebuild a future timeline

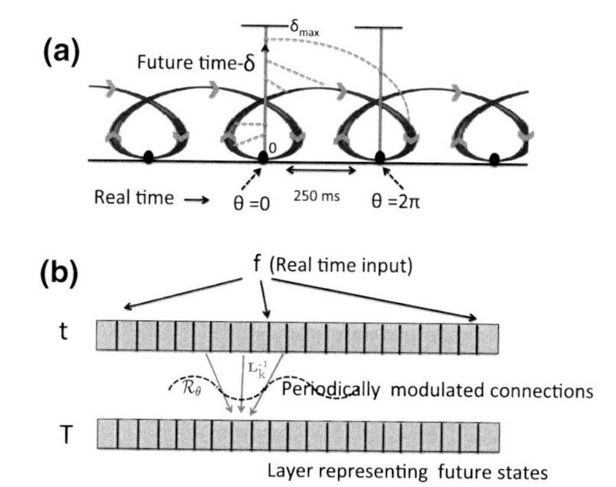

Fig. 1 **a** Schematic of future timeline construction. **b** Two layer network with theta-modulated connections

at the next cycle. To anticipate the mapping onto hippocampal neurophysiology, we assume that the Laplace variable s changes systematically along the dorsoventral axis of the hippocampus (Fig. 1b). Local theta phase changes along the dorsoventral position such that theta appears as a traveling wave [7, 12]. The central hypothesis here is that changing synaptic weights as a function of local phase within each theta cycle [18] implements the desired translation.

2 Time-Translating a Memory Representation

The model is implemented as a two-layer feed forward network (Fig. 1b) where the **t** layer holds the Laplace transformed past memory and the **T** layer reconstructs a temporally fuzzy estimate of past events. Let's start by describing the untranslated temporal representation following [13, 14]. The stimulus at time τ is denoted as $\mathbf{f}(\tau)$. The **t** layer nodes are leaky integrators activated by the stimulus. The nodes within the **t** layer are parametrized by their decay rate s. The nodes in the **T** layer are in one to one correspondence with the nodes in the **t** layer. The activity of the two layers is described by

$$\frac{d}{d\tau}\mathbf{t}(\tau, s) = -s\mathbf{t}(\tau, s) + \mathbf{f}(\tau) \tag{1}$$

$$\mathbf{T}(\tau, s) = \mathbf{L}_{\mathrm{k}}^{-1}\ \mathbf{t}(\tau, s) \tag{2}$$

By integrating Eq. 1, note that at time τ, the **t** layer nodes encode the Laplace transform of the entire past of the stimulus function leading up to time τ. The connection between the **t** and **T** layers is given by a constant linear operator $\mathbf{L}_{\mathrm{k}}^{-1}$ that constructs the activity of each **T** node by linearly combining the activities of k neighboring **t** nodes on either side. The connection strengths are derived to ensure that $\mathbf{L}_{\mathrm{k}}^{-1}$ approximates the inverse Laplace transform operation, with increasing accuracy for larger k [13]. An explicit instantiation of the connectivity is given in the appendix. The **T** layer inverts the Laplace transformed stimulus history in **t** so as to represent a fuzzy and approximate reconstruction of the stimulus history at every instant τ, $\mathbf{T}(\tau, s) \simeq \mathbf{f}(\tau - k/s)$. In other words, the various nodes in **T** layer (with different s) represent the stimulus history from different past moments. The error in the representation of the stimulus history $\mathbf{f}(\tau' < \tau)$ in the **T** layer increases as τ' recedes into past, and this error is precisely scale-invariant [13].

We extend this framework to construct a future timeline. The critical step is to swiftly time-translate the current memory state in the **T** layer to mimic a distant future state without changing the memory representation in the **t** layer. To achieve this, we hypothesize that the connection strengths between the two layers are periodically modulated in sync with the local phase of the theta oscillations at any location on the dorsoventral axis. For a translation δ, the time-translated memory state $\mathbf{T}(\tau + \delta, s)$ could be obtained if we had access to the time-translated Laplace transform $\mathbf{t}(\tau + \delta, s)$, which is simply

$$\mathbf{t}(\tau + \delta, s) = e^{-s\delta}\mathbf{t}(\tau, s) \tag{3}$$

Noting that the factor $\exp(-s\delta)$ can be absorbed as a part of the connection strength, a time-translated state of **T** can be obtained from the current state of **t** by extending Eq. 2 as

$$\mathbf{T}_\theta(\tau, s) = \left[\mathbf{L}_\mathrm{k}^{-1} \cdot \mathcal{R}_\theta\right] \, \mathbf{t}(\tau, s) \tag{4}$$

where $\mathcal{R}_\theta$ is the translation operator—larger translation for larger θ; $-\pi < \theta < \pi$. $\mathcal{R}_\theta$ can be understood as a modulation to the fixed synaptic weights in $\mathbf{L}_\mathrm{k}^{-1}$. To translate by an amount δ, $\mathcal{R}_\theta$ would simply be a diagonal operator with entries $\exp(-s\delta)$, where δ is monotonically related to θ.[1]

We hypothesize that the anatomical gradient of s values is arranged along the dorsoventral axis of the hippocampus and denote the local phase at any s as θ_s. It is known that the local theta phase changes along the dorsoventral axis such that theta oscillation appears as a traveling wave with a phase offset of π from one end to the other [7, 12]. As a reference, we denote the phase at the very first node as θ_o. Let there be a total of $N + 1$ nodes whose s values are given by s_o, s_1..., s_N. We shall interchangeably use the subscripts s and n because they are monotonically related. Figure 2 shows a schematic to visualize the theta oscillations. Note from Fig. 2 that the relationship between the phase at any two values of s is given by

$$\theta_s/\pi = \theta_o/\pi - n/N \tag{5}$$

[1]The phase θ deterministically varies in real time τ with a periodicity of 250 ms. However, in the context of representing memory from much larger timescales, it is convenient to treat θ and τ in Eq. 4 as independent. Within each theta cycle, treating τ as a constant is a fair approximation.

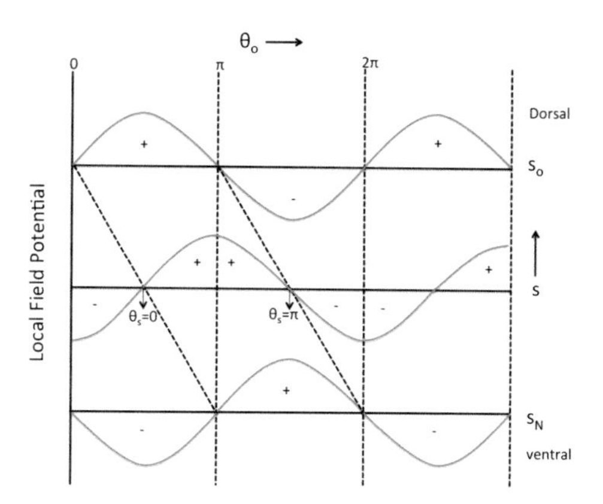

Fig. 2 Traveling theta wave the along the s axis. The two *diagonal lines* indicate the positions where the local phases are 0 and π

for $0 < \theta_s < \pi$. To keep the notation simple, we let θ_s take values between $-\pi$ and $+\pi$, but take the reference phase θ_o to vary between 0 and 2π, with the implicit understanding that the regime $(-\pi, 0)$ is mapped on to $(\pi, 2\pi)$. Following [3, 4] we assume that the negative phase of theta does not contribute to prediction.

Since the synaptic modulations only depend on the local theta phase, $\mathcal{R}_\theta$ should be dependent only on the local phase at each s, the diagonal entries of $\exp(-s\delta)$ in the $\mathcal{R}_\theta$ operator should only depend on s and its current phase: $\exp(-\Phi_s(\theta_s))$. Since the different nodes in the **T** layer represent memory from various timescales in a scale-invariant fashion, it would be ideal to preserve this scale-invariance even in the time-translated state. To this end we pick all the functions Φ_s to be identical; so that $\mathcal{R}_\theta$ is a diagonal operator with entries $\exp(-\Phi(\theta_s))$, where Φ is a monotonic function of the local phase at any location.

Finally, we impose the constraint that all nodes in the **T** layer in the positive phase of theta should be coherently time-translated to the same point in the future. Thus, instantaneous read-out of all nodes at any instant within the theta cycle provides a coherent estimate of a specific future moment δ. However, the $\mathcal{R}_\theta$ operator time-translates the activity of different **T**-nodes by different amounts given by $\Phi(\theta_s)/s$. Requiring coherence in time-translation for the nodes in the positive half cycle implies that $\Phi(\theta_s)/s\ (= \delta)$ should be a constant for all nodes with $0 < \theta_s < \pi$:

$$\Delta\left(\Phi\left(\theta_s\right)/s\right) = \Delta\left(\Phi\left(\theta_o - \pi n/N\right)/s_n\right) = 0. \tag{6}$$

For this to hold at all values of θ_o between 0 and 2π, $\Phi(\theta_s)$ must be an exponential function so that θ_o can be functionally decoupled from n; consequently s_n should also have an exponential dependence on n. With the above criterion when $0 < \theta_s < \pi$, the general solution can be written as

$$\Phi(\theta_s) = \Phi_o \exp[b\theta_s] \tag{7}$$

$$s_n = s_o(1 + c)^{-n} \tag{8}$$

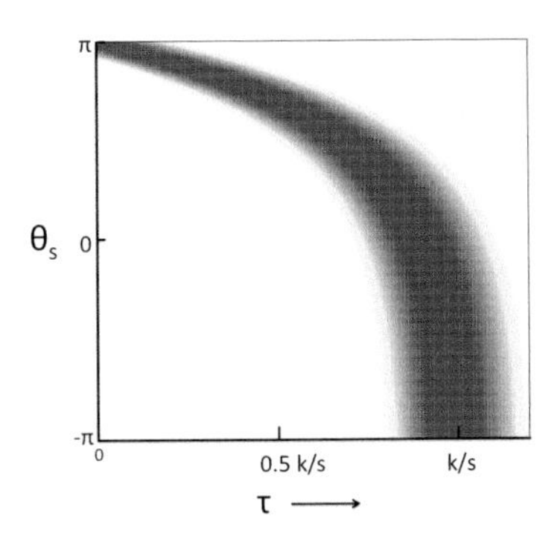

Fig. 3 Activity of a node in the **T** layer described by Eq. 10 as a function of τ and local phase θ_s. ($k = 10$, $\Phi_{\max} = 10$). Activity less than half the maximum is thresholded to zero. Note that only values of $\theta_s > 0$ contribute to prediction

where c is a positive number. That is, a coherent time-translation across the memory nodes requires the values s_n to lie on a Weber-Fechner scale.

The maximum value that Φ can take (at $\theta_s = \pi$) is $\Phi_{\max} = \Phi_o \exp[b\pi]$, such that $\Phi_{\max}/\Phi_o = s_o/s_N$ and $b = (1/\pi)\log(\Phi_{\max}/\Phi_o)$.[2] These considerations lead to a monotonic relationship between the magnitude of translation δ of the nodes in the positive cycle as a function of the reference phase θ_o:

$$\delta(\theta_o) = (\Phi_o/s_o)\exp[b\theta_o] \tag{9}$$

Hence, in addition to a Weber-Fechner distribution of the s_n, the assumption of coherent time-translation also leads to an exponential acceleration in δ as theta phase is swept through; δ also lies on a Weber-Fechner scale within the theta cycle. This solution is scale-invariant, with each value of s contributing to the time-translation for the same fraction of the theta cycle.

The linearity of Eqs. 1 and 4 implies that linearly superposing different stimulus functions would simply result in a linearly superposed $\mathbf{T}_\theta$ representation. For brevity, we just take the stimulus to be a single delta function at $\tau = 0$. Then $\mathbf{T}_\theta(\tau, s)$ can be approximately evaluated as

$$\mathbf{T}_\theta(\tau, s) \simeq \frac{s}{k!}\left[s\tau + \Phi\left(\theta_s\right)\right]^k e^{-[s\tau+\Phi(\theta_s)]} \tag{10}$$

Fig. 3 shows the activity of a node in the **T** layer (any s) as a function of time and local theta phase. In the negative half cycle ($< -\pi < \theta_s < 0$) the node is active only around the time k/s, while in the positive cycle it gets activated earlier at later phases.

---

[2] To ensure continuity around $\theta_s = 0$, we take the Eq. 7 to hold true even when $\theta_s \in (-\pi, 0)$. However, since notationally θ_s makes a jump from $+\pi$ to $-\pi$, $\Phi(\theta_s)$ must make a quick transition from $\Phi_{\max}$ ($\theta_s = \pi$) to $\Phi_{\min} = \Phi_o^2/\Phi_{\max}$ ($\theta_s = -\pi$).

3 Timeline of Future Prediction

At any time τ, the δ-translated **T** state can be used to predict the stimuli expected at a time δ in the future. As δ is swept through within a theta cycle a timeline of future predictions can be simulated. While the details of this future timeline would depend on a number of factors that are not specified by the biology we can nonetheless describe some basic properties.

First, to generate well-timed predictions, it is necessary to learn the temporal relationships between various stimuli. Because $\mathbf{T}_0$ (the untranslated state) contains information about the timing of previously-presented stimuli, associating $\mathbf{T}_0$ with the currently-available stimulus will store temporal relationships between stimuli separated in time. A great many learning rules could be used to affect this learning. Here we make a minimal assumption that simple associative (Hebbian) learning takes place between $\mathbf{T}_0$ and the stimulus. A particular state of activity in the **T** layer will predict a stimulus to the extent it resembles the states in which that stimulus was encoded.

To evaluate the construction of future timeline, consider a thought experiment where stimulus A is presented at $\tau = 0$ and stimulus B is presented after a delay of τ_o much larger than the period of a theta oscillation. Stimulus B is associated with the state of $\mathbf{T}_0$ at time τ_o, which holds the memory of A a time τ_o in the past. Later when A is repeated (at a time $\tau = 0$), the subsequent activity in the **T** nodes can be used to generate predictions for the future occurrence of B. The future prediction as a function of θ_o and τ is obtained as the sum of activity of each **T** node multiplied by the learned association strength:

$$\mathbf{p}(\theta_o, \tau) = \sum_{n=\ell}^{N} \mathbf{T}_{\theta_o}\left(\tau, s_n\right)\ \mathbf{T}_0\left(\tau_o, s_n\right)/s_n^w, \tag{11}$$

The factor s_n^w (for any w) allows for differential association strengths for the different nodes, while still preserving the scale invariance property in $\mathbf{p}$. Rewriting θ_o in terms of δ (Eq. 9), analytic approximation leads to

$$\mathbf{p}(\delta, \tau) \simeq \frac{\tau_o^{w-2}}{k!^2} \frac{(\tau/\tau_o + \delta/\tau_o)^k}{(1 + \tau/\tau_o + \delta/\tau_o)^{2k+2-w}} \times \Gamma\left[2k + 2 - w, \Phi_{\max}(1 + \tau/\delta + \tau_o/\delta)\right] \tag{12}$$

The prediction generated within each theta cycle unfolds as θ_o is swept from 0 to 2π. The predictions for a more distant future time corresponding to a larger value of δ will peak at a larger θ_o. Figure 4a shows Eq. 12 for two different values of τ_o. At the instant $\tau = 0$, the predictions for the two stimuli are plotted in the left panel of Fig. 4a. For larger τ_o, the peak of the prediction occurs at a larger δ. So, the future prediction at any given moment is ordered within the theta cycle. The right panel of Fig. 4a shows how the prediction for one stimulus with a given τ_o would shift the peak

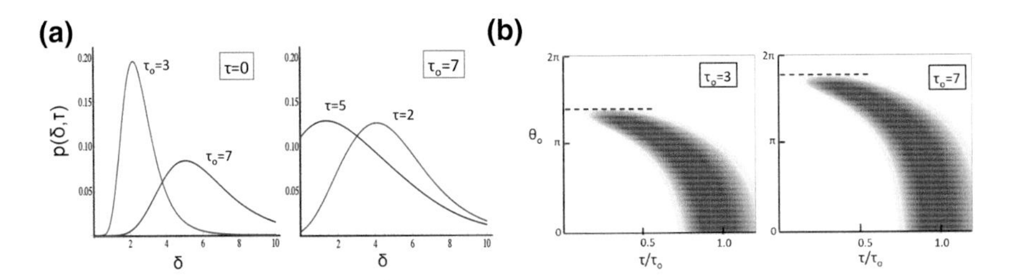

Fig. 4 **a** Equation 12 is plotted as a function of δ for different τ and τ_o. Here $\Phi_{\max} = 10$, $\Phi_o = 1, k = 10, s_o = 10, w = 1$. **b** Equation 12 is plotted as a function of θ_o and τ for the same parameters. Activity less than half the maximum is thresholded to zero

to lower values of δ as we evolve forward in real time (τ). An important consequence of this construction of future timeline is that the activity of the nodes representing future prediction will exhibit phase precession with respect to the hippocampal theta phase θ_o, as shown in Fig. 4b. Neurons in the ventral striatum (known to code for prediction) indeed phase precesses relative to hippocampal theta over large portions of the environment terminating in a reward location [17]. Because δ is monotonically related to θ_o, when τ_o is small the prediction nodes should precess from π to 0, but when τ_o is large they will precess from $\theta_o = 2\pi$ to 0.

4 Discussion

This hypothesis is quite different in mechanism from existing models of phase precession [2, 4, 6, 9]. It makes several distinctive predictions about theta oscillations and phase precession that could be tested with existing technology. Cognitively, the function of this mechanism is to generate a timeline $\mathbf{p}(\delta, \tau)$ of future predictions where the timing of predicted stimuli is reflected in the phase of firing. Neurons coding for distant events should start firing when the predicting stimulus is presented and they should gradually increase their firing as the predicted stimulus comes closer in time [17].

Irrespective of the neural implementation, this computational mechanism for time-translating the memory state constructs a Weber-Fechner timeline on which future events can be represented. The use of a two-layer network allows future states of memory to be constructed on the **T** layer without destroying the current state of memory on the **t** layer. Moreover, because the **t** layer holds the Laplace transform of the past (memory), time-translation can be accomplished in parallel across all nodes without having to sequentially visit intermediate states. Consequently the physical time taken for the translation is decoupled from the magnitude of translation, providing a huge computational advantage. A neuromorphic chip implementing this mechanism could be very useful for AI agents foraging and processing natural world information on the fly.

Here we have assumed an equivalence between the temporal memory and one-dimensional spatial memory because Eq. 1 can be extended to the spatial domain by modulating the decay rates s with the velocity of motion [5]. Two-dimensional place cells that phase precess are observed when a rat manuevers through an open field [16]. Equation 3 readily generalizes to two dimensions, with the understanding that the translation is itself two-dimensional. In two dimensions, this method for translation also allows for the exploration of multiple distinct paths in parallel.

Acknowledgments Supported by NSF PHY 1444389 and the Initiative for the Physics and Mathematics of Neural Systems.

Appendix

This appendix provides an explicit recipe for computing $\mathbf{L}_{\mathrm{k}}^{-1}$ for the distribution of s_n derived in the text and $k = 2$. A more general derivation can be found in [14]. When $s_n = s_o(1 + c)^{-n}$ the connection strengths in the $\mathbf{L}_{\mathrm{k}}^{-1}$ operator takes a special form– for every n, the local connectivity from the $(2k + 1)$ **t**-nodes to the n-th **T**-node has an identical form multiplied by s_n. For example, with $k = 2$, the connection strengths to the n-th **T** node from the **t** nodes in the local neighborhood are given by

$$\mathbf{t}_{n+2} \rightarrow s_n \frac{(c+1)^5}{c^2(c+2)^2}$$

$$\mathbf{t}_{n+1} \rightarrow s_n \frac{-(c+1)^2}{c}$$

$$\mathbf{t}_{n} \rightarrow s_n \frac{c^4 + 3c^3 + c^2 - 4c - 2}{c^2(c+2)^2}$$

$$\mathbf{t}_{n-1} \rightarrow s_n \frac{1}{c^2 + c}$$

$$\mathbf{t}_{n-2} \rightarrow s_n \frac{1}{c^2(c+1)(c+2)^2}$$

The factor s_n can be treated as a post-synaptic weight and the rest (which only depend on c) can be treated as pre-synaptic weight which are constant along the dorsoventral axis.

References

1. Battaglia, F.P., Sutherland, G.R., McNaughton, B.L.: Local sensory cues and place cell directionality: additional evidence of prospective coding in the hippocampus. J. Neurosci. **24**(19), 4541–4550 (2004)
2. Burgess, N., Barry, C., O'Keefe, J.: An oscillatory interference model of grid cell firing. Hippocampus **17**(9), 801–12 (2007)
3. Hasselmo, M.E., Bodelón, C., Wyble, B.P.: A proposed function for hippocampal theta rhythm: separate phases of encoding and retrieval enhance reversal of prior learning. Neural Comput. **14**, 793–817 (2002)
4. Hasselmo, M.E.: How We Remember: Brain Mechanisms of Episodic Memory. MIT Press, Cambridge (2012)
5. Howard, M.W., MacDonald, C.J., Tiganj, Z., Shankar, K.H., Du, Q., Hasselmo, M.E., Eichenbaum, H.: A unified mathematical framework for coding time, space, and sequences in the hippocampal region. J. Neurosci. **34**(13), 4692–4707 (2014)
6. Lisman, J.E., Jensen, O.: The theta-gamma neural code. Neuron **77**(6), 1002–16 (2013)
7. Lubenov, E.V., Siapas, A.G.: Hippocampal theta oscillations are travelling waves. Nature **459**(7246), 534–9 (2009)
8. MacDonald, C.J., Lepage, K.Q., Eden, U.T., Eichenbaum, H.: Hippocampal "time cells" bridge the gap in memory for discontiguous events. Neuron **71**, 737–749 (2011)
9. Mehta, M.R., Lee, A.K., Wilson, M.A.: Role of experience and oscillations in transforming a rate code into a temporal code. Nature **417**(6890), 741–6 (2002)
10. O'Keefe, J., Recce, M.L.: Phase relationship between hippocampal place units and the EEG theta rhythm. Hippocampus **3**, 317–30 (1993)
11. Pastalkova, E., Itskov, V., Amarasingham, A., Buzsaki, G.: Internally generated cell assembly sequences in the rat hippocampus. Science **321**(5894), 1322–7 (2008)
12. Patel, J., Fujisawa, S., Berényi, A., Royer, S., Buzsáki, G.: Traveling theta waves along the entire septotemporal axis of the hippocampus. Neuron **75**(3), 410–7 (2012)
13. Shankar, K.H., Howard, M.W.: A scale-invariant representation of time. Neural Comput. **24**, 134–193 (2012)
14. Shankar, K.H., Howard, M.W.: Optimally fuzzy scale-free memory. J. Mach. Learn. Res. **14**, 3753–3780 (2013)
15. Shankar, K.H.: Generic construction of scale-invariantly coarse grained memory. In: Chalup, S.K., Blair, A.D., Randall, M. (eds.) ACALCI 2015. LNCS, vol. 8955, pp. 175–184. Springer, Heidelberg (2015)
16. Skaggs, W.E., McNaughton, B.L., Wilson, M.A., Barnes, C.A.: Theta phase precession in hippocampal neuronal populations and the compression of temporal sequences. Hippocampus **6**, 149–172 (1996)
17. van der Meer, M.A.A., Redish, A.D.: Theta phase precession in rat ventral striatum links place and reward information. J. Neurosci. **31**(8), 2843–54 (2011)
18. Wyble, B.P., Linster, C., Hasselmo, M.E.: Size of CA1-evoked synaptic potentials is related to theta rhythm phase in rat hippocampus. J. Neurophysiol. **83**(4), 2138–44 (2000)

Assessment of Vaccination Strategies Using Fuzzy Multi-criteria Decision Making

Daphne Lopez and M. Gunasekaran

Abstract Alternative selection of vaccination strategy has become a challenging task for the public health, and it is considered as a complex decision making problem. Decision makers often use linguistic variables to rate the alternatives. The objective of this research is to use fuzzy logic based VIKOR method for evaluating H1N1 Influenza vaccination strategies. The experimental design of the proposed decision making model is illustrated with a case study in Vellore, Tamil Nadu, India. The alternative of a vaccination strategy considered in this study includes the combination of "people", "spatial" and "temporal".

Keywords Multi criteria decision making • Fuzzy VIKOR • Vaccination strategy • H1N1 influenza

1 Introduction

Decision making process plays a vital role in selecting and organizing the health care resources efficiently that improves human health. Recently, health care systems face more challenges towards resources and expenditures. Thus, professionals from health care organization are trying to provide high-quality health care delivery using the limited resources. This creates awareness on using decision making process in health care organization [1]. Multi criteria decision making is one of the popular decision making processes used in many organizations. It is defined as a set of processes that aid decision-making, where decisions are selected based on multiple criteria. There are a number of decision making methods available to choose the

D. Lopez (✉) · M. Gunasekaran
School of Information Technology and Engineering, VIT University, Vellore, Tamil Nadu, India
e-mail: daphnelopez@vit.ac.in

M. Gunasekaran
e-mail: gunavit@gmail.com

V. Ravi et al. (eds.), *Proceedings of the Fifth International Conference on Fuzzy and Neuro Computing (FANCCO - 2015)*, Advances in Intelligent Systems and Computing 415, DOI 10.1007/978-3-319-27212-2_16

"most preferred alternative". Fuzzy VIKOR-based MCDM method is used in this paper to select the best vaccination strategy to protect people from the H1N1 influenza epidemic. The paper is organized in the following way: Sect. 3 explains the Fuzzy VIKOR based MCDM method in detail, a case study in detail is analyzed in Sect. 4 and results are discussed in Sect. 5.

2 Related Work

The field of decision making is the complete study of how decisions are really made and how they can be made better or more effectively. Many of the researchers are trying to make an efficient decision making model to prevent and control the spreading of epidemics [2, 3, 4]. There are a number of decision making approaches already available such as Analytic Hierarchy Process (AHP), Grey System Theory (GST), Analytic Network Process (ANP) and so on. Analytic Hierarchy Process (AHP) is the familiar type of multi criteria decision making method that was initially identified by Thomas L. Saaty. AHP is used to make decisions in contractor prequalification [5]. ANP and AHP models predict the decision in same way but the difference is that AHP makes decisions into a hierarchy with an end whereas the ANP structures it as a network [6]. Many of the organizations used linear programming based Data Envelopment Analysis (DEA) approach to monitor the performance. For example, it is used to measure the bridge maintenance performance in some countries [7]. Some decision making approaches like Technique for Order of Preference by Similarity to Ideal Solution (TOPSIS) create the decisions depending upon the geometric distance between the events [8]. Differential equations are used to evaluate decisions in Grey System Theory (GST) [9]. Decision making researchers have been merging the above techniques and have produced hybrid technique [10]. In addition to that, VIKOR (VIsekriterijumska optimizacija iKOmpromisno Resenje) first developed by Opricovic to solve a distinct decision problem with contradictory criteria. VIKOR method concentrates on selecting, ranking the best solution from multiple alternatives and also identifies the compromise solution for decision makers to reach an absolute decision [11]. VIKOR method is most often used in MCDM problem, as it finds the ranking index based on the exact measure of "closeness" to the ideal solution, identified compromise solution gives a more group utility for the "majority" and a less individual regret for the "opponent" [12].

Though number of decision making methods has been identified in the past decades, more techniques and tools are required to deal with uncertainty especially in medicine and biology. Identifying the best decision involves more levels of imprecision and uncertainty, that has most often happened in epidemiological studies [13]. For example, depending upon the immunization level a single disease may have different impact on different people. In addition to that, a particular symptom may be indicative of different diseases, and the occurrence of numerous diseases in a single patient may disturb the predictable symptom. In the study of vaccination strategy design, the following criteria should be considered, population for the immunization programme, the proportion of susceptible to be vaccinated, the spatio temporal parameters of the epidemics, history and social habits of the people, and the nature of the strategy. Thus, epidemiological studies have high impact of imprecision and uncertainty in the measures in analysis. Fuzzy VIKOR-based MCDM method is used in this paper to assess the best vaccination strategy to prevent and control the H1N1 influenza epidemic.

3 Fuzzy VIKOR-Based MCDM Method

Combining fuzzy logic with VIKOR MCDM method has been developed to solve uncertainty in decision making process. It provides two main solution such as best solution and compromise solution. Those solutions can be used to resolve a fuzzy MCDM problem [14]. Fuzzy VIKOR-based MCDM algorithm has a sequence of steps as described below:

Step 1 Define the objectives of the decision making and identify the problem scope

Step 2 Identify p alternative methods called $A = \{A_1, A_2, \ldots, A_P\}$

Step 3 Identify q selection criteria called $C = \{C_1, C_2, \ldots, C_q\}$

Step 4 Identify the number of decision makers to select an alternative, let there be set of r decision makers $DM = \{DM_1, DM_1, \ldots, DM_r\}$

Step 5 Identify the appropriate linguistic variables, let a triangular fuzzy number $\tilde{t}$ as (t_1, t_2, t_3) and its membership function $\mu_{\tilde{t}}(x)$ can be defined as [15]:

$$\mu_{\tilde{t}}(x) = \begin{cases} 0, & x < t1, \\ \frac{x - t1}{t2 - t1}, & t1 \le x \le t2, \\ \frac{t3 - x}{t3 - t2}, & t2 \le x \le t3, \\ 0, & x > t3, \end{cases} \quad (1)$$

Step 6 Get the fuzzy alternatives with respect to each criteria and fuzzy criteria weights by detailed questionnaire answered by all decision makers.

Step 7 Construct a fuzzy decision matrix based on aggregated fuzzy weights of each criteria and aggregated ratings fuzzy alternatives, let a set of fuzzy ratings of alternatives $A_i(i = 1, 2, \ldots, p)$ with respect to criteria $C_j(j = 1, 2, \ldots, q)$, called $X = \{x_{ij}, i = 1, 2, \ldots, p, j = 1, 2, \ldots, q\}$ Calculate the aggregated decision maker's fuzzy assessments of alternatives $\tilde{x}_{ij}^k = \left(\tilde{x}_{ij1}^k, \tilde{x}_{ij2}^k, \tilde{x}_{ij3}^k\right)$ and aggregated fuzzy weights of criteria $\tilde{w}_{ij}^k = \left(\tilde{x}_{ij1}^k, \tilde{x}_{ij2}^k, \tilde{x}_{ij3}^k\right)$ as defined as follows [16]:

$$\tilde{x}_{ij} = \sum_{l=1}^{q} \vartheta_1 \tilde{x}_{ijl} \quad (2)$$

$$\tilde{w}_j = \sum_{l=1}^{q} \vartheta_1 \tilde{w}_{jl} \tag{3}$$

Where $\vartheta_1 \in [0, 1]$ denotes assigned weights to the 1th decision maker, and $\sum_{l=1}^{q} \vartheta_l = 1$.

Let the matrix for vaccination alternative selection problem be defined as:

$$\tilde{D} = \begin{bmatrix} \tilde{x}_{11} & \tilde{x}_{12} & \cdots & \tilde{x}_{1q} \\ \tilde{x}_{21} & \tilde{x}_{22} & & \tilde{x}_{2q} \\ \vdots & \vdots & \cdots & \vdots \\ \tilde{x}_{p1} & \tilde{x}_{p2} & \cdots & \tilde{x}_{pq} \end{bmatrix}, \tilde{w} = \left(\tilde{w}_1, \tilde{w}_2, \ldots, \tilde{w}_q\right)^T \tag{4}$$

Step 8 Identify the fuzzy best $\tilde{f}_j^*$ and fuzzy worst $\tilde{f}_j^-$ values of all criteria. $i = 1, 2, \ldots, p, j = 1, 2, \ldots, q$

$$\tilde{f}_j^* = \begin{Bmatrix} \max_i \tilde{x}_{ij}, & \text{for benifit criteria} \\ \min_i \tilde{x}_{ij}, & \text{for cost criteria} \end{Bmatrix} \tag{5}$$

$$\tilde{f}_j^- = \begin{Bmatrix} \min_i \tilde{x}_{ij}, & \text{for benifit criteria} \\ \max_i \tilde{x}_{ij}, & \text{for cost criteria} \end{Bmatrix} \tag{6}$$

Step 9 Calculate the normalized fuzzy distance $\tilde{d}_{ij}, (i = 1, 2, \ldots, p, j = 1, 2, \ldots, q)$

$$\bar{d}_{ij} = \frac{d\left(\tilde{f}_j^*, \tilde{x}_{ij}\right)}{d\left(\tilde{f}_j^*, \tilde{f}_j^-\right)} \tag{7}$$

The Euclidean distance between two triangular fuzzy numbers $A_i = (A_1, A_2, A_3)$ and $B_i = (B_1, B_2, B_3)$ as defined by [14]:

$$D(A_i, B_i) = \sqrt{\frac{1}{6}\left[(a_1 - b_1)^2 + 4(a_2 - b_2)^2 + (a_3 - b_3)^2\right]} \tag{8}$$

Step 10 Graded mean integration based deffuzzifying the fuzzy weight can be defined as [17]:

$$\bar{w}_j = \frac{\bar{w}_j + 4 \times \bar{w}_j + \bar{w}_j}{6}, j = 1, 2, \ldots, q \tag{9}$$

Step 11 Calculate the group utility values S_i and individual regret values R_i, $i = 1, 2, \ldots, p$, by the following equation:

$$S_i = \sum_{j=1}^{n} \frac{\bar{w}_j.\bar{d}_{ij}}{\sum_{j=1}^{n} \bar{w}_j} \quad (10)$$

$$R_i = \max_j \left(\frac{\bar{w}_j.\bar{d}_{ij}}{\sum_{j=1}^{n} \bar{w}_j} \right) \quad (11)$$

Step 12 Calculate the compromise measure Q_i, $i = 1, 2, \ldots, p$, by the following equation:

$$Q_i = \vartheta \frac{S_i - S^*}{S^- - S^*} + (1 - \vartheta) \frac{R_i - R^*}{R^- - R^*} \quad (12)$$

where $S^* = min_i S_i$, $S^- = max_i S_i$, $R^* = min_i R_i$, $R^- = max_i R_i$, ϑ = weight for maximum group utility value and $(1-\vartheta)$ = weight for individual regret value. In this study the threshold value of $\vartheta = 0.5$

Step 13 Rank the alternatives by sorting S_i, R_i and Q_i ($i = 1, 2, \ldots, p$) by increasing order.

Step 14 The alternative $(A^{(1)})$ is the best ranked by Q_i (minimum), if below statistics are satisfied:

C1. Acceptable advantage:

$$Q\left(A^{(2)}\right) - Q\left(A^{(1)}\right) \geq \frac{1}{p-1} \quad (13)$$

where $(A^{(2)})$ is the alternative with second position in the ranking list by Q_i

C2. Acceptable stability: The alternative $(A^{(1)})$ should be the best ranked by S_i or/and R_i.

If these two conditions are not satisfied go to next step.

Step 15 Propose a compromise solutions:

All the alternatives $A^{(i)}(i = 1, 2, \ldots, p)$ are the compromise solutions, if C1 is not satisfied, then explore the maximum value of N according to the equation:

$$Q\left(A^{(N)}\right) - Q\left(A^{(1)}\right) < \frac{1}{p-1} \quad (14)$$

- Alternatives $(A^{(1)})$ and $(A^{(2)})$ are the compromise solutions, if the condition C2 is not satisfied.

4 Case Study: Assessment of Vaccination Alternative Selection Process

H1N1 was reported first in Mexico 2009 and then it spread worldwide [18]. A flu vaccine is used to protect and control the people from influenza epidemics. It takes about 14 days after vaccination to produce antibodies that provide protection from H1N1. Two types of vaccines have been developed by Center for Disease Control and prevention (CDC), "flu shot" and "nasal spray flu vaccine". A 2009 H1N1 "flu shot" is an inactivated vaccine that contains killed virus. This type of vaccination is injected using a needle, most often it is given in the arm. The 2009 H1N1 "flu shot" vaccination is given for people who in the age group of 6 months and older, healthy people, pregnant women and people with chronic diseases. The 2009 H1N1 nasal spray flu vaccine contains live and weakened viruses that give protection against the flu. LAIV vaccine is approved for healthy people and the individuals with the age between 2 and 49 years and pregnant women [19].

All above priorities come under one of the special characteristics of people, spatial and temporal. Health care organization while selecting the best vaccination alternative to protect the people from H1N1 influenza should consider the above three characteristics (people, spatial and temporal). Hence, multiple alternatives are presented and comparison is done to select one best alternative (Fig. 1). However, in sensible situations, majority of the selection criteria are indefinable and it is not easy to show preference with precise numerical value and to create evaluations which exactly express the feeling for decision makers. As a result, the decision makers frequently apply linguistic terms to express their judgments and to evaluate the alternatives for diverse subjective criteria and the weights of the criteria.

Step 1–4 Components of VIKOR-based MCDM as shown in Tables 1, 2, 3.
Step 5 Linguistic variables defined based on the Eq. (1) and shown in Table 4.
Step 6 Fuzzy ratings of alternatives and weights with respect to criteria is shown in Tables 5 and 6 respectively.
Step 7 Aggregated fuzzy ratings of alternatives and weights with respect to criteria are calculated based on the Eqs. (2) and (3) respectively, the results as shown in Table 7.

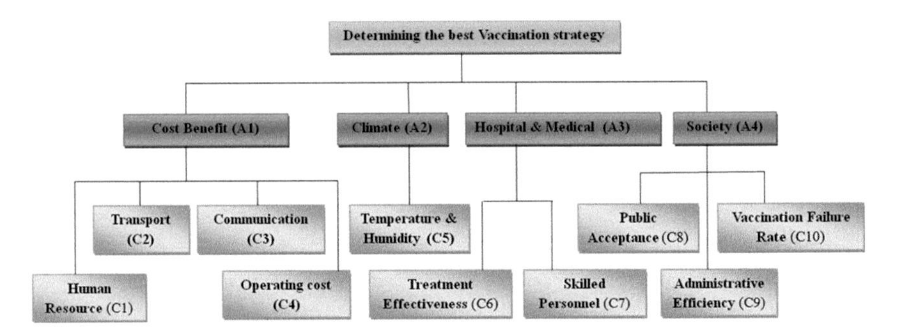

Fig. 1 Hierarchical structure of the problem

Table 1 Alternatives

Notation	Alternatives	Description
A_1	People	Age < 9 & Age > 65, population affected with Asthma, Diabetes, Heart disease, Liver disorders, Blood disorders, Endocrine disorders, Kidney disorders and pregnant women, people who are morbidly obese, people who live with or care for children younger than 6 months of age [20]
A_2	Spatial	Cities where the hygienic level is very low, schools, college, hospitals and high mobility places [20]
A_3	Temporal	Time when vaccination has to be given (H1N1 vaccination has to be given 4 weeks once, before October is the best time to vaccinate against H1N1, second dose should be given within 21 days after the first dose [21]
A_4	People & Spatial	People working in health care departments, people who are living in the city where the hygienic level is very low [20]
A_5	People & Temporal	People who are less than 9 years old should be vaccinated within 1 month after the first dose [21].
A_6	Spatial & Temporal	H1N1 is positively correlations with rainfall and wind speed, and negatively correlated with temperature and humidity [18].
A_7	People, Spatial & Temporal	Months of October/November, people who are living in the city where the hygienic level is very low.

Table 2 Criteria

Notation	Criteria	Description
C_1	Unskilled personnel (Labourers)	People involving in vaccination distribution
C_2	Transport	Ambulance and other vehicles
C_3	Communication cost	Getting survey from the patient about his/her social habits, health condition, last vaccination details and announcing vaccination venue & schedule
C_4	Operating cost	Cost for the infrastructure, medical equipment, staff salary, and other resources
C_5	Temperature & Humidity	Considering climatic conditions and seasonal attributes
C_6	Treatment effectiveness	Percentage of recovery/prevented from the disease after vaccination
C_7	Skilled personnel	Doctors, nurses and medical assistant
C_8	Public acceptance	Public satisfaction rate
C_9	Administrative efficiency	Percentage of people vaccinated
C_{10}	Vaccination failure rate	Percentage of vaccinated infected people

Table 3 Decision Makers

DM_n	Decision makers
DM_1	Public Health Officer
DM_2	Doctor
DM_3	Paramedical Force
DM_4	PRO Leading Office
DM_5	Public Survey

Table 4 Linguistic variables for rating the Alternatives and Criteria Weights

Alternatives		Criteria Weights	
Linguistic variable	Fuzzy numbers	Linguistic variable	Fuzzy numbers
Very Low (VL)	(0,0,1)	Very Low (VL)	(0,0,0.25)
Low (L)	(0,1,3)	Low (L)	(0,0.25,0.5)
Medium Low (ML)	(1,3,5)	Medium (M)	(0.25,0.5,0.75)
Medium (M)	(3,5,7)	High (H)	(0.5,0.75,1)
Medium High (MH)	(5,7,9)	Very High (VH)	(0.75,1,1)
High (H)	(7,9,10)		
Very High (VH)	(9,10,10)		

Table 5 Fuzzy rating of criteria weights

Decision makers	Criteria									
	C_1	C_2	C_3	C_4	C_5	C_6	C_7	C_8	C_9	C_{10}
DM_1	H	L	M	H	H	VH	H	H	H	VH
DM_2	M	L	M	M	M	H	H	M	H	VH
DM_3	L	M	M	M	M	VH	H	H	VH	VH
DM_4	M	L	L	H	M	H	H	H	H	VH
DM_5	M	L	L	M	H	VH	H	VH	VH	VH

Step 8 Determination of fuzzy best $\tilde{f}_j^*$ $(j = 1, 2, \ldots, q)$ and fuzzy worst $\tilde{f}_j (j=1,2,\ldots,q)$ values of all criteria are identified based on the Eqs. (5) and (6) respectively, the results are as follows:

$$\tilde{f}_1^* = (0, 0.6, 2.2), \tilde{f}_2^* = (1.2, 2.6, 4.6), \tilde{f}_3^* = (0, 0.4, 1.8), \tilde{f}_4^* = (0, 0.2, 1.4),$$
$$\tilde{f}_5^* = (8.2, 9.6, 10), \tilde{f}_6^* = (8.2, 9.6, 10), \tilde{f}_7^* = (0.6, 1, 3.4), \tilde{f}_8^* = (5.8, 7.8, 9.4), \tilde{f}_9^* = (7.8, 9.2, 9.8),$$
$$\tilde{f}_{10}^* = (0.6, 1.2, 2.6), \tilde{f}_1^- = (7, 9, 10), \tilde{f}_2^- = (6.6, 8.6, 9.8), \tilde{f}_3^- = (6.6, 8.2, 9.2), \tilde{f}_4^- = (7, 9, 10),$$
$$\tilde{f}_5^- = (2.2, 3.8, 5.6), \tilde{f}_6^- = (2.6, 4.2, 6.6), \tilde{f}_7^- = (7.4, 9.2, 10), \tilde{f}_8^- = (2.2, 4.2, 6.2),$$
$$\tilde{f}_9^- = (0.8, 2.6, 4.6), \tilde{f}_{10}^- = (5.6, 7.2, 8.4)$$

Table 6 Fuzzy rating of alternatives with respect to criteria

Decision makers	Alternatives	Criteria									
		C_1	C_2	C_3	C_4	C_5	C_6	C_7	C_8	C_9	C_{10}
DM_1	A_1	L	H	VH	ML	L	MH	VH	VH	MH	L
	A_2	VL	L	ML	VL	H	ML	ML	M	M	H
	A_3	VL	MH	VL	H	VH	M	L	MH	ML	MH
	A_4	ML	H	VH	MH	MH	MH	H	H	H	ML
	A_5	L	ML	ML	VL	VH	H	MH	MH	VH	VL
	A_6	L	MH	VL	VL	VH	H	VL	H	H	MH
	A_7	H	MH	M	H	MH	VH	M	VH	VH	M
DM_2	A_1	MH	H	H	VH	M	MH	H	M	H	H
	A_2	L	MH	M	L	H	ML	ML	L	MH	L
	A_3	L	MH	L	H	M	ML	M	M	ML	ML
	A_4	H	MH	M	M	ML	M	M	ML	H	M
	A_5	MH	M	H	VL	MH	H	MH	H	MH	M
	A_6	M	M	L	L	VH	H	L	MH	H	M
	A_7	H	M	MH	H	M	VH	M	VH	H	ML
DM_3	A_1	ML	H	MH	VH	ML	ML	H	ML	VH	H
	A_2	L	M	ML	VL	H	M	M	ML	M	MH
	A_3	L	M	VL	H	H	M	L	M	ML	M
	A_4	M	H	MH	ML	M	MH	H	H	MH	H
	A_5	VL	M	ML	L	H	H	MH	M	H	L
	A_6	L	M	L	H	H	L	MH	MH	H	M
	A_7	H	M	ML	H	M	VH	M	H	H	M
DM_4	A_1	L	H	MH	M	L	M	H	H	M	VL
	A_2	VL	L	ML	VL	H	M	M	MH	M	H
	A_3	L	MH	L	H	VH	M	L	MH	L	M
	A_4	M	MH	H	M	H	H	MH	H	H	M
	A_5	ML	ML	M	L	H	MH	H	MH	VH	VL
	A_6	L	MH	L	L	H	MH	VL	H	H	MH
	A_7	H	M	MH	H	M	H	M	H	H	M
DM_5	A_1	L	MH	H	ML	H	M	H	VH	H	ML
	A_2	L	ML	VL	VL	M	MH	M	ML	ML	VH
	A_3	VL	M	MH	H	VH	M	VL	MH	ML	M
	A_4	M	H	VH	MH	M	ML	MH	H	VL	ML
	A_5	ML	L	L	L	VH	H	MH	H	VH	VL
	A_6	L	MH	VL	L	VH	H	L	MH	H	M
	A_7	H	M	MH	H	MH	H	M	H	H	ML

Table 7 Aggregated fuzzy ratings of alternatives and aggregated fuzzy ratings of criteria weights

C_n	Alternatives							Weights
	A_1	A_2	A_3	A_4	A_5	A_6	A_7	
C_1	(1.2,2.6,4.6)	(0,0.6,2.2)	(0,0.6,2.2)	(3.4,5.4,7.2)	(1.4,4.2,4.6)	(0.6,1.8,3.8)	(7,9,10)	(0.25,0.5,0.75)
C_2	(6.6,8.6,9.8)	(1.2,2.6,4.6)	(4.2,6.2,8.2)	(6.2,8.2,9.6)	(1.6,3.4,5.4)	(4.2,6.2,8.2)	(3.4,5.4,7.4)	(0.05,0.3,0.55)
C_3	(4.6,8.4,9.6)	(1.2,2.8,4.6)	(0,0.4,1.8)	(6.6,8.2,9.2)	(2.4,4.2,6)	(0,0.6,2.2)	(3.8,5.8,7.8)	(0.15,0.4,0.65)
C_4	(4.6,6.2,7.4)	(0,0.2,1.4)	(7,9,10)	(3.4,7.4,9)	(0,0.6,2.2)	(1.4,2.4,4)	(7,9,10)	(0.35,0.6,0.85)
C_5	(2.2,3.8,5.6)	(6.2,8.2,9.4)	(7.4,8.8,9.4)	(3.8,5.8,7.6)	(7.4,8.6,9.8)	(8.2,9.6,10)	(3.8,5.8,7.8)	(0.35,0.6,0.85)
C_6	(3.4,5.4,7.4)	(2.6,4.6,6.6)	(2.6,4.2,6.6)	(4.2,6.2,8)	(6.6,8.6,9.8)	(5.2,7,8.4)	(8.2,9.6,10)	(0,65,0.9,1)
C_7	(7.4,9.2,10)	(2.2,4.2,6.2)	(0.6,1,3.4)	(5.4,7.4,9)	(5.4,7.4,9.2)	(1,1.8,3.4)	(3,5,7)	(0.5,0.75,1)
C_8	(5.8,6.2,8.4)	(2.4,4.2,6.2)	(4.2,6.2,8.2)	(5.8,7.8,9)	(5.4,7.4,8.6)	(5.8,7.8,9.4)	(7.8,9.6,10)	(0.5,0.75,0.95)
C_9	(6.2,8,9.2)	(3,5,7)	(0.8,2.6,4.6)	(5.2,6.8,8)	(7.8,9.2,9.8)	(7,9,10)	(7.4,9.2,10)	(0.45,0.85,1)
C_{10}	(3,4.4,5.8)	(5.6,7.2,8.4)	(3,5,7)	(3,5,6.8)	(0.6,1.2,2.6)	(4.2,6.2,8.2)	(2.2,4.2,6.2)	(0.75,1,1)

Table 8 Fuzzy distance for all alternatives and crisp values of fuzzy weights

C_n	Alternatives							Weights
	A_1	A_2	A_3	A_4	A_5	A_6	A_7	
C_1	0.2432	0	0	0.5729	0.3897	0.1487	1	0.5
C_2	1	0	0.6071	0.9361	0.2266	0.6071	0.4690	0.3
C_3	0.9946	0.3075	0	1	0.4873	0.0306	0.6992	0.4
C_4	0.9087	0	1	0.7996	0.0543	0.2647	1	0.6
C_5	1	0.2290	0.1369	0.6606	0.8831	0	0.6569	0.6
C_6	0.7930	0.9464	0.6443	0.6443	0.1712	0.4918	0	0.875
C_7	1	0.3783	0.7799	0.7799	0.7840	0.0870	0.4801	0.75
C_8	0.3906	1	0.0466	0.0466	0.1398	0	0.4849	0.7416
C_9	0.1862	0.6974	0.3633	0.3633	0	0.0579	0.0282	0.8083
C_{10}	0.5301	1	0.6328	0.6328	0	0.8439	0.5520	0.9583

Table 9 S, R and Q values for all alternatives

Criteria	Alternatives						
	A_1	A_2	A_3	A_4	A_5	A_6	A_7
S	0.6668	0.5565	0.5556	0.6010	0.2849	0.2722	0.4876
R	0.1147	0.1462	0.1339	0.0928	0.0900	0.1237	0.0918
Q	0.7197	0.8602	0.7217	0.4415	0.0160	0.2998	0.2889

Step 9 Normalized fuzzy distance $\tilde{d}_{ij}$ $(i = 1, 2, \ldots, p; j = 1, 2, \ldots, q)$ are calculated based on the Eq. (7) and the results are depicted in Table 8.

Step 10 Graded mean integration based defuzzification, the fuzzy weights are calculated based on the Eq. (9), shown in the last column of Table 8.

Step 11 Group utility values S_i $(i = 1, 2, \ldots, p)$ and individual regret values R_i $(i = 1, 2, \ldots, p)$ are calculated based on Eqs. (10) and (11) respectively, the calculated values are shown in the first two rows of Table 9.

Step 12 Compromise measure Q_i $(i = 1, 2, \ldots, p)$ values are calculated based on Eq. (12) is given in the last row of Table 9.

Step 13 The ranking of seven above mentioned alternatives by sorting S_i, R_i and Q_i $(i = 1, 2, \ldots, p)$ in increasing order is given in Table 10.

Step 14 As we see in Table 10, vaccination alternative A_5 is apparently the best strategy in accordance with the value of $Q_i (i = 1, 2, \ldots, p)$. Also, the conditions $\mathbf{C_1}$ and $\mathbf{C_2}$ are satisfied:

Table 10 The ranking of the seven alternatives by S, R and Q in increasing order

Criteria	Alternatives						
	A_1	A_2	A_3	A_4	A_5	A_6	A_7
S	7	5	4	6	2	1	3
R	4	7	6	3	1	5	2
Q	5	7	6	4	1	3	2

Table 11 Ranking of alternatives

Rank	Notation	Alternatives
1	A_5	People & Temporal
2	A_7	People & Spatial & Temporal
3	A_6	Spatial & Temporal
4	A_4	People & Spatial
5	A_1	People
6	A_3	Temporal
7	A_2	Spatial

$$\mathbf{C_1}: Q\left(A^{(2)}\right) - Q\left(A^{(1)}\right) \geq \frac{1}{p-1}$$

Where, Alternative ($A^{(1)}$) is best ranked by the compromise measure Q_i (minimum)($i = 1, 2, \ldots, p$), which is Q_5

Alternative ($A^{(2)}$) is the second position in the ranking list by compromise measure Q_i (minimum) ($i = 1, 2, \ldots, p$), which is Q_7

$$0.2889 - 0.016 \geq \frac{1}{7-1}$$

$0.2727 \geq 0.1666$, hence the condition $\mathbf{C_1}$ is satisfied.

$\mathbf{C_2}$: $R_5 < R_7 < R_4 < R_1 < R_6 < R_3 < R_2 < R_5$, hence the condition $\mathbf{C_2}$ is satisfied. Thus, people & temporal based strategy A_5 is the most suitable vaccination method for protecting people from H1N1 influenza epidemic (Table 11).

5 Result and Conclusion

Choosing the best vaccination strategy has been a significant problem worldwide due to mismanagement of public health organization and lack of vaccination awareness among people and supporting staff. Selecting the suitable vaccination strategy for protecting people from H1N1 influenza epidemic is a complicated MCDM problem, for it requires consideration of multiple alternative solutions and several tangible and intangible criteria. During the vaccination alternative selection process, it is often difficult to provide exact and crisp values for the selection parameters and make evaluations due to uncertainty of information and the vagueness of human recognition. In this paper, fuzzy VIKOR is presented to process both tangible and intangible criteria. The applicability of the proposed approach has been illustrated using a case study of vaccination alternative selection problem to prevent and control from H1N1 in Vellore, Tamil Nadu, and India. The ranking order of all the vaccination alternative methods is depicted in Table 9 on the basis of cost benefit, climate, medical providers and their related sub-criteria. According to the final score, the fifth alternative, people & temporal based strategy is the most suitable vaccination method for protecting people from H1N1 influenza epidemic.

Acknowledgment This work is fully supported by the Indian Council of Medical Research, Government of India under Award Number 32/1/2010-ECD-I.

References

1. Cromwell, I., Peacock, S., Mitton, C.: 'Real-world' health care priority setting using explicit decision criteria: a systematic review of the literature. BMC Health. Serv. Res. **15**, 1–11 (2015)
2. Phuong, N., Kreinovich, V.: Fuzzy logic and its applications in medicine. Int. J. Med. Inform. **62**, 165–173 (2001)
3. De, S., Biswas, R., Roy, A.: An application of intuitionistic fuzzy sets in medical diagnosis. Fuzzy Sets Syst. **117**, 209–213 (2001)
4. Massad, E., Burattini, M., Ortega, N.: Fuzzy logic and measles vaccination: designing a control strategy. Int. J. Epidemiol. **28**, 550–557 (1999)
5. Al-Harbi, K.: Application of the AHP in project management. Int. J. Project Manage. **19**, 19–27 (2001)
6. Lu, S.T., Lin, C.W., Ko, P.H.: Application of analytic network process (ANP) in assessing construction risk of urban bridge project. In: Proceedings of the Second International Conference on Innovative Computing, Information and Control, pp. 169–169 (2007)
7. Ozbek, M., de la Garza, J., Triantis, K.: Efficiency measurement of bridge maintenance using data envelopment analysis. J. Infrastruct. Syst. **16**, 31–39 (2010)
8. Simsek, B., İç, Y., Simsek, E.: A TOPSIS-based Taguchi optimization to determine optimal mixture proportions of the high strength self-compacting concrete. Chemometr. Intell. Lab. Syst. **125**, 18–32 (2013)
9. Wang, J., Xu, Y., Li, Z.: Research on project selection system of pre-evaluation of engineering design project bidding. Int. J. Project Manage. **27**, 584–599 (2009)
10. Jato-Espino, D., Castillo-Lopez, E., Rodriguez-Hernandez, J., Canteras-Jordana, J.: A review of application of multi-criteria decision making methods in construction. Autom. Constr. **45**, 151–162 (2014)
11. Opricovic, S.: Fuzzy VIKOR with an application to water resources planning. Expert Syst. Appl. **38**, 12983–12990 (2011)
12. Opricovic, S., Tzeng, G.: Extended VIKOR method in comparison with outranking methods. Eur. J. Oper. Res. **178**, 514–529 (2007)
13. Broekhuizen, H., Groothuis-Oudshoorn, C., van Til, J., Hummel, J., IJzerman, M.: A review and classification of approaches for dealing with uncertainty in multi-criteria decision analysis for healthcare decisions. Pharmaco Econ. **33**, 445–455 (2015)
14. Chen, T.Y., Ku, T.C., Tsui, C.W.: Determining attribute importance based on triangular and trapezoidal fuzzy numbers in (z) fuzzy measures. In: The 19th International Conference on Multiple Criteria Decision Making, pp. 75–76 (2008)
15. Bellman, R.E., Zadeh, L.A.: Decision-making in a fuzzy environment. Manage. Sci. **17**, 141–164 (1970)
16. Dursun, M., Karsak, E., Karadayi, M.: Assessment of health-care waste treatment alternatives using fuzzy multi-criteria decision making approaches. Resour. Conserv. Recycl. **57**, 98–107 (2011)
17. Chou, C.: The canonical representation of multiplication operation on triangular fuzzy numbers. Comput. Math Appl. **45**, 1601–1610 (2003)

18. Lopez, D., Gunasekaran, M., Murugan, B.S., Kaur, H., Abbas, K.M.: (2014). Spatial big data analytics of influenza epidemic in Vellore, India. In: 2014 IEEE International Conference on Big Data, pp. 19–24 (2014)
19. Centers for Disease Control and Prevention. http://www.cdc.gov/h1n1flu/
20. Centers for Disease Control and Prevention: People at High Risk of Developing Flu–Related Complications. http://www.cdc.gov/flu/about/disease/high_risk.htm
21. Centers for Disease Control and Prevention: Vaccine against 2009 H1N1 Influenza Virus. http://www.cdc.gov/h1n1flu/vaccination/public/vaccination_qa_pub.htm

Ranking and Dimensionality Reduction Using Biclustering

V. Hema Madhuri and T. Sobha Rani

Abstract Organizing and searching the data tries to detect groups where objects exhibit similar properties. As the dimensionality d increases, the space in which data is represented increases rapidly therefore the available data becomes sparse. When d is high, all objects appear to be sparse and dissimilar in many ways. Here, a study is made to reduce the number of dimensions using biclustering method to rank the features/dimensions. Classification rate is used as validation criteria for the selection of appropriate dimensions. Ranking algorithms such as Relief F, Symmetrical uncertainty and Information gain are compared with the proposed ranking using biclustering. It is found that for large data sets with large number of dimensions, ranking using biclustering achieves classification rates with less number of features than the other ranking algorithms.

Keywords Biclustering · Dimensionality reduction · Ranking algorithms

1 Introduction

Dimensionality reduction deals with the reduction of features and is an important issue in datamining and pattern recognition fields. In order to eliminate the confusion that may result from considering irrelevant and redundant features, dimensionality reduction is attempted.

1.1 Dimensionality Reduction

Dimensionality reduction can be considered as a method which can be used in bringing down the number of dimensions/attributes/features of data to a lower number of

---

V. Hema Madhuri · T. Sobha Rani (✉)
School of Computer and Information Sciences, University of Hyderabad, Hyderabad, India
e-mail: tsris@uohyd.ernet.in

V. Ravi et al. (eds.), *Proceedings of the Fifth International Conference on Fuzzy and Neuro Computing (FANCCO - 2015)*, Advances in Intelligent Systems and Computing 415, DOI 10.1007/978-3-319-27212-2_17

dimensions/attributes/features. The resulting set of features may retain more useful information since redundant and irrelevant dimensions are eliminated without much loss to the information [5]. Features are said to be redundant if their values are correlated [12] and they are said to be irrelevant if they do not contribute to the understand the data. Dimensionality reduction can be attempted differently depending on the way data is treated.

- Feature Selection: Feature selection selects a subset of features for the construction of the model. These subset of features may help better in understanding, easier to interpret and takes less time to train the model and avoids over-fitting. This process may involve removal of redundant and irrelevant features. Feature elimination could be done using certain measures. The measures chosen for selection of features should be computable quickly and also retain the usefulness. Some of the measures that can be used to decide the relevance of a feature are the popular measures like Information Gain, Mutual Information [12] and Pearson product-moment correlation coefficient [8, 10]. For each of the class or feature combinations, scores of significance test is done [9, 12].
- Feature Extraction: Feature extraction is the other way of reducing the number of features by considering combinations of two or more features. A transformation function is applied to data to project the existing data on to a new feature space with lower dimensions. Most popularly used methods are Principal Component Analysis (PCA) and Nonlinear dimensionality reduction, partial least squares and Isomap.

Curse of dimensionality [3] phenomena [3] arises when analysing and organizing data in high-dimensional spaces which does not not occur in low-dimensional settings. When the dimensionality increases, the volume of the space increases so fast that the available data becomes sparse. Performance of a classifier degrades above a certain maximum number of features.

2 Literature Survey

Some of the most frequently used approaches in dimensionality reduction are:

2.1 Principal Component Analysis

In PCA, x new features are forming linear combinations of n orthogonal attributes where $x < n$. In order to do PCA, first covariance matrix of the features is calculated. Then the eigen vectors and eigen values of the covariance matrix are computed. Eigen vector with highest eigen value is taken as the first principal component of the data set. New feature vectors are formed by calculating other components. PCA can be used for large data sets. But the drawback of PCA is that it is not applicable to non-linear data sets.

2.2 *Linear Discriminant Analysis*

Unlike PCA, LDA separates classes of objects by finding linear combination of features. By taking linear combination, it results in dimensionality reduction. It minimizes intra variance and maximizes inter variance. This guarantees maximum separability. Similar to PCA, LDA looks for linear combinations of variables in data.

2.3 *Existing Heuristics for Dimensionality Reduction*

PCA and LDA are dependent on matrix manipulations which may become slow when large dimensions are involved. Some of the heuristics methods used for dimensionality reduction are listed here.

1. *Dimensionality reduction using Laplacian eigenmaps*: It is a computationally efficient approach to nonlinear dimensionality reduction that has locality-preserving properties and a natural connection to clustering [2].
2. *Using local Fisher discriminant analysis*: LFDA combines the ideas of FDA (Fisher discriminant analysis) and LPP(locality-preserving projection) [26]. LFDA has an analytic form of the embedding transformation and the solution can be easily computed by solving a generalized eigenvalue problem.
3. *Using rough and fuzzy-rough based approaches*: [16] This approach looks at the problem of selecting input features that are most useful in predicting an outcome while retaining the underlying semantics of the data. Several rough and fuzzy based methods are used for feature selection and a rough set based feature grouping is provided.
4. *Using Fast correlation-based filter selection*: This method [28] uses a fast filter method to identify the relevant features and redundancy among relevant features without using pairwise correlation. Here, predefined threshold is used to specify the number of features that will be considered.

3 Measures Used for Validation

In order to validate the reduction in dimensions, classification can be used as one verification method. If the same or more accurate classification rates are achieved using the subset of features selected using a dimensionality reduction method, it is one way of validation.

Table 1 Confusion matrix

	Predicted as positives	Predicted as negatives
Actual positives (P)	True positives (TP)	False negatives (FN)
Actual negatives (N)	False positives (FP)	True negatives (TN)

The measures used for dimensionality reduction in the experiment are based on confusion matrix obtained in a classification task. The two-class confusion matrix shown in the Table 1 illustrates the outcomes that are possible. If the entire data is used to build the model, estimating the accuracy of the classifier becomes difficult since it leads to overoptimistic predictions of accuracy of the model. Therefore, often the available data is randomized and divided into train, validation and test data sets to determine the accuracy of the model.

1. **A**ccuracy: Accuracy is the ability of the classifier to determine the class of a randomly selected object.

$$Accuracy = \frac{(TP + TN)}{(P + N)}$$

2. **G**-mean: This measure is based on recall of both the positive and negative classes. G-mean value is very low when there is a high bias in the classifier towards one particular class. This means g-mean is zero when none of the positive objects are identified.

$$gmean = \sqrt{specificity \times sensitivity}$$

3. **A**UC: Area Under the Curve is useful in evaluating the better model. The value of AUC is 1 if the model is perfect or it is near to 0.5 if it performs random guessing.

$$AUC = \frac{(1 + TPR - FPR)}{2}$$

4 Proposed Approach

Since there is no way one can identify the desired set of features that are the best in describing a data set, most of the dimensionality reduction methods depend on the ranking of the features. To which particular value of the rank one needs to consider the features is dependent on the measure that one needs to optimize. It means that the best one could achieve with the least number of features. In the proposed approach, biclustering is used to rank the features, and then a greedy method is proposed to obtain the best set of features for the classification task.

4.1 Biclustering

A data set may consists of rows and columns which correspond to objects and attributes or the samples/instances and features respectively. It is of interest to find homogeneous groups such as market segments, meaningful patterns of text or co regulated genes in case of market analysis, textmining and DNA analysis respectively. There may be different set of interests. For example, if a product like a bike is to be sold, there may be a group of customers who will be interested in economy, cost and service cost and some other group may be interested in sportiness and horse power.

Groups can be formed by only a subset of attributes or objects in a data set. Standard clustering algorithms do not consider a subset of features in clustering the objects. A solution was proposed by Hartigan [14] for this problem which was later came to be known as biclustering [1]. Biclustering also known as two-mode clustering [21], co-clustering [11] is simultaneous clustering of objects and attributes of data a set. Biclustering attempts to find submatrices, which are a subset of objects and attributes where the objects exhibit highly correlated activities for every attribute. In recent times biclustering concept is being used in gene expression data analysis [6]. There are several ways of doing a biclustering.

4.1.1 Types of Biclusters

A bicluster is formed by a subset of objects and attributes of the data matrix. These objects and attributes follow a consistent pattern [19] in the biclusters.

- **Biclusters with constant values**
  A biclustering algorithm tries to find sub-matrices consisting of similar values in rows and columns. Evaluating the data using variance can be helpful for noisy data.
- **Biclusters with constant values on rows or columns**
  A bicluster having constant value on each row but different rows may have different values. Normalization technique can be used for preprocessing.
- **Biclusters with coherent values**
  In these biclusters, with respect to columns the rows change in a synchronized way and vice versa.
- **Biclusters with coherent evolutions**
  In applications like gene expression matrix rather than looking for exact values, it is more interesting to observe up or down regulated changes across genes or conditions.

4.2 *Proposed Dimensionality Reduction Method Using Biclustering*

1. Data is divided into train, validation and test data sets.
2. The model is trained on the train data set.
3. Using biclustering algorithm, ranking of the features is done. Fewer number of times the feature is selected in biclusters higher is the rank of that feature. For each (feature in the decreasing order of rank) Repeat steps 4,5 and 6
4. Accuracy, gmean and AUC are calculated for the training data.
5. Same measures are calculated for the validation data set by the model trained on the training set.
6. Feature is added the subset of features already considered.
7. For each of the measures, the feature having highest value for the validation set is determined.
8. Thus selected set of features are used to evaluate the test data set. Consensus of average is taken from the five folds and the features are ranked. This is explained in greater depth in Sect. 5.

5 Experiments and Results

5.1 *Data Sets*

Data sets used in the current paper are listed in the Table 2. Data sets chosen for this experimentation are two-class problems and multi-class problems where one class

Table 2 Data sets used

Data set	Instances	Features	Positive class	Negative class
Pima	768	8	268-Diabetic	500-Non diabetic
Wisconsin	699	9	241-Malignant	458-Benign
Spambase	4601	57	1813-Spam	2188-Non spam
Ozone	2563	73	128-Ozone day	2335-Normal day
Musk2	6598	166	1017-Musk	5581-Non musk
Arrythmia	279	452	207-Arrythmia	245-Non arrythmia

is treated as positive and all the other classes are treated as negative. Arrythmia is a multi-class problem in which *Normal* is treated as positive and all other classes are put under *Non-Arrythmia*.

5.2 Selection of Biclustering Algorithm

A few biclustering algorithms are available in literature. Bimax [23], Plaid model [18], Iterative signature algorithm [4] and Cheng and Church [6] algorithms are more often used. Cheng and Church method is used for experimentation in this work. According to them biclustering follow an additive model which searches bicluster with constant values. Two parameters, delta and alpha values, are needed to be determined in Cheng and Church biclustering algorithm.

Cheng and Church use greedy iterative search to minimize the mean square residue (MSR). The resulting biclusters are as large as possible such that the H score is below a threshold I.

$$H(I,J) = \frac{1}{\|I\|\|J\|} \sum_{i \in I, j \in J} (a_{ij} - a_{iJ} - a_{Ij} + a_{IJ})^2$$

Here, a_{iJ} is the mean of row i, a_{Ij} is the mean of column j and a_{IJ} is the overall mean. The algorithm searches for a subgroup where the H score is below a delta value and above a alpha fraction of the overall score.

1. **D**elta Parameter: Delta value is used as threshold in the score function H. The quality of the clusters selected, is determined by this delta value. Smaller value of delta gives better clusters. If the delta value is very small it results in very small biclusters which leads to loss of information. Hence, a suitable value has to be determined for delta.
2. **A**lpha Parameter: The speed at which the algorithm clusters the data is determined by the alpha value. This value is used in the deletion phase of Cheng and Church algorithm. Greater the alpha value, faster is the deletion.

α and δ Determination The process of determining delta and alpha values [24] is described here.

1. Since α determines the speed at which algorithm clusters, experiments were conducted to determine the possible values for α.
2. Keeping one of the α constant, the δ values are taken on x axis and sub matrix size (the number of rows and columns of the bicluster selected) are taken on y axis. Experimenting with various δ values on the Cheng and Church algorithm the sub matrix size is determined. Figure 1.
3. x value where the slope changes is taken as δ.
4. Now this δ value is kept constant. α values are taken on x axis and time taken for clustering is taken on y axis. Biclustering is done by varying the α values. Time taken is plotted for each of the delta values as shown in Fig. 2.

Table 3 Values of δ and α

Data set	Delta value	Alpha value
Wisconsin	0.6	1.2
Pima	0.6	1.2
Spambase	0.5	2
Ozone	5.2	4.6
Musk2	50	100
Arrythmia	0.37	0.7

5. x value where the slope changes is taken as α. If the slope remains almost the same, some average value is taken as α.
6. The values chosen for δ and α are shown in Table 3.

5.3 Method

Naive Bayes and J48 classifiers are used for validation purpose. For each data set and for each of the classifiers AUC, gmean and accuracy values are calculated. Experimental results for Wisconsin data set using Naive Bayes classifier is as shown in Fig. 3 and for Pima data set using Naive Bayes classifier is shown in Fig. 4. In order to carry out the experiments, following packages and tools are used.

- **R**: A language and environment for statistical computing. R Foundation for Statistical Computing, Vienna, Austria. URL http://www.R-project.org [25].
- **Package "biclust"**: A toolbox for bicluster analysis in R. [17].
- **Package: RWeka** RWeka interfaces Weka's functionality to R [15].

1. The data sets are taken from "UCI Machine learning repository". The values of data setare randomized and stored in .csv format.
2. The data setis partitioned into three sets namely train data set (40 %), validation data set (10 %) and test data set (50 %).
3. The experiment is done as five folds.
4. For every fold, the biclustering is done on train data.
5. The result of biclustering algorithm is the set of biclusters. The results are extracted using "biclust package".
6. For every feature, the feature count (the number of times the feature is selected in all the biclusters) is determined. The features selected using biclustering are selected from first 50 biclusters the algorithm gives or lesser (if algorithm gives lesser biclusters). If more number of biclusters are selected, it results in smaller biclusters which do not give better information of features (Fig. 1).

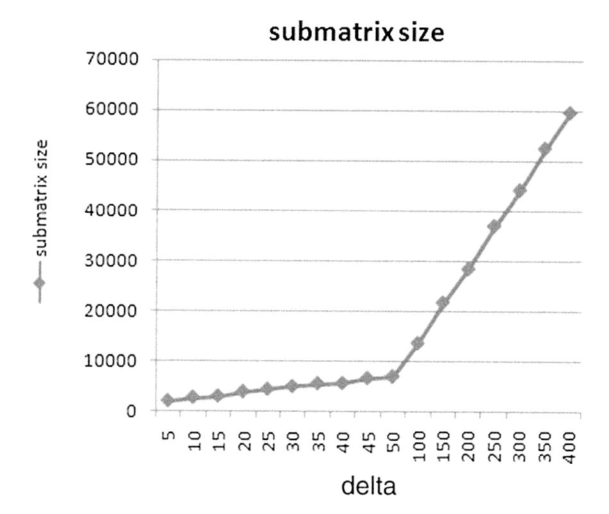

Fig. 1 Determination of α and δ values

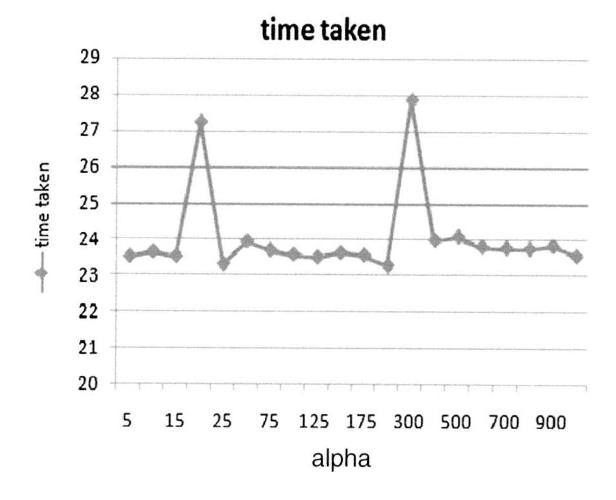

Fig. 2 Musk: Selection of α and δ values

7. Lower the feature count, higher is the rank of the feature. Because if a feature is selected in more biclusters, it is less unique, which do not contribute much to dimensionality reduction.
8. The feature whose rank is higher, better contributes to dimensionality reduction. Therefore, the features are arranged in the decreasing order of rank.
9. Thus the order in which the features are selected are listed in the column 1. An illustration is given how the features appear for each fold. One of the fold results of Wisconsin and Pima are shown in Figs. 3 and 4 respectively. Here, column 1 shows features in decreasing order of rank.
10. Each feature is considered in iteration to calculate the AUC, gmean and Accuaracy for train data. The model is trained using train data. The same feature set is used on validation data and all the measures are calculated. For example in Fig. 3, the decreasing order of ranking is 4, 6, 8, 7, 1, 5, 9, 3, 2. Considering third

		AUC		gmean		Accuracy	
Feature	Feature set	train	validation	train	validation	Train	validation
2	2, 9	0.681	0.7137	0.6577	0.6918	71.7105	82.89
5	5, 2, 9	0.6824	0.7	0.6564	0.6708	72.0395	81.58
4	4, 5, 2, 9	0.6838	0.7	0.6597	0.6708	72.0395	81.58
3	3, 4, 5, 2, 9	0.6852	0.693	0.6672	0.6574	71.7105	80.26
6	6, 3, 4, 5, 2, 9	0.7087	0.7396	0.6981	0.7181	73.3553	84.21
8	8, 6, 3, 4, 5, 2, 9	0.7198	0.7069	0.7128	0.6759	74.0132	81.58
7	7, 8, 6, 3, 4, 5, 2, 9	0.717	0.7223	0.7104	0.6961	73.6842	82.89
1	1, 7, 8, 6, 3, 4, 5, 2, 9	0.717	0.6902	0.7104	0.6487	73.6842	78.95
The selected features are:		2	3	4	5	6	

Measure	Train	Test
AUC	0.7087	0.7331
gmean	0.6981	0.7309
Accuracy	73.3553	0.7577

Fig. 3 One fold of Wisconsin NB

		AUC		gmean		Accuracy	
Feature	Feature set	train	validation	train	validation	Train	validation
4	4, 10	0.8246	0.8631	0.8121	0.8626	87.5	88.24
6	6, 4, 10	0.9155	0.9183	0.914	0.9172	93.3824	94.12
8	8, 6, 4, 10	0.9333	0.9183	0.9333	0.9172	93.3824	94.12
7	7, 8, 6, 4, 10	0.9279	0.9444	0.9279	0.9428	92.6471	97.06
1	1, 7, 8, 6, 4, 10	0.9506	0.9444	0.9502	0.9428	94.1176	97.06
5	5, 1, 7, 8, 6, 4, 10	0.9506	0.9444	0.9502	0.9428	94.1176	97.06
9	9, 5, 1, 7, 8, 6, 4, 10	0.9615	0.9444	0.9613	0.9428	95.5882	97.06
3	3, 9, 5, 1, 7, 8, 6, 4, 10	0.9615	0.9444	0.9613	0.9428	95.5882	97.06
2	2, 3, 9, 5, 1, 7, 8, 6, 4, 10	0.9615	0.9444	0.9613	0.9428	95.5882	97.06
The selected features are:		4	6	7	8		

Measure	Train	test
AUC	0.9279	0.9478
gmean	0.9279	0.9473
Accuracy	92.6471	0.9535

Fig. 4 One fold of Pima NB

row, the feature set here is 4, 6, 8 and in 10 column the class variable. The model is trained on features 4, 6, 8 using train data and validated on validation data for each of the measures.

11. For each of the validation results column, the feature with highest value in each column is selected. In Fig. 3, feature set till the feature 7 has highest value in fourth column (AUC), sixth column (gmean) and eighth column (Accuracy).

12. The features with equal and higher rank than this feature are chosen for all the measures. In Fig. 3, in all the three columns the features 4, 6, 8, 7 are chosen.
13. Consensus of feature is taken for all the three measures. The feature is selected if it is taken in more than two columns. Here in 3, 4, 6, 8, 7 are chosen.
14. So these features are the set of selected features.
15. Thus if a feature is selected in more that two folds it contributes to dimensionality reduction.
16. These features are used to evaluate the test data set and the results are displayed.
17. This experiment is carried out on all the three data sets and the "Feature Selection" is made.

An illustration is given for Pima data setin Fig. 4. Similarly, the values are calculated for Pima and Musk2 data sets.

5.4 Classificaion Results Using Naive Bayes

Naive Bayes classifier is used for validation method for dimensionality reduction. For Musk2 data set AUC, gmean and accuracy graphs for one fold are shown in respectively. Number of features are taken on x axis and measure (AUC or gmean or Accuracy) is taken on y axis. The graphs are plotted for train and validation data.

5.5 Classification Results Using J48

J48 classifier is used for validation method for dimensionality reduction. Here also Musk2 data set AUC, gmean and Accuracy graphs for one fold are shown in respectively. Number of features are taken on x axis and measure (AUC or gmean or Accuracy) is taken on y axis. The graphs are plotted for train and validation data. In the case of Musk2 data set, feature set till the first occurrence of the maximum value of the measures is used.

6 Discussion

6.1 Comparison of Features Selected Using Biclustering and Other Ranking Algorithms

In order to compare the results of dimensionality reduction using biclustering algorithm, three ranking algorithms are chosen. The ranking algorithms considered for analysis are Relief F, Symmetrical Uncertainty and Gain Ratio.

Table 4 Average number of features selected

Data Set	Biclustering NB	Biclustering J48	Ranking (J48) Symm Uncer	Ranking (J48) Relief F	Ranking (J48) Gain ratio
Wisconsin	3	4	4	3	2
Pima	2	6	6	7	6
Musk2	27	13	49	52	58

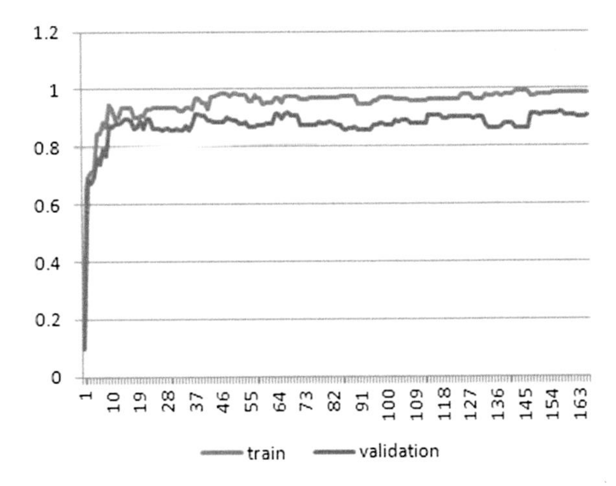

Fig. 5 One fold of Musk2 using bicluster ranking

Relief F [20]: This is a ranking algorithm which estimates how well an attribute can distinguish between instances that are closer to each other. It is not dependent on heuristics. It requires linear time with respect to the number of given features and training instances.

Gain Ratio: Information gain ratio is a ratio of information gain to the intrinsic information in decision tree learning. The bias towards multi-valued attributes is reduced in this method.

Symmetrical Uncertainty [13]: Evaluates the worth of a set attributes by measuring the symmetrical uncertainty with respect to another set of attributes. A ranking can be obtained using symmetrical uncertainty. Symmetrical uncertainty takes care of the information gain's bias towards features with more values.

Comparison results are given in Table 4. Procedure followed for this experimentation is as given in Sect. 5 from points 10 to 17. The only difference is, here ranking is done using ranking algorithms. The classifier used is J48. Classification results (gmean) are presented in Figs. 5, 6, 7, and 8 using ranking algorithms bicluster, symmetric uncertainty, gain ratio, Relief F respectively. It can be observed that classification rates increase almost gradually with the all three ranking algorithms gain ratio, Relief F and symmetric uncertainty whereas bicluster achieves the maximum at a faster rate with less number of features there after there are oscillations. Similar behaviour is observed in all the folds (Table 5).

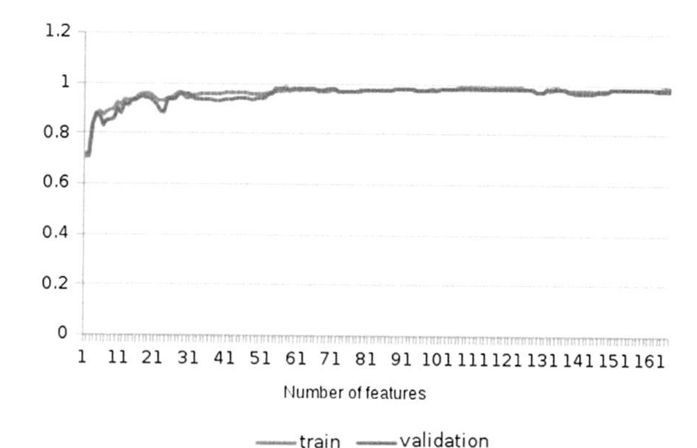

Fig. 6 One fold of Musk2 using Symmetric Uncertainty

Table 5 Average number of features selected for other data sets

Data set	Biclustering NB	Biclustering J48
Spambase	15	11
Ozone	28	8
Arrythmia	47	47

Table 6 Wisconsin—Test data set results

	Biclustering NB	Biclustering J48	Ranking (J48) Symm Uncer	Ranking (J48) Relief F	Ranking (J48) Gain ratio
AUC	0.93374	0.89352	0.92678	0.92034	0.90348
gmean	0.93334	0.8897	0.92562	0.91836	0.9009
Accuracy	93.818	91.14092	93.36528	93.3502	91.8386

Table 7 Pima—Test data set results

	Biclustering NB	Biclustering J48	Ranking (J48) Symm Uncer	Ranking (J48) Relief F	Ranking (J48) Gain ratio
AUC	0.7299	0.7019	0.70898	0.7059	0.70898
gmean	0.72786	0.69064	0.69034	0.68756	0.69034
Accuracy	75.25	73.38798	75.27184	74.91102	75.27184

Various comparisons are made on the basis of number of features selected Table 4, the time taken for the process of feature selection Table 9 and also the test results of different algorithms with respect to the given data sets. Table 4 shows the number of features chosen for each of the data sets. Tables 6, 7 and 8 present the average classification results for the test data set for the data sets Wisconsin, Pima and Musk2

Table 8 Musk 2—Test data set results

	Biclustering NB	Biclustering J48	Ranking (J48) Symm Uncer	Ranking (J48) Relief F	Ranking (J48) Gain ratio
AUC	0.709	0.8974	0.91304	0.90008	0.90088
gmean	0.66742	0.8947	0.91072	0.89692	0.89756
Accuracy	83.74	94.46898	95.56092	94.88262	95.0159

Table 9 Total time taken for classification for all the folds including the ranking

Algorithm	Classifier	Wisconsin	Pima	Musk
BC–CC	NB	11.43 s (3)	34.63 s (2)	39.71 min (27)
BC–CC	J48	7.54 s (4)	24.33 s (6)	53.43 min (13)
Symm Unc	J48	7.23 s (4)	14.58 s (6)	36.39 min (49)
Relief F	J48	7.40 s (3)	14.75 s (7)	29.22 min (52)
Gain ratio	J48	7.03 s (2)	2.01 min (6)	35.56 min (58)

Table 10 Ozone—Test data set results

Data Set	Biclustering NB	Biclustering J48
AUC	0.565425	0.57016
gmean	0.273625	0.4042
Accuracy	91.4065	0.81578

Table 11 Arrythmia—Test data set results

Data Set	Biclustering NB	Biclustering J48
AUC	0.73622	0.78202
gmean	0.72886	0.78088
Accuracy	75.20398	0.7806

respectively. For Spambase, Ozone and Arrythmia data sets only the biclustering results are provided in Tables 12, 10 and 11 respectively.

Table 14 gives the results for feature selection using cooperative game theory (CGFS-mRMR and CGFS-SU) [27] and also feature reduction using mRMR [22] and Relief F methods for Musk2 data set. They have obtained slightly better accuracies than the proposed method but with higher number of features (Table 13).

Table 12 Spambase—Test data set results

Data Set	Biclustering NB	Biclustering J48
AUC	0.82128	0.87826
gmean	0.81604	0.87804
Accuracy	0.82006	87.98672

Table 13 Total time taken for classification for all the folds including the ranking (in minutes)

Algorithm	Classifier	Email Spam	Ozone	Arrythmia
BC–CC	NB	2.6	1.539	22.98
BC–CC	J48	4.14	1.812	25.09

Table 14 Comparison of Musk2 with other methods

Classifier	CGFS-mRMR	mRMR	CGFS-SU	SU	Relief F
SVM	95.56 (28)	95.21 (26)	95.92 (28)	96.01 (28)	94.92 (7)
NB	92.83 (24)	92.51 (19)	91.54 (10)	91.72 (21)	89 (7)

Table 15 Pima and Wisconsin results obtained with other methods

Data set	EFS-RPS	FS-SSGA	FPS-SSGA	FS-RST	EIS-RFS
Pima	73.35 (4)	67.7 (3)	72.65 (8)	73.83 (8)	74.49 (8)
Wisconsin	95.85 (5)	95.14 (5)	96.13 (6)	95.86 (9)	96.86 (9)

Table 15 provides the results obtained by evolutionary feature selection on top of the fuzzy rough set based prototype selection for Pima and Wisconsin data sets [7]. They both are providing accuracies.

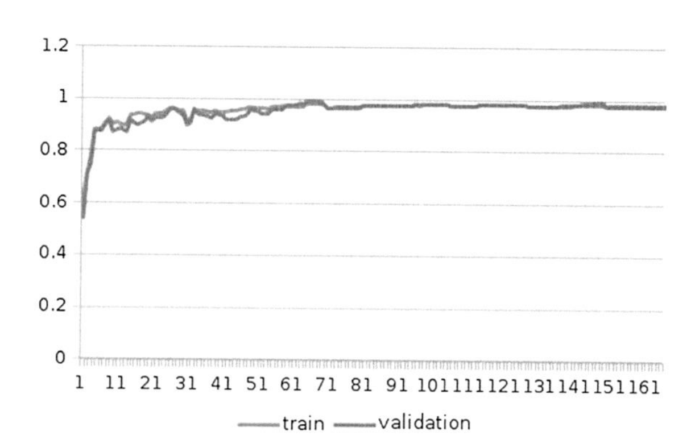

Fig. 7 One fold of Musk2 using Gain ratio

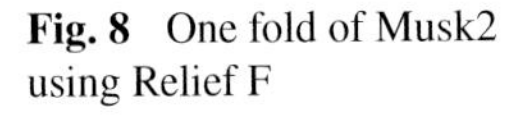

Fig. 8 One fold of Musk2 using Relief F

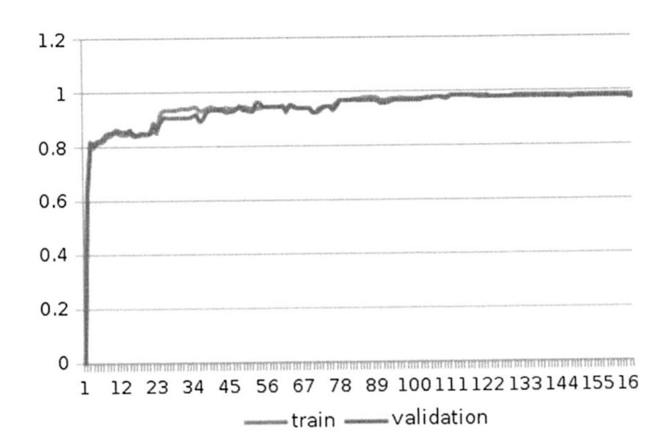

7 Conclusions

In this work, biclustering is used to rank the features and dimensionality reduction is achieved by adding features in the ranking order. Cheng and Church biclustering is used as the ranking algorithm. Experimentation is done on six data sets. From Wisconsin data set results (Table 6), it can be inferred that Naive Bayes classifier is giving performance measures similar to symmetrical Uncertainty and Relief F algorithms. From Pima data set results (Table 7), it is evident that Naive Bayes is giving better performance measures than the other ranking algorithms. From Musk2 data set results (Table 8), other algorithms are doing better than Bicluster. But the number of features are less than the other methods. Ozone results are a classical case of imbalanced data sets, where the results obtained for the majority/negative class dominates the positives/minority class results. One way of handling this is to do data balancing. Then the results would improve. Table 9 gives the time taken by Naive Bayes and J48 and shows that they are closer for less number of features whereas J48 is taking greater time to train in case of Musk2.

It is very important to note that training set is only 40 % of the entire data. With less training data, we are able to select fewer number of relevant features. In some of the algorithms entire data set is considered for selection of the features. From Table 4, number of features selected by NB and J48 for Wisconsin data set are similar to other ranking algorithms. For Pima data set NB is giving lesser number of features while J48 is similar to other ranking algorithms. Whereas in Musk data set the time taken is high for training but the number of features selected are less. If there is a way to specify a threshold, like top x % features, that would reduce the training time drastically.

References

1. Bacelar-Nicolau, H.: Mathematical Classification and Clustering. Kluwer Academic Publishers, New York (1996)
2. Belkin, M., Niyogi, P.: Laplacian eigenmaps for dimensionality reduction and data representation. Neural Comput. **15**, 1373–1396 (2003)
3. Bellman, R.E.: Some new techniques in the dynamic-programming solution of variational problems. Q. Appl. Math. **23**, 295–305 (1958)
4. Bergmann, S., Ihmels, J., Barkai, N.: Iterative signature algorithm for the analysis of large-scale gene expression data. Phys. Rev. E **67**, 031902 (2003)
5. Bermingham, M.L., Pong-Wong, R., Spiliopoulou, A., Hayward, C., Rudan, I., Campbell, H., Wright, A.F., Wilson, J.F., Agakov, F., Navarro, P.: Application of high-dimensional feature selection: evaluation for genomic prediction. Sci. Rep. **5** (2015)
6. Cheng, Y., Church, G.M.: Biclustering of expression data. ISMB **8**, 93–103 (2000)
7. Derrac, J., Verbiest, N., Garca, S., Cornelis, C., Herrera, F.: On the use of evolutionary feature selection for improving fuzzy rough set based prototype selection. Soft Comput. **17**, 223238 (2013)
8. Fisher, R.A.: Frequency distribution of the values of the correlation coefficient in samples from an indefinitely large population. Biometrika, 507–521 (1915)
9. Forman, G.: An extensive empirical study of feature selection metrics for text classification. J. Mach. Learn. Res. **3**, 1289–1305 (2003)
10. Gayen, A.K.: The frequency distribution of the product-moment correlation coefficient in random samples of any size drawn from non-normal universes. Biometrika **38**, 219–247 (1951)
11. Govaert, G., Nadif, M.: Co-clustering. Wiley, New York (2013)
12. Guyon, I., Elisseeff, A.: An introduction to variable and feature selection. J. Mach. Learn. Res. **3**, 1157–1182 (2003)
13. Hall, M.A.: Correlation based feature selection for machine learning. Thesis Report, University of Waikato, April 1999
14. Hartigan, J.A.: Direct clustering of a data matrix. J. Am. Stat. Assoc. **67**, 123–129 (1972)
15. Hornik, K., Buchta, C., Zeileis, A.: Open-source machine learning: R meets weka. Comput. Stat. **24**, 225–232 (2009)
16. Jensen, R., Shen, Q.: Semantics-preserving dimensionality reduction: rough and fuzzy-rough-based approaches. IEEE Trans. Knowl. Data Eng. **16**, 1457–1471 (2004)
17. Kaiser, S., Leisch, F.: A toolbox for bicluster analysis in r (2008)
18. Lazzeroni, L., Owen, A.: Plaid models for gene expression data. Stat. Sinica **12**, 61–86 (2002)
19. Madeira, S.C., Oliveira, A.L.: Biclustering algorithms for biological data analysis: a survey. IEEE/ACM Trans. Comput. Biol. Bioinformatics (TCBB) **1**, 24–45 (2004)
20. Marko, R.S., Igor, K.: Theoretical and empirical analysis of Relief and ReliefF. Mach. Learn. **53**, 23–69 (2003)
21. Mechelen, I.V., Bock, H., Boeck, P.: Two-mode clustering methods: a structured overview. Stat. Methods Med. Res. **13**, 363–394 (2004)
22. Peng, H., Long, M.F., Ding, C.: Feature selection based on mutual information: criteria of max-dependency, max-relevance, and min-redundancy. IEEE Trans. Pattern Anal. Mach. Intell. **27** (2005)
23. Prelic, A., Bleuler, S., Zimmermann, P., Wille, A., Bhlmann, P., Gruissem, W., Hennig, L., Thiele, L., Zitzler, E.: A systematic comparison and evaluation of biclustering methods for gene expression data. Bioinformatics **22**, 1122–1129 (2006)
24. Qu, H., Wang, L., Liang, Y., Wu, C.: An improved biclustering algorithm and its application to gene expression spectrum analysis. Genomics, Proteomics Bioinformatics **3**, 189–193 (2005)
25. R Core Team, R: A language and environment for statistical computing. In: R Foundation for Statistical Computing, Vienna, Austria, 2012 (2014)
26. Sugiyama, M.: Dimensionality reduction of multimodal labeled data by local fisher discriminant analysis. J. Mach. Learn. Res. **8**, 1027–1061 (2007)

27. Suna, X., Liua, Y., Lic, J., Zhua, J., Liua, X., Chena, H.: Using cooperative game theory to optimize the feature selection problem. Neurocomputing **97**, 8693 (2012)
28. Yu, L., Liu, H.: Feature selection for high-dimensional data: a fast correlation-based filter solution. ICML **3**, 856–863 (2003)

Effect of Fractional Order in Pole Motion

Sucheta Moharir and Narhari Patil

Abstract The frequency response methods are most powerful in conventional control system. The impulse response characteristics are related to the location of poles of F(s). In this paper, we discuss impulse response and frequency response of $\{1/(s^2 + as + b)^q\}$ for different fractional values of q where $0 < q < 1$, $q = 1$, $1 < q < 2$ in pole motion. The different characters of the impulse response and frequency response are shown in numerical examples. The numbers of figures are presented to explain the concepts.

Keywords Transfer function · Impulse response · Frequency response · MAT-LAB

1 Introduction

The transfer function G(s) is given by

$$G(s) = \frac{L_o}{L_i} \tag{1}$$

where L denotes the Laplace transform. The frequency response function and the transfer function are interchangeable by the substitution $s = j\omega$ [1].

S. Moharir (✉)
St. Vincent Pallotti College of Engineering and Technology, Nagpur, Maharashtra, India
e-mail: sucheta.moharir@gmail.com

N. Patil
Shri Sant Gajanan Maharaj College of Engineering, Shegaon, Maharashtra, India
e-mail: napshegaon@rediffmail.com

V. Ravi et al. (eds.), *Proceedings of the Fifth International Conference on Fuzzy and Neuro Computing (FANCCO - 2015)*, Advances in Intelligent Systems and Computing 415, DOI 10.1007/978-3-319-27212-2_18

The frequency response function $G(j\omega)$ is

$$G(s) = \frac{F_o}{F_i} \tag{2}$$

where F denotes the Fourier transform.

TF has been used in many applications. One important application among them is monitoring the mechanical integrity of transformer windings (during testing and while in service). Mechanical deformation arises mainly due to short circuit forces, unskilled handling and rough transportation. Information related to winding deformation is embedded in the TF. Hence the first step should ensure accurate diagnosis, a correct interpretation of TF [2]. TF can be used to describe a variety of filter or to express solution of linear differential equation accurately [3]. The TF of system is analyzed and response curves are simulated [4]. The location of poles and zeros gives idea regarding response characteristics of a system. Transfer function is mainly used in control system and signal processing.

2 System Stability

If poles are in LHP, the system is stable; if poles are in RHP, the system is unstable and poles on imaginary axis then system is marginally stable or limitedly stable.

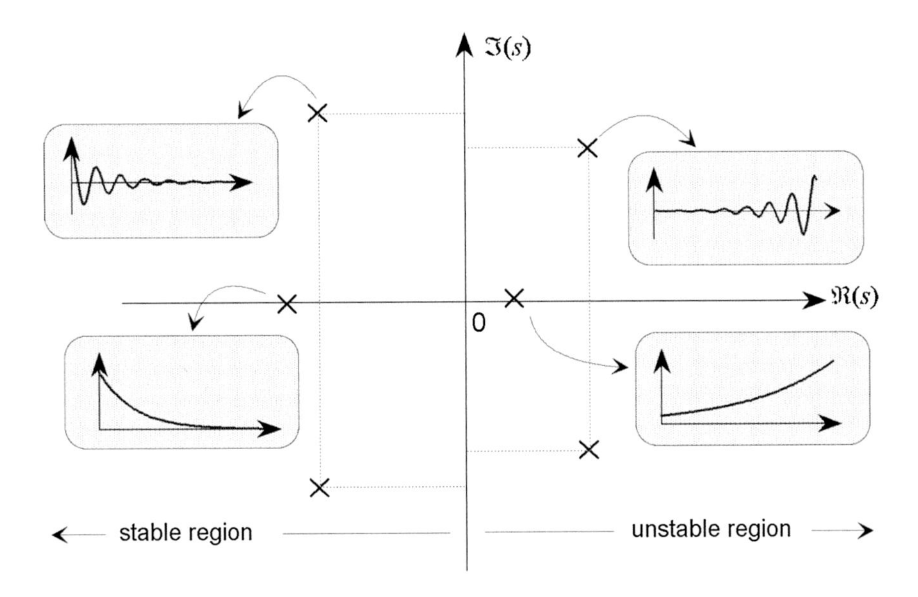

Fig. 1 Stable and unstable region according to pole position

3 Impulse Response and Frequency Response Analysis

Impulse signals and their responses, Frequency response are commonly used in control systems analysis and design. Both responses are a function of frequency and apply only to the steady state sinusoidal response of the system [5]. One of the most useful representations of a transfer function is a logarithmic plot which consists of two graphs, one giving magnitude and the other phase angle both plotted against frequency in logarithmic scale i.e. Bode plots [6, 7].

Theorem 1 For $F(s) = \frac{1}{(s^2+as+b)^q}$ we have

Case (i): When both of the poles lie in the open LHP, $\lim_{t\to\infty} f_1(t) = 0$ with $q > 0$;
Case (ii): When both the poles lie on the imaginary axis, $f_1(t)$ has damped oscillation with $0 < q < 1$; has undamped oscillation with $q = 1$; and has diverging oscillation with $q > 1$;
Case (iii): When both of the poles lie in the open RHP, $\lim_{t\to\infty} f_1(t) = \infty$ with $q > 0$ [8].

Consider second order transfer function with different fractional order then movement of poles in LHP, on imaginary axis, in RHP and MAT-LAB based evaluation of impulse responses and frequency esponses are given.

Numerical example 1: To observe the impulse response and frequency response for Case (i), Case (ii), Case (iii) for the following values of a and b for $q = 0.7$, 1.0, 1.3 [9, 10].

Case (i): When $a_1 = 1$, $b_1 = 1$; Case (ii): When $a_2 = 2$, $b_2 = 2$; Case (iii): When $a_3 = 3$, $b_3 = 3$ and the impulse response and frequency response of $\lim_{t\to\infty} f_1(t) = 0$ is demonstrated in Figs. 2, 3, 4, 5, 6 and 7.

When poles moves away from origin in LHP then impulse response tends to zero as time tends to infinity and peak time, rise time goes on decreasing. But in Case (i), Case (ii), Case (iii) peak time decreases and rise time increases as q changes from 0.7,1, and 1.3. Frequency response for all values i.e. $q = 0.7$, 1, 1.3 are same. As poles move away from origin, frequency responses of Case (i) for all values, magnitude coincide with zero line and then decreasing with negative slopes and phase angle coincide with zero line then decreases and again coincide at $-180°$ line. Case (ii) and Case (iii) magnitude coincide with different straight lines below zero line and then decreases with same negative slope and phase angle coincide with zero line then decreases with different slopes and again coincide at $-180°$ line.

Numerical example 2: To observe the impulse response and frequency response for Case (iv), Case (v), Case (vi) for the following values of a and b for $q = 0.7$, 1.0, 1.3. Case (iv): When $a_4 = 0$, $b_4 = 1$; Case (v): When $a_5 = 0$, $b_5 = 4$; Case (vi): When $a_6 = 0$, $b_6 = 9$ and the impulse response and frequency response of $F(s) = \frac{1}{(s^2+as+b)^q}$ in Figs. 8, 9, 10, 11,12 and 13.

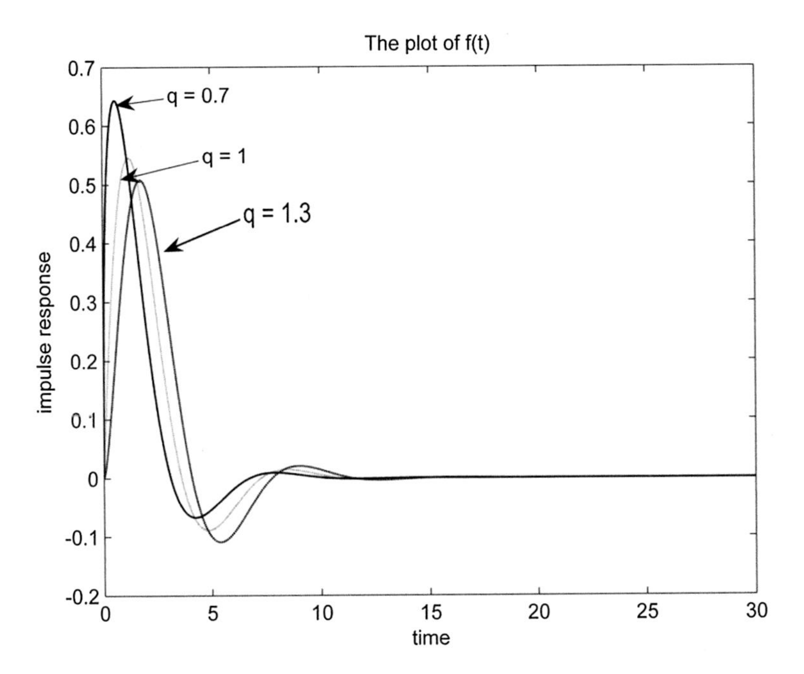

Fig. 2 Impulse response of case(i)

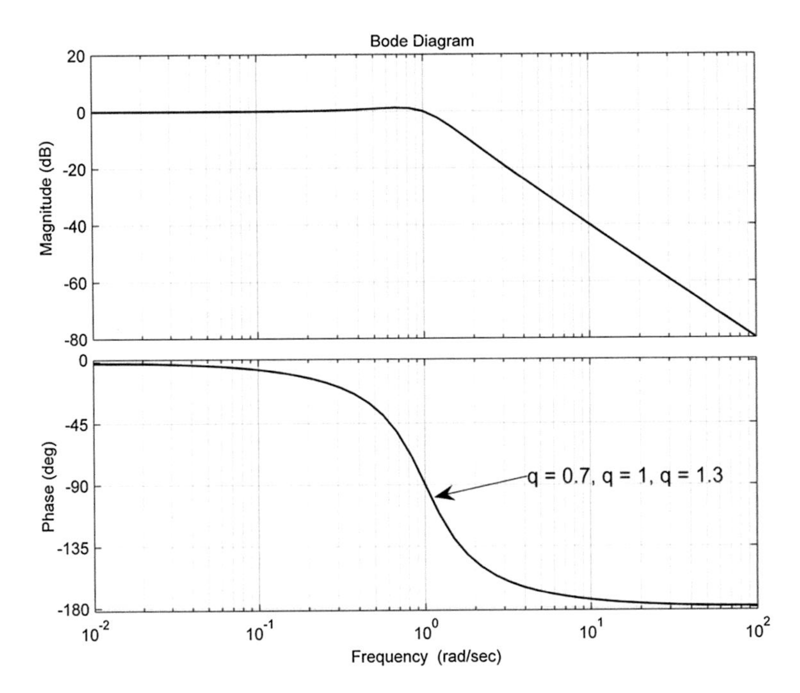

Fig. 3 Frequency response of Case (i)

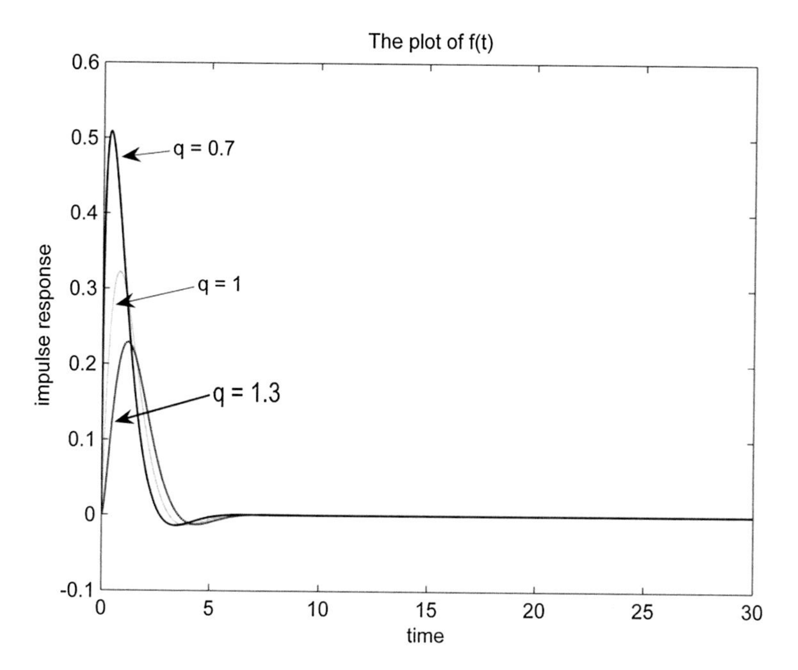

Fig. 4 Impulse response of Case (ii)

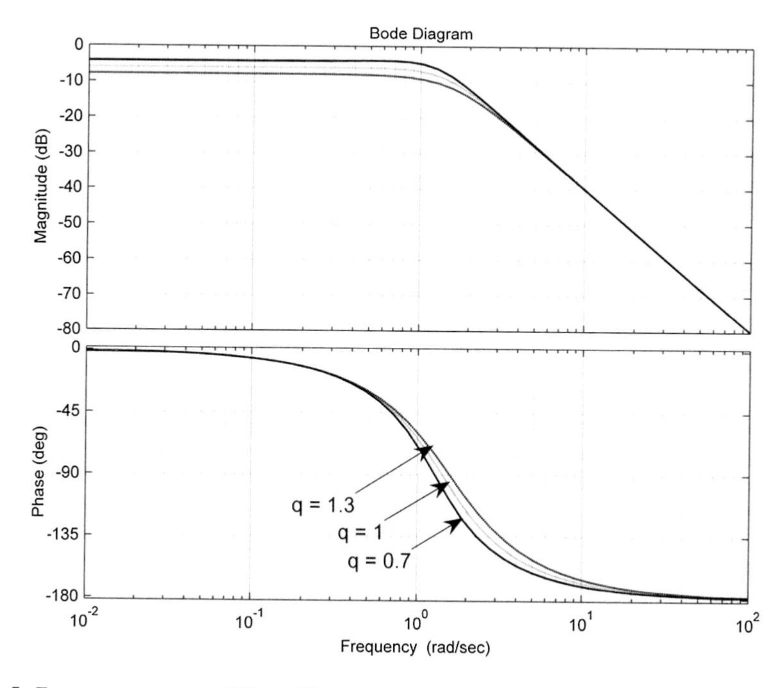

Fig. 5 Frequency response of Case (ii)

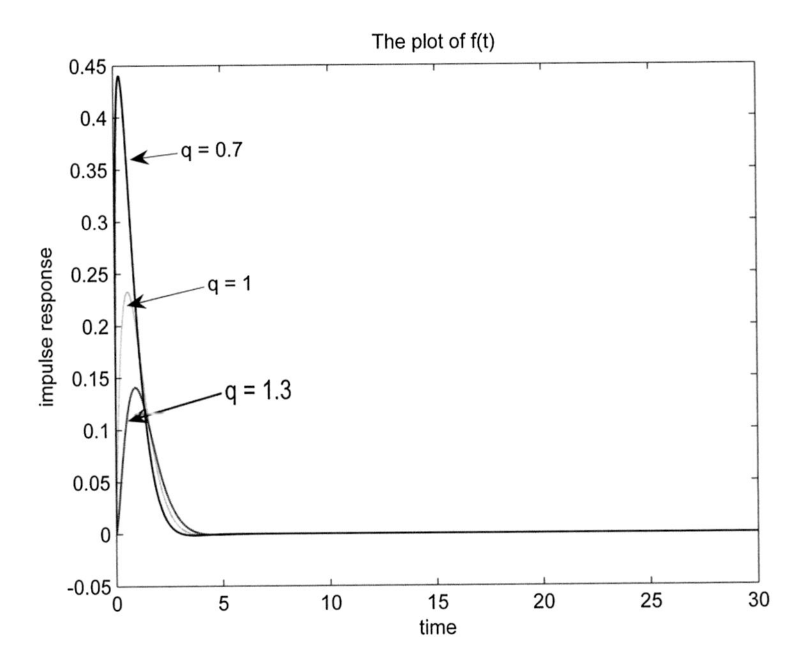

Fig. 6 Impulse response of Case (iii)

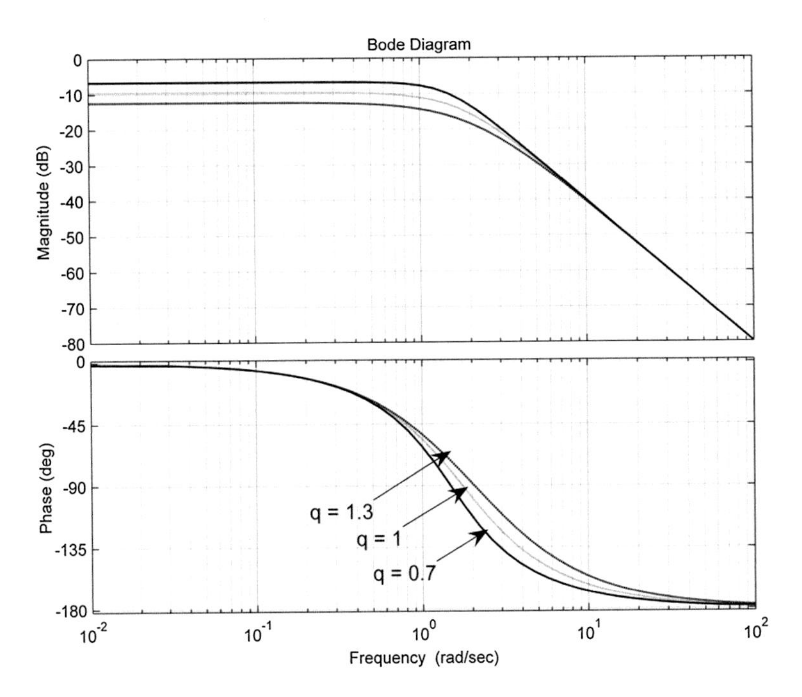

Fig. 7 Frequency response of Case (iii)

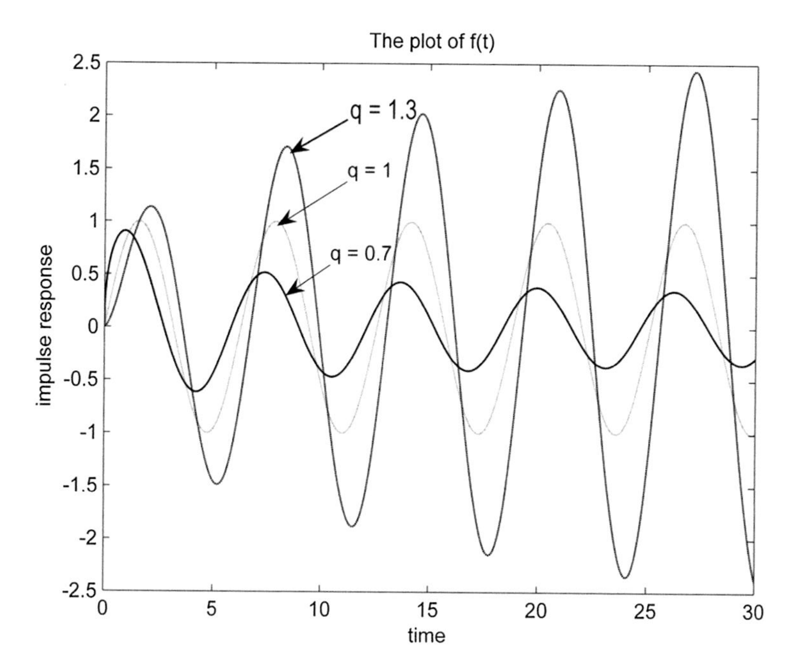

Fig. 8 Impulse response of Case (iv)

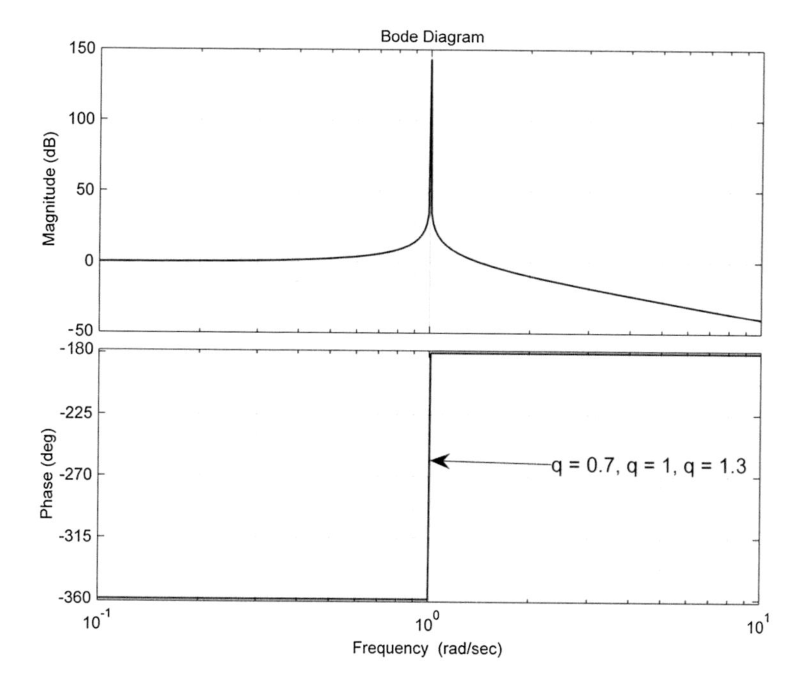

Fig. 9 Frequency response of Case (iv)

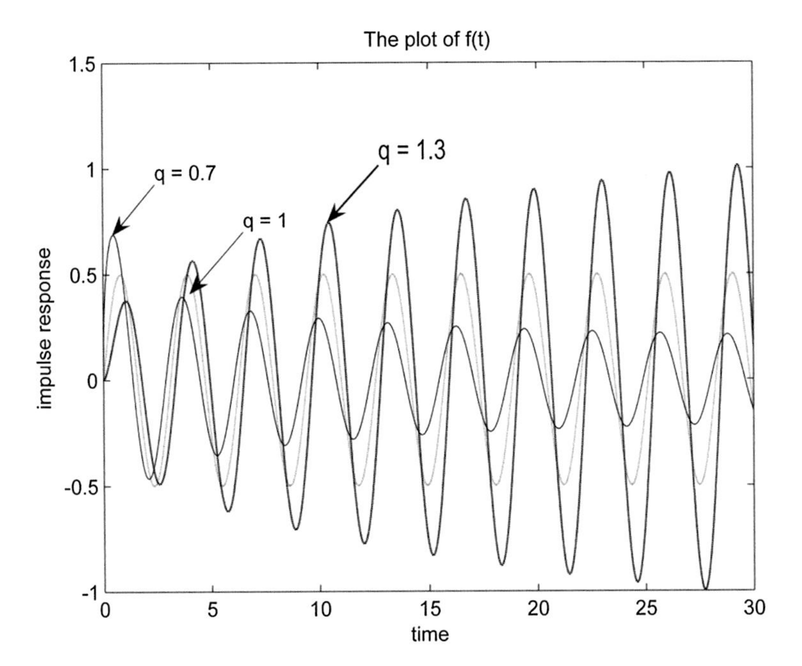

Fig. 10 Impulse response of Case (v)

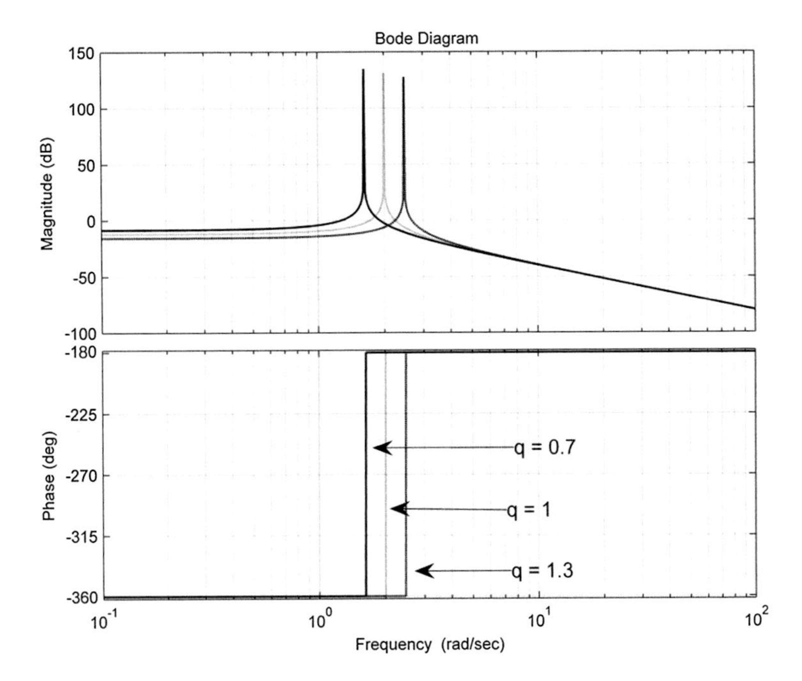

Fig. 11 Frequency response Case (v)

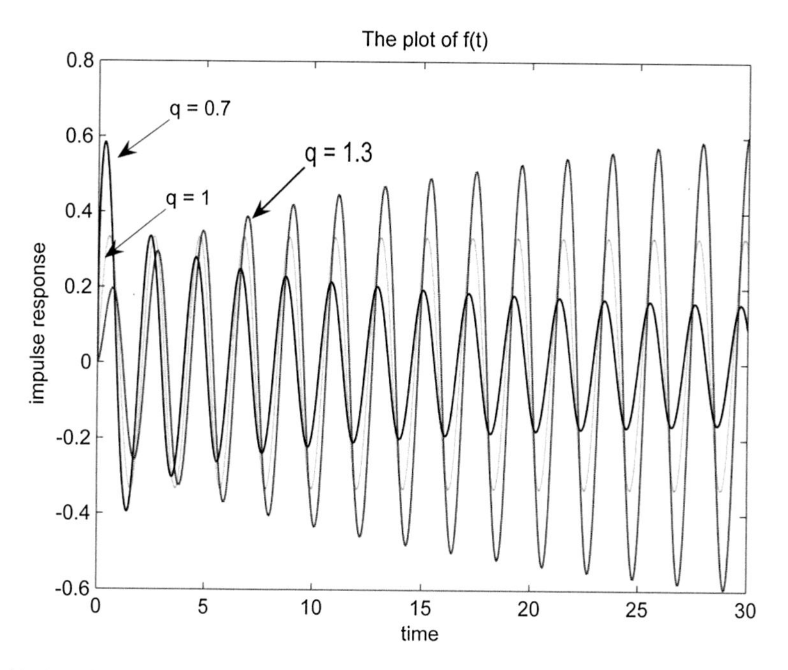

Fig. 12 Impulse response of Case (vi)

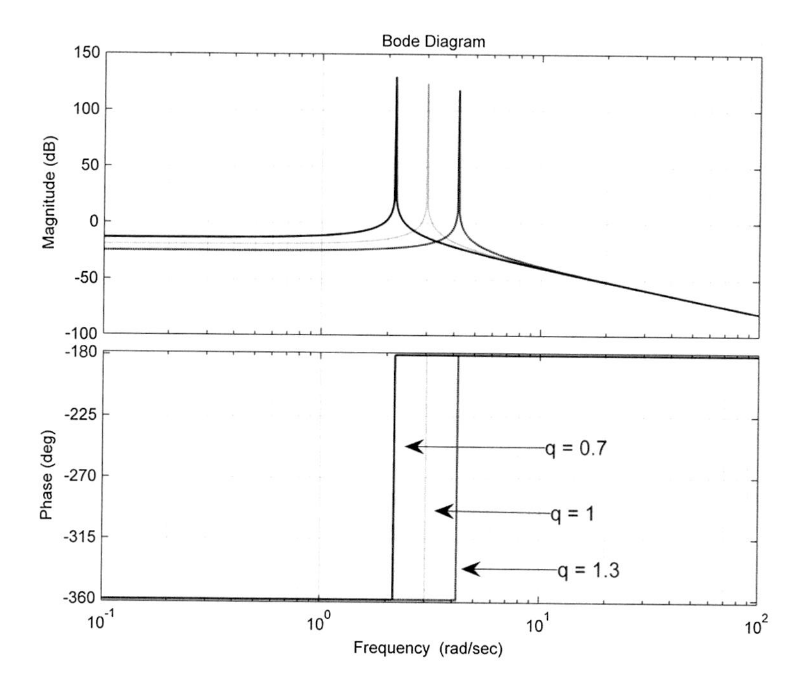

Fig. 13 Frequency response of Case (vi)

When poles moves away from origin on imaginary line (vertical direction) then impulse response of all values oscillations goes on increasing and Frequency response is same for Case (iv) for all values i.e. q = 0.7, 1, 1.3. Case (v) and Case (vi) magnitude coincide with different straight lines below zero line, increases at different frequency and then decreases with same negative slope and phase angle coincide with −360° line then increases with different slopes and again coincide at −180° line.

Numerical example 3: To observe the impulse response and frequency response for Case (vii), Case (viii), Case (ix) for the following values of a and b for q = 0.7, 1.0, 1.3. Case (vii): When a7 = −1, b7 = 1; Case (viii): When a8 = −2, b8 = 2; Case (ix): When a9 = −3, b9 = 3 and the impulse response and frequency response of $F(s) = \frac{1}{(s^2 + as + b)^q}$ is demonstrated in Figs. 14, 15, 16, 17, 18 and 19.

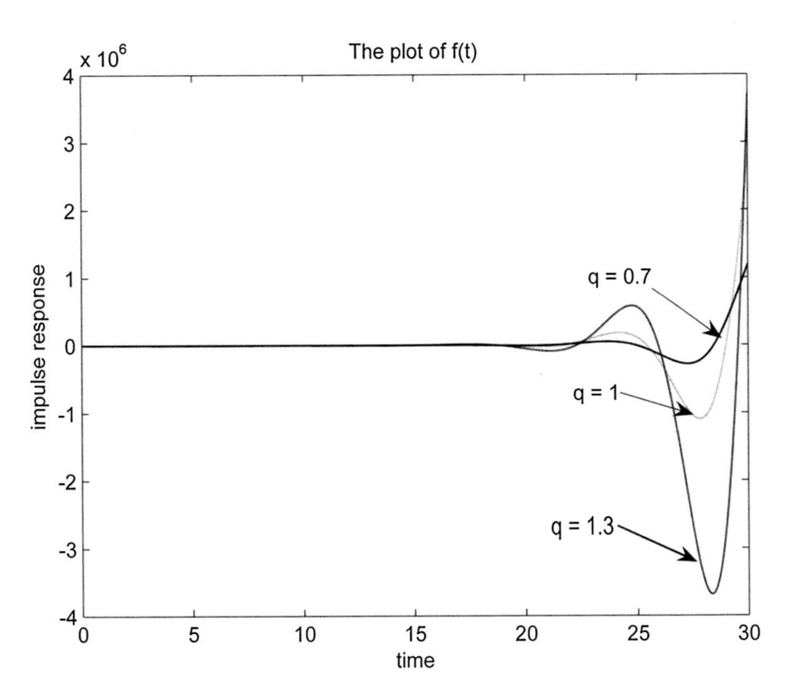

Fig. 14 Impulse response of Case (vii)

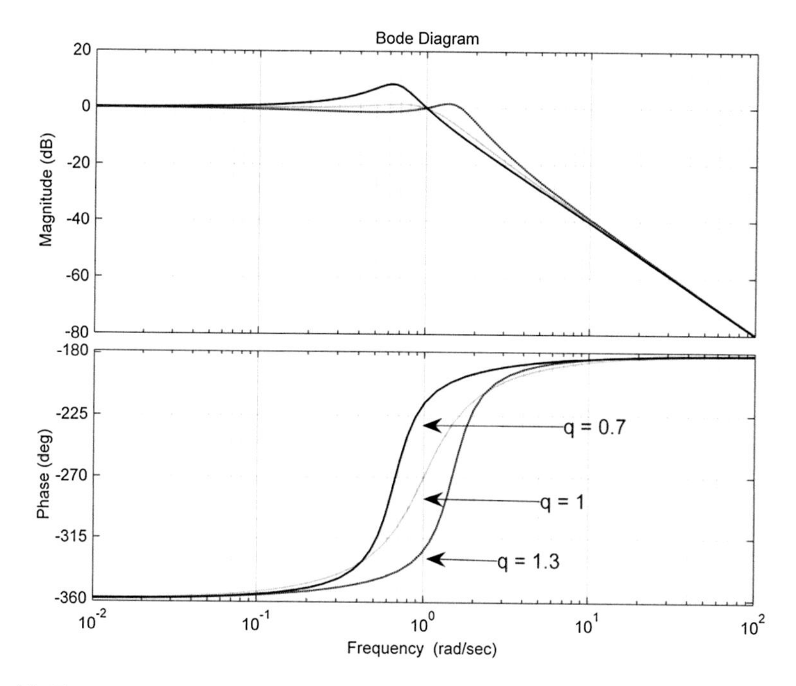

Fig. 15 Frequency response of Case (vii)

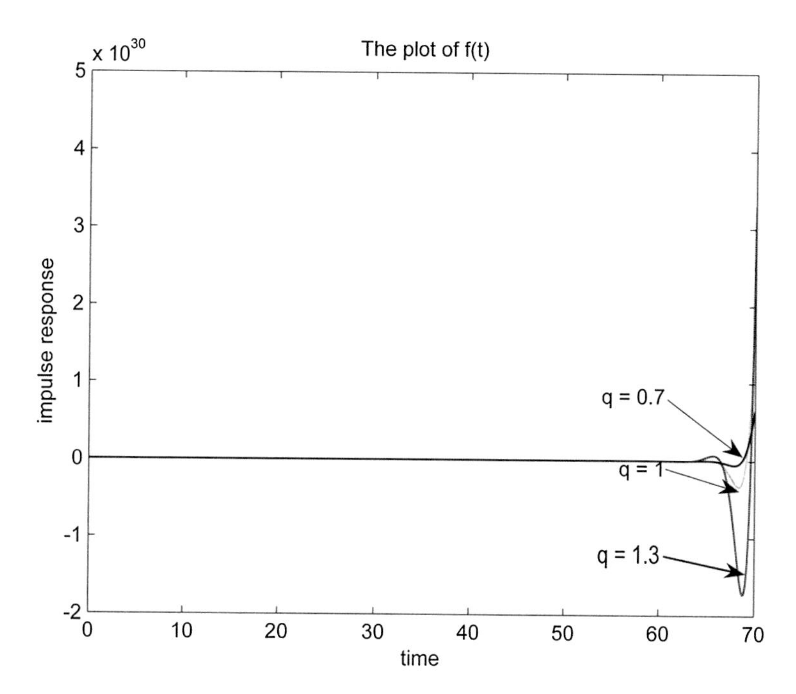

Fig. 16 Impulse response of Case (viii)

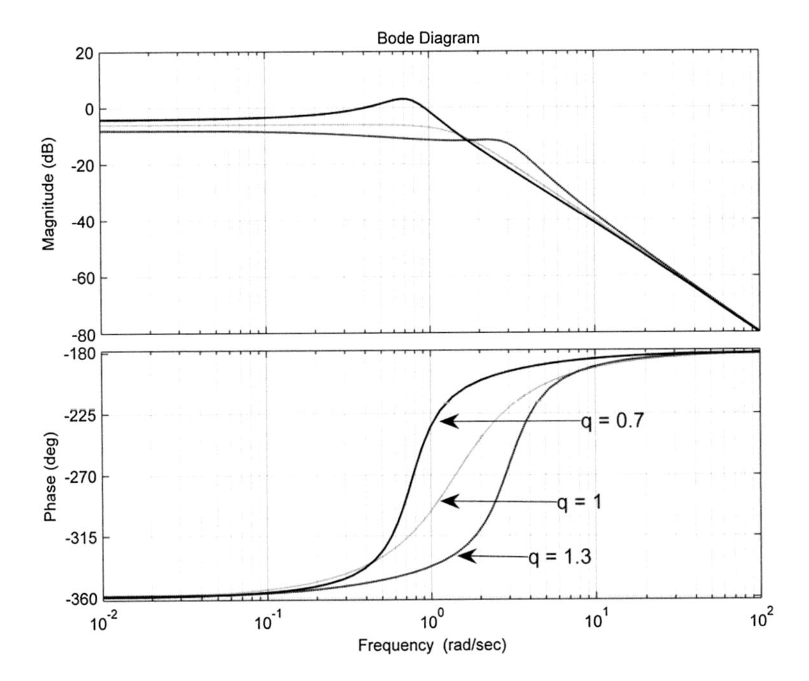

Fig. 17 Frequency response of Case (viii)

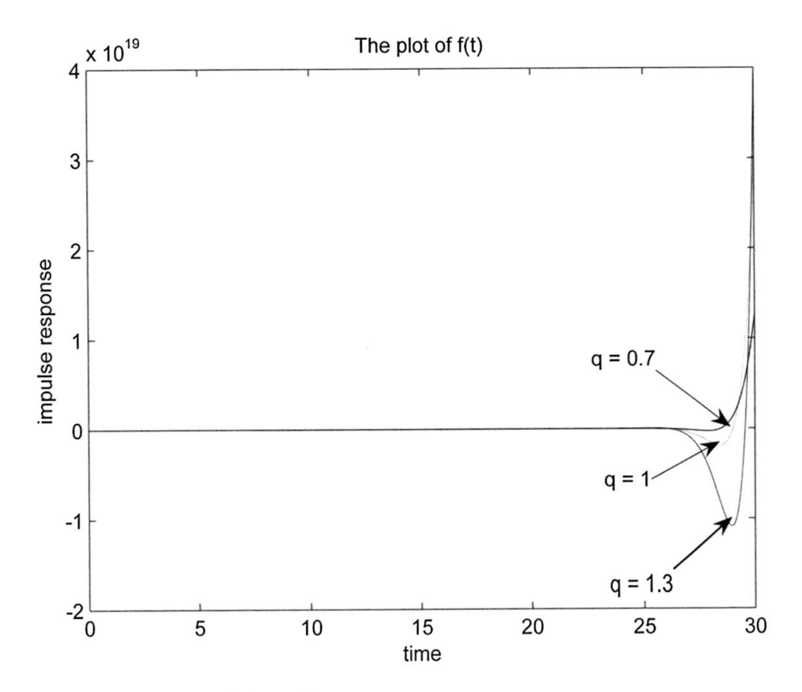

Fig. 18 Impulse response of Case (ix)

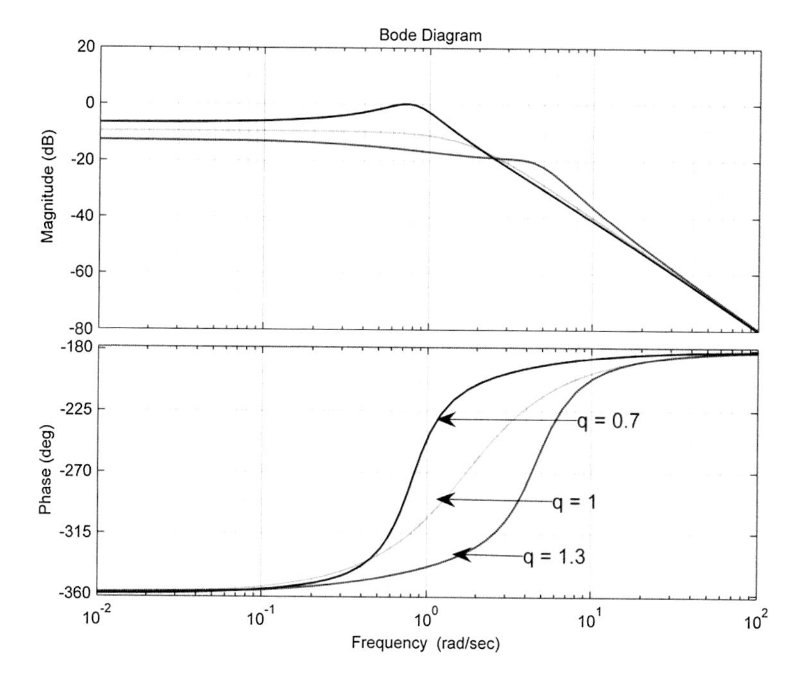

Fig. 19 Frequency response of Case (ix)

When poles moves away from origin in RHP then impulse response tends to infinity as time increases. As poles move away from origin, magnitude coincide with zero line for Case (vii) and in Case (viii) and Case (ix), magnitude goes down zero linechanges with different slopes and again coincide and phase angle coincide with $-360°$ line then increases and again coincide at $-180°$ line.

4 Conclusion

When poles moves away from origin in LHP then impulse response tends to zero as time tends to infinity and peak time, rise time goes on decreasing. But in Case (i), Case (ii), Case (iii) peak time decreases and rise time increases as q changes from 0.7,1,1.3 Frequency response for all values i.e. $q = 0.7$, 1, 1.3 are same. As poles move away from origin, frequency responses of Case (i) for all values, magnitude coincide with zero line and then decreasing with negative slopes and phase angle coincide with zero line then decreases and again coincide at $-180°$ line. Case (ii) and Case (iii) magnitude coincide with different straight lines below zero line and then decreases with same negative slope and phase angle coincide with zero line then decreases with different slopes and again coincide at $-180°$ line. When poles moves away from origin on imaginary line (vertical direction) then impulse response of all values oscillations goes on increasing and Frequency response is same for Case (iv) for all values i.e. q = 0.7, 1, 1.3. Case (v) and Case (vi) magnitude coincide with different straight lines below zero line, increases at different frequency and then decreases with same negative slope and phase angle coincide with −360° line then increases with different slopes and again coincide at −180° line. When poles moves away from origin in RHP then impulse response tends to infinity as time increases. As poles move away from origin, magnitude coincide with zero line for Case (vii) and in Case (viii) and Case (ix) magnitude goes down zero line changes with different slopes and again coincide and phase angle coincide with −360° line then increases and again coincide at −180° line.

References

1. Can, S., Unal, A.: Transfer functions for nonlinear systems via Fourier-Borel transforms. IEEE (1988)
2. Satish, L.: An effort to understand what factors affect the transfer function of a two- winding transformer. IEEE Trans. Power Deliv. **20**(2) (2005)
3. Wang, S.Y., Lu, G.Y., Li, B.: Discuss about linear transfer function and optical transfer function. Appl. Mech. Mater. 275–277 (2013)
4. Wang, J., Wang, T., Wang, J.: Application of π equivalent circuit in mathematic modeling and simulation of gas pipe. Appl. Mech. Mater. 496–500 (2014)
5. Nise, N.S.: (Pomana): Control Systems Engineering. Wiley, New York (2010)
6. Nagrath, I.J., Gopal, M.: Control Systems Engineering. New Age International Publishers Book (2006)
7. Moharir, S.S., Patil, N.A.: Effect of order of pole and zero on frequency response. In: Proceeding of the ICMS 2014 International Conference on Mathematical Sciences Published by Elsevier in association with University of Central Florida, The Institute of Mathematical Sciences Chennai, Chennai, pp. 98–101 (2014)
8. Jiao, Z., Chen, Y.Q.: Impulse response of a generalized fractional second order filter. In: Proceedings of the ASME 2011 International Design Engineering Technical Conference and Computers and Information in Engineering conference IDETC/CIE, pp. 303–310 (2011)
9. Sheng, H., Li, Y., Chen, Y. Q.: Application of numerical inverse Laplace transform algorithms in fractional calculus. In: The 4th IFAC Workshop Fractional Differentiation and its Applications (2010)
10. Li, Y., Sheng, H., Chen, Y.Q.: Analytical impulse response of a fractional second order filter and its impulse response invariant discretization. Elsevier Signal Process. **91** (2011)

Sewage Water Quality Index of Sewage Treatment Plant Using Fuzzy MCDM Approach

Purushottam S. Dange and Ravindra K. Lad

Abstract Sewage produced from the human activity leads to pollute an environment. A tremendous environmental degradation because of less focus on pollution due to sewage. Therefore, it is required to revise pollution control strategies. This problem is taken into consideration and a feasible method is proposed for ranking of Sewage Treatment Plants (STPs) based on their environmental pollution potential. The sensitivity analysis is also includes in this study to see variation in the pollution potential of STPs. This analysis can be used for making policy of paying tax on the basis of pollution potential to control the pollution. The framed method explained in the paper using Fuzzy Multi Criteria Decision Making (FMCDM) for ranking the STPs is one of the important methods in decision problems. The case study is also incorporated which relates to three Sewage Treatment Plants located in Pimpri Chinchwad Municipal Corporation, Pimpri, Pune, Maharashtra, India.

Keywords Sewage water quality index · Fuzzy multi criteria decision making · Linguistic variable and ranking of sewage treatment plant

1 Introduction

Ever increasing pollution levels are primarily responsible for impairing natural environment. The permissible limits set by the pollution control authorities permit to discharge treated sewage of Sewage Treatment Plant in the receiving water bodies (e.g. rivers, streams and alike). The water flow is depleting in water bodies

P.S. Dange (✉) · R.K. Lad (✉)
Department of Civil Engineering, D.Y.P.I.E.T, Pimpri, Pune 411018 MH, India
e-mail: dange.purushottam@gmail.com

R.K. Lad
e-mail: ravindraklad@yahoo.com

P.S. Dange · R.K. Lad
Savitribai Phule Pune University, Pimpri, Pune 18 MH, India

V. Ravi et al. (eds.), *Proceedings of the Fifth International Conference on Fuzzy and Neuro Computing (FANCCO - 2015)*, Advances in Intelligent Systems and Computing 415, DOI 10.1007/978-3-319-27212-2_19

over a period of time and the ever increasing pollution load in the natural bodies are the matters of serious concern. In the absence of adequate dilution, discharging wastewater within the permissible limit also leads to degradation of natural environment. For example, as considering BOD as a parameter, the Central Pollution Control Board allows discharging sewage up to 20 mg/L. It means, if BOD_5 at 20 °C is up to 19 mg/L there is no any legal offence, but if it is 21 mg/L then it violates the norms. But, as receiving water bodies are already environmentally degraded, so even if BOD_5 at 20 °C is up to 19 or 5 mg/L there is a possibility of degradation of environment up to a certain degree because of the environmental quality of rivers or streams are already degraded, therefore, it is required to change the concept of crisp (Permissible limit OR Not Permissible limit value) to fuzzy values (Permissible limit AND Not Permissible limit value).

Limited work is reported on Sewage Water Quality Index (SWQI). Therefore, there is a need to look into this aspect of the SWQI, as the same would be useful in considering its impact on the receiving body at the time of final disposal. So FMCDM method has been developed for ranking STPs, based on their water pollution potential.

1.1 Fuzzy Multi-criteria Decision Making (FMCDM) Studies

Many authors have developed different methodologies for ranking alternatives using fuzzy logic during the last four decades. Jain [7, 8] developed a method to rank alternatives on the basis of maximizing set. The membership level concept was proposed by Bass and Kwakernaak [3]. But from the above two papers Baldwin and Guild [2] were concluded that these two methods are suffered from some difficulties for comparing the alternatives and having some disadvantages. The overall review of ranking methods on the basis of fuzzy numbers has explained by Bortolan et al. [4]. Raj et al. [1] were developed a methodology for ranking alternatives using a concept of maximizing and minimizing set. Singh et al. [12] developed a methodology for contractor selection using fuzzy logic. Lad et al. [11] developed Combined Water and Air Pollution Potential Index for ranking of different types of industries.

Zadeh [13] is a pioneer who developed the fuzzy set theory. It is generally agreed that an important point in the evaluation of the modern concept of uncertainty was the publication of a seminal paper by Lotfi A. Zadeh. In his paper, Zadeh introduced a theory whose objects–fuzzy sets–are sets with boundaries that are not precise. He introduced that the membership is not a matter of affirmation or denial in fuzzy, but it is a matter of a certain degree. Hipel et al. [5] suggested that a decision making problems are very complex, if multiple criteria, uncertainty, risk and vagueness are involved in decision making issues.

2 Methodology

FMCDM approach is developed for the determination of SWQI. This approach is used for the determination of SWQI of existing STPs. For ranking of STPs, the Fuzzy decision framework is as shown in Fig. 1.

Linguistic terms used for this study are: Very Important (VI), Important (I), Average (A), Least Important (LI) and Very Least Important (VLI). Table 1 shows the linguistic terms and fuzzy numbers used in this study. Figure 2 shows fuzzy sets for the linguistic terms.

The first step is the identification for defining quality of treated wastewater of Sewage Treatment Plant as per the consent of Central Pollution Control Board are: pH, BOD, COD, SS, TDS, Temperature, Oil & grease, Phenolic Compounds, Total residual chlorine, Chlorides, Sulphates and Ammonical Nitrogen. An importance weightage for each of the sub criteria of water pollution is developed on the basis of opinion of experts who are involved in the field of Environmental Engineering. Table 2 shows experts' opinion.

An expert has taken into account the exposure levels and the effects of each parametric value on receiving body along with its effect on human health while giving the weightage to each parameter for the determination of SWQI.

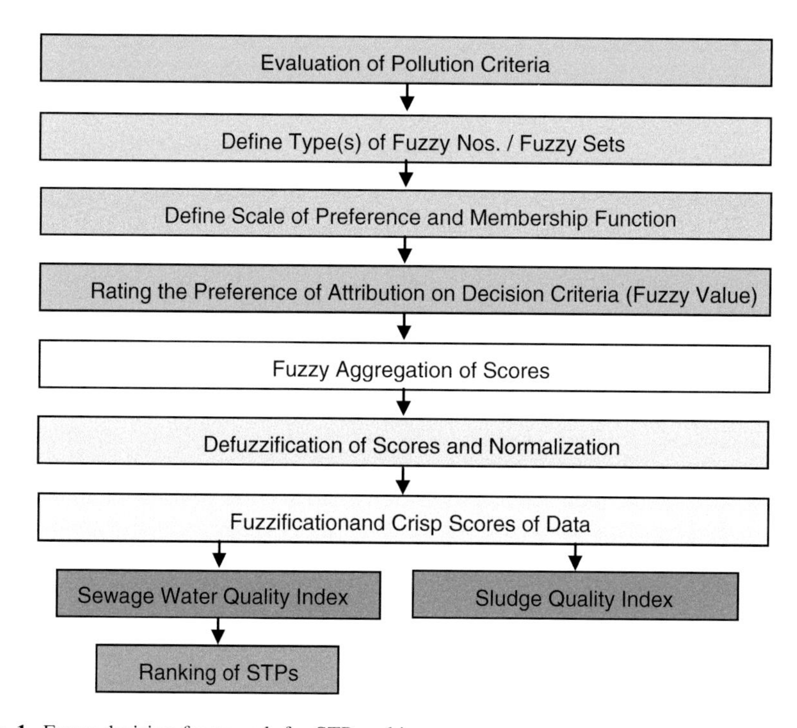

Fig. 1 Fuzzy decision framework for STP ranking

Table 1 Linguistic terms and fuzzy numbers

Linguistic terms	Fuzzy number
VI (Very Important)	(0.777,0.888,1.000,1.000)
I (Important)	(0.555,0.666,0.777,0.888)
A (Average)	(0.333,0.444,0.555,0.666)
LI (Least Important)	(0.111,0.222,0.333,0.444)
VLI (Very Least Important)	(0.000,0.000,0.111,0.222)

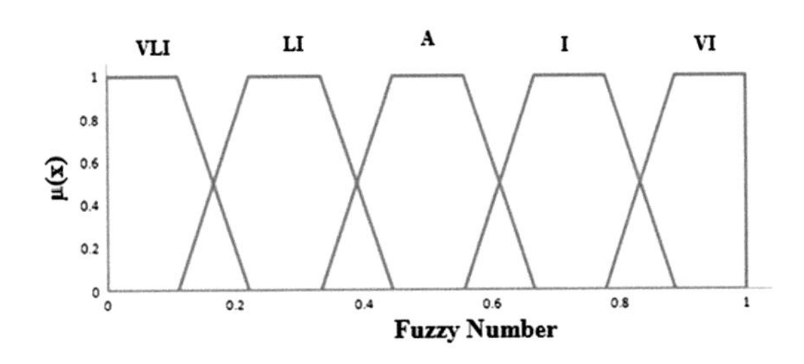

Fig. 2 Fuzzy sets for the linguistic terms

Table 2 Experts' opinion

Sub criteria	Experts' opinion									
	Academicians					Professionals				
	E1	E2	E3	E4	E5	E1	E2	E3	E4	E5
pH	I	A	I	I	I	VI	I	I	I	I
BOD, mg/L	VI	VI	VI	VI	VI	VI	VI	VI	VI	VI
COD, mg/L	I	I	VI	I	VI	VI	I	I	I	I
SS, mg/L	I	I	I	A	A	I	A	I	A	A
TDS, mg/L	I	I	I	I	I	I	I	I	I	I
Temperature, °C	A	I	A	A	I	A	A	A	I	A
Oil and grease, mg/L	A	A	A	A	A	A	A	A	A	A
Phenolic Compounds, mg/L	LI	LI	LI	LI	A	A	LI	LI	A	A
Total residual chlorine, mg/L	A	LI	A	LI	A	LI	LI	LI	A	LI
Chlorides, mg/L	A	A	A	A	A	A	A	A	A	A
Sulphates, mg/L	LI	A	LI	A	LI	LI	LI	A	LI	A
Ammonical Nitrogen, mg/L	LI	A	A	A	A	LI	LI	A	A	A

E Environmental expert

Using Eq. (1) the linguistic terms given by experts can be further simplified to calculate the Average Fuzzy Number (AFN). The linguistic terms as assigned by the experts for each sub criterion of treated sewage of Sewage Treatment Plant can be converted to fuzzy numbers used in the Eq. (1) through Table 1 and Fig. 2.

$$A_{ij}^{k} = (1/p) \times \left(a_{i1}^{k} + a_{i2}^{k} + \cdots + a_{ip}^{k}\right) for\, i = 1, 2, \ldots, n\, and\, j = 1, 2, \ldots, p \quad (1)$$

where a_{ij}^{k} is the fuzzy number assigned to a sub criterion by expert for the criterion, p is the number of experts and n is the number of fuzzy numbers.

The next step is defuzzification. The defuzzification means produces a crisp value which represents the degree of satisfaction of the aggregated fuzzy number. In this study, trapezoidal sets were developed to represent the experts' opinion because an importance of parameter was considered as a range value but not with specific value. Let a fuzzy numbers of trapezoidal set be parameterized by x_1, x_2, x_3 and x_4 as shown in the Fig. 2, then 'e' for the sub criteria as defuzzified value (crisp score) can be obtained by using Eq. (2)

$$e = (x_1 + x_2 + x_3 + x_4)/4 \tag{2}$$

Zimmerman [14] and Kaufmann and Gupta [9] given in details about membership functions, different types of fuzzy numbers, aggregation and defuzzification methods. Interested researchers may refer these papers.

The normalized weight for each sub criterion of water pollution can be calculated by dividing the crisp score of each sub criterion (C_{mk}) by the sum total of crisp score of all sub criteria $(\sum C_{mk})$ where m is the criterion and k is the sub criterion.

After this the next is to convert the values of effluent wastewater to the fuzzy numbers based on permissible limits set by pollution control board. For example, if COD of a given sample is 95 mg/L, the membership function of that sample would then be 0.38 (see Fig. 3) as the permissible limit of COD is 250 mg/L.

Figure 3 shows fuzzy set for the parameter Chemical Oxygen Demand (COD). Similarly, fuzzy sets for other parameters of treated sewage can be developed. These fuzzy sets form basic part of the fuzzy model which is based on permissible limit of pollution parameter.

Fuzzy decision matrix for sub criterion (COD) C can be written as

$$X_C = \begin{array}{c} \\ \end{array}\begin{matrix} \mu_1 & \mu_2 & \mu_3 \cdots\cdots & \mu_n \end{matrix}$$
$$X_C = \begin{bmatrix} x_1 & x_2 & x_3 \cdots\cdots & x_n \\ y_1 & y_2 & y_3 \cdots\cdots & y_n \\ z_1 & z_2 & z_3 \cdots\cdots & z_n \end{bmatrix} \begin{vmatrix} 1 \\ 2 \\ 3 \end{vmatrix}$$

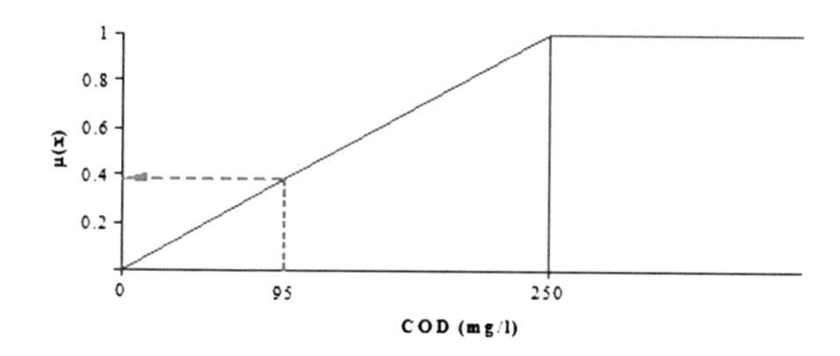

Fig. 3 Fuzzy Set for not acceptable (Pollution Parameter COD)

where $x_1, x_2, x_3 \ldots x_n, y_1, y_2, y_3 \ldots y_n$ and $z_1, z_2, z_3 \ldots z_n$ are fuzzy values of COD for Sewage Treatment Plants STP_1, STP_2 and STP_3 respectively.

The crisp scores of the sub criterion COD for each Sewage Treatment Plant (STP) can be obtained by using following equations:

$$STP_1 = (x_1 + x_2 + x_3 + \cdots + x_n)/n$$
$$STP_2 = (y_1 + y_2 + y_3 + \cdots + y_n)/n$$
$$STP_3 = (z_1 + z_2 + z_3 + \cdots + z_n)/n$$

Similarly crisp scores can be computed for the other sub criteria of treated sewage. The total score (TS_{mi}), for each STP, of treated sewage criteria can be calculated separately using [6] Eq. (3).

$$TS_{mi} = \sum [X_{mk} \cdot W(C_{mk})] \text{ for } k = 1, 2, \ldots n \tag{3}$$

where,

TS_{mi} total score of the STP i against the criterion m

X_{mk} crisp score of the STP data against sub criterion k of the criterion m and

$W(C_{mk})$ weight (importance value) of sub criterion k of the criterion m

3 Case Study

The available data of treated sewage from three STPs located in Pimpri Chinchwad Municipal Corporation (PCMC), Pimpri, Pune, in the State of Maharashtra, India were collected for this study. The data were collected for all three seasons. The collected data is of 120 samples per season.

3.1 Evaluation of Sewage Water Quality Index (SWQI) for Sewage Treatment Plants

For the determination of SWQI, Fuzzy Multi Criteria Decision Making approach was used. The normalized weight for each sub criterion of treated sewage was calculated as below: Experts' opinions (linguistic terms) or perceptions for eight sub criteria (parameters) of treated sewage have been taken, from five academicians and five professionals, who are involved in the field of Environmental Engineering. The selection of experts from different domain helped in understanding the significance of individuals subjectively and studies its influence on the SWQI and ranking of treatment plants. Table 2 shows the importance weights in terms of linguistic term as assigned to each of sub criteria of treated sewage by experts (academicians and professionals).

The experts' opinions (linguistic terms) or perceptions of each sub criterion were converted to fuzzy numbers using Table 1 and Fig. 2 and then using Eqs. (1) and (2), the Average Fuzzy Numbers (AFNs) and crisp score (defuzzified value) respectively, for each sub criterion of treated sewage of Sewage Treatment Plants were calculated.

As a sample calculation, the fuzzy decision matrix (F), the average fuzzy score matrix (AF) and crisp score for sub criteria of treated sewage of Sewage Treatment Plant are shown as below.

Fuzzy decision matrix (Academicians)

$$F_{C1} = \begin{bmatrix} (0.555, 0.666, 0.777, 0.888)(0.333, 0.444, 0.555, 0.666)(0.555, 0.666, 0.777, 0.888)(0.555, 0.666, 0.777, 0.888)(0.555, 0.666, 0.777, 0.888) \\ (0.777, 0.888, 1.000, 1.000)(0.777, 0.888, 1.000, 1.000)(0.777, 0.888, 1.000, 1.000)(0.777, 0.888, 1.000, 1.000)(0.777, 0.888, 1.000, 1.000) \\ (0.555, 0.666, 0.777, 0.888)(0.555, 0.666, 0.777, 0.888)(0.777, 0.888, 1.000, 1.000)(0.555, 0.666, 0.777, 0.888)(0.777, 0.888, 1.000, 1.000) \\ (0.555, 0.666, 0.777, 0.888)(0.555, 0.666, 0.777, 0.888)(0.555, 0.666, 0.777, 0.888)(0.333, 0.444, 0.555, 0.666)(0.333, 0.444, 0.555, 0.666) \\ (0.555, 0.666, 0.777, 0.888)(0.555, 0.666, 0.777, 0.888)(0.555, 0.666, 0.777, 0.888)(0.555, 0.666, 0.777, 0.888)(0.555, 0.666, 0.777, 0.888) \\ (0.333, 0.444, 0.555, 0.666)(0.555, 0.666, 0.777, 0.888)(0.333, 0.444, 0.555, 0.666)(0.333, 0.444, 0.555, 0.666)(0.555, 0.666, 0.777, 0.888) \\ (0.333, 0.444, 0.555, 0.666)(0.333, 0.444, 0.555, 0.666)(0.333, 0.444, 0.555, 0.666)(0.333, 0.444, 0.555, 0.666)(0.333, 0.444, 0.555, 0.666) \\ (0.111, 0.222, 0.333, 0.444)(0.111, 0.222, 0.333, 0.444)(0.111, 0.222, 0.333, 0.444)(0.111, 0.222, 0.333, 0.444)(0.333, 0.444, 0.555, 0.666) \\ (0.333, 0.444, 0.555, 0.666)(0.111, 0.222, 0.333, 0.444)(0.333, 0.444, 0.555, 0.666)(0.111, 0.222, 0.333, 0.444)(0.333, 0.444, 0.555, 0.666) \\ (0.333, 0.444, 0.555, 0.666)(0.333, 0.444, 0.555, 0.666)(0.333, 0.444, 0.555, 0.666)(0.333, 0.444, 0.555, 0.666)(0.333, 0.444, 0.555, 0.666) \\ (0.111, 0.222, 0.333, 0.444)(0.333, 0.444, 0.555, 0.666)(0.111, 0.222, 0.333, 0.444)(0.333, 0.444, 0.555, 0.666)(0.111, 0.222, 0.333, 0.444) \\ (0.111, 0.222, 0.333, 0.444)(0.333, 0.444, 0.555, 0.666)(0.333, 0.444, 0.555, 0.666)(0.333, 0.444, 0.555, 0.666)(0.333, 0.444, 0.555, 0.666) \end{bmatrix}$$

Average fuzzy score matrix (Academicians):

$$AFCS = \begin{bmatrix} 0.511,\ 0.622,\ 0.733,\ 0.844 \\ 0.777,\ 0.888,\ 1.000,\ 0.916 \\ 0.644,\ 0.755,\ 0.866,\ 0.933 \\ 0.466,\ 0.577,\ 0.688,\ 0.799 \\ 0.555,\ 0.666,\ 0.777,\ 0.888 \\ 0.422,\ 0.533,\ 0.644,\ 0.755 \\ 0.333,\ 0.444,\ 0.555,\ 0.666 \\ 0.155,\ 0.266,\ 0.377,\ 0.488 \\ 0.244,\ 0.355,\ 0.466,\ 0.577 \\ 0.333,\ 0.444,\ 0.555,\ 0.666 \\ 0.200,\ 0.311,\ 0.422,\ 0.533 \\ 0.289,\ 0.400,\ 0.511,\ 0.622 \end{bmatrix} \begin{bmatrix} pH \\ BOD \\ COD \\ SS \\ TDS \\ Temperature \\ Oil\,\&\,Grease \\ Phenolic\ Compounds \\ Total\ \text{Residual}\ Chlorine \\ Chlorides \\ Sulphates \\ Ammonical\ Nitrogen \end{bmatrix}$$

Crisp scores:

$$Criteria, C_1(pH) = \frac{(0.511 + 0.622 + 0.733 + 0.844)}{4} = 0.677$$

$$Criteria, C_2(BOD) = \frac{(0.777 + 0.888 + 1.000 + 0.916)}{4} = 0.916$$

$$Criteria, C_3(COD) = \frac{(0.644 + 0.755 + 0.866 + 0.933)}{4} = 0.799$$

$$Criteria, C_4(SS) = \frac{(0.466 + 0.577 + 0.688 + 0.799)}{4} = 0.633$$

$$Criteria, C_5(TDS) = \frac{(0.555 + 0.666 + 0.777 + 0.888)}{4} = 0.722$$

$$Criteria, C_6(Temperature) = \frac{(0.422 + 0.533 + 0.644 + 0.755)}{4} = 0.588$$

$$Criteria, C_7(Oil\,\&\,Grease) = \frac{(0.333 + 0.444 + 0.555 + 0.666)}{4} = 0.500$$

$$Criteria, C_8(Phenolic\,Compounds) = \frac{(0.155 + 0.266 + 0.377 + 0.488)}{4} = 0.322$$

$$Criteria, C_9(Total\,Residual\,Chlorine) = \frac{(0.244 + 0.355 + 0.466 + 0.577)}{4} = 0.411$$

$$Criteria, C_{10}(Chlorides) = \frac{(0.333 + 0.444 + 0.555 + 0.666)}{4} = 0.500$$

$$Criteria, C_{11}(Sulphates) = \frac{(0.200 + 0.311 + 0.422 + 0.533)}{4} = 0.366$$

$$Criteria, C_{12}(Ammonical\,Nitrogen) = \frac{(0.289 + 0.400 + 0.511 + 0.622)}{4} = 0.455$$

Average Fuzzy Numbers (AFNs) and crisp score respectively, for each sub criterion of treated sewage of Sewage Treatment Plant are as shown in Table 3.

Then, the normalized weight for each sub criterion of treated sewage was calculated by dividing the score of each sub criterion (C_{mk}) by the sum total of all sub criteria $(\sum C_{mk})$ for Sewage Treatment Plant. The weights for each sub criteria depend upon the characteristics of wastewater discharges. The normalized weight for each sub criterion of treated sewage of Sewage Treatment Plant is as shown in Table 4.

Table 3 Average fuzzy number (Academicians)

Sub criteria	AFN1	AFN2	AFN3	AFN4	AVG(C_k)
Water pollution					
pH	0.511	0.622	0.733	0.844	0.677
BOD, mg/L	0.777	0.888	1.000	1.000	0.916
COD, mg/L	0.644	0.755	0.866	0.933	0.799
SS, mg/L	0.466	0.577	0.688	0.799	0.633
TDS, mg/L	0.555	0.666	0.777	0.888	0.722
Temperature °C	0.422	0.533	0.644	0.755	0.588
Oil and grease, mg/L	0.333	0.444	0.555	0.666	0.500
Phenolic Compounds, mg/L	0.155	0.266	0.377	0.488	0.322
Total residual chlorine, mg/L	0.244	0.355	0.466	0.577	0.411
Chlorides, mg/L	0.333	0.444	0.555	0.666	0.500
Sulphates, mg/L	0.200	0.311	0.422	0.533	0.366
Ammonical Nitrogen, mg/L	0.289	0.400	0.511	0.622	0.455
$\sum C_k$					**6.888**

Table 4 Normalized weight (Academicians)

Sub criteria	Weight
Water Pollution	
pH	0.100
BOD, mg/L	0.135
COD, mg/L	0.118
SS, mg/L	0.093
TDS, mg/L	0.106
Temperature, °C	0.087
Oil and grease, mg/L	0.073
Phenolic Compounds, mg/L	0.047
Total residual chlorine, mg/L	0.060
Chlorides, mg/L	0.073
Sulphates, mg/L	0.054
Ammonical Nitrogen, mg/L	0.067

The parametric data of treated sewage were converted to the fuzzy numbers based on the permissible limits of parameter (see Fig. 3). The next step was to determine the total score. To obtain the total score the fuzzy crisp scores of data and the normalized weight of sub criteria were operated by a matrix as shown below.

Total Score Matrix for Sewage Treatment Plants (Academicians)

$$
TS_{\mathrm{water}} = \begin{bmatrix} 0.824 & 0.844 & 0.834 \\ 0.875 & 0.290 & 0.757 \\ 0.323 & 0.129 & 0.276 \\ 0.325 & 0.101 & 0.288 \\ 0.000 & 0.000 & 0.000 \\ 0.000 & 0.000 & 0.000 \\ 0.000 & 0.000 & 0.000 \\ 0.000 & 0.000 & 0.000 \\ 0.000 & 0.000 & 0.000 \\ 0.000 & 0.000 & 0.000 \\ 0.000 & 0.000 & 0.000 \\ 0.000 & 0.000 & 0.000 \end{bmatrix}_{STP1\ STP2\ STP3} \begin{bmatrix} 0.100 \\ 0.135 \\ 0.118 \\ 0.093 \\ 0.106 \\ 0.087 \\ 0.073 \\ 0.047 \\ 0.060 \\ 0.073 \\ 0.054 \\ 0.067 \end{bmatrix}_{Wck} \begin{bmatrix} \mathrm{pH} \\ \mathrm{BOD} \\ \mathrm{COD} \\ \mathrm{SS} \\ \mathrm{TDS} \\ \mathrm{Temperature} \\ \mathrm{Oil\&Grease} \\ \mathrm{Phenolic\ Compounds} \\ \mathrm{Total\ Residual\ Chlorine} \\ \mathrm{Chlorides} \\ \mathrm{Sulphates} \\ \mathrm{Ammonical\ Nitrogen} \end{bmatrix}
$$

The total score (TS), for each Sewage Treatment Plant of treated sewage were calculated separately using Eq. (3) [6].

As a sample calculation, the total score for sub criteria of treated sewage due to Sewage Treatment Plant1 (Academicians) is as shown below.

$$TS_{STP1} = (0.824 \times 0.100 + 0.875 \times 0.135 + 0.323 \times 0.118 + 0.325 \times 0.093)$$
$$TS_{STP1} = 0.268$$

Total score as SWQI for sub criteria of treated sewage of all Sewage Treatment Plants are as shown in Table 5.

Similarly, SWQI was calculated using perception of professionals and the same is shown in Table 6.

Table 5 SWQI (Academicians)

STP No.	SWQI for period of					
	June 14 to Sept. 14	Rank	Oct. 14 to Jan. 15	Rank	Feb. 15 to Mar. 15 and April 14 to May 15	Rank
1	0.268	1	0.207	2	0.227	2
2	0.148	3	0.157	3	0.144	3
3	0.244	2	0.248	1	0.233	1

Table 6 SWQI (Professionals)

STP No.	SWQI For Period of					
	June 14 to Sept. 14	Rank	Oct. 14 to Jan. 15	Rank	Feb. 15 to Mar. 15 and April 14 to May 15	Rank
1	0.275	1	0.215	2	0.234	2
2	0.157	3	0.166	3	0.153	3
3	0.251	2	0.254	1	0.239	1

4 Sensitivity Analysis

The SWQI was also developed by considering stringent discharge norms for BOD. For this, the permissible limit for BOD was reduced to 10 mg/L from 20 mg/L. Tables 7 and 8 show the SWQI with considering the stringent norms.

5 Discussions

From the normalized weightage method for the determination of SWQI, following points were observed:

(i) The final ranking of STPs did not change with the linguistic opinion of the experts (Academicians and Professionals). However, SWQI for STP marginally changed. This was mainly due to the change in the normalized weightage factors derived on the basis of the linguistic term assignment by the experts.

Table 7 Sensitivity Analysis SWQI (Academicians)

STP No.	SWQI for period of					
	June 14 to Sept. 14	Rank	Oct. 14 to Jan. 15	Rank	Feb. 15 to Mar. 15 and April 14 to May 15	Rank
1	0.504	1	0.366	2	0.412	2
2	0.226	3	0.255	3	0.223	3
3	0.448	2	0.421	1	0.414	1

Table 8 Sensitivity Analysis SWQI (Professionals)

STP No.	SWQI For Period of					
	June 14 to Sept. 14	Rank	Oct. 14 to Jan. 15	Rank	Feb. 15 to Mar. 15 and April 14 to May 15	Rank
1	0.510	1	0.374	2	0.418	2
2	0.235	3	0.263	3	0.232	3
3	0.455	2	0.427	1	0.421	1

(ii) The SWQI of $\mathbf{STP_1}$, $\mathbf{STP_2}$ and $\mathbf{STP_3}$ was obtained as **0.207 to 0.275, 0.144 to 0.157 and 0.233 to 0.254** respectively (See Tables 5 and 6).
(iii) STP_1 ranked first out of three STPs and it showed higher pollution potential while STP_2 ranked number three with lowest pollution potential compared to other STPs during rainy season (June 14 to Sept. 14).
(iv) STP_3 ranked first out of three STPs and it showed higher pollution potential while STP_2 ranked number three with lowest pollution potential compared to other STPs during rainy season (Oct. 14 to Jan. 15).
(v) STP_3 ranked first out of three STPs and it showed higher pollution potential while STP_2 ranked number three with lowest pollution potential compared to other STPs during rainy season (Feb. 15 to Mar. 15 and April 14 to May 15).

From the sensitivity analysis, it is observed that, the rank of the STP does not change but the SWQI increases.

6 Conclusions

The SWQI model using Fuzzy Multiple Criteria Decision Making technique has been developed in the present study to evaluate SWQI to rank the Sewage Treatment Plants on the basis of their wastewater discharges. From the study and analysis of the results of the present study following conclusions have been drawn:

(i) Application of fuzzy approach to the standards of Pollution Control is found to be more appropriate compared to the current crisp approach.
(ii) The fuzzy approach, used in this study, quantifies the Pollution Potential of Sewage Treatment Plants based on an integrated index, taking into consideration the wastewater discharges.
(iii) When the comparison is made among the linguistic terms assignments by Academicians and Professionals the index value changes marginally, but the final ranking of the Sewage Treatment Plants does not change.
(iv) It is possible to rank different types of Sewage Treatment Plants based on their pollution potential and it is possible to encourage concern authorities to bring pollution potential at minimum level, by controlling pollution.
(v) From the sensitivity analysis study, it is concluded that the pollution potential is increases however; it retains their present rankings. Because of increase in the pollution level, it is suggested that the concept of pollution tax can be linked with this study by decision makers to control the pollution levels of developing countries.

References

1. Anand Raj, P., Nagesh Kumar, D.: Ranking alternatives with fuzzy weights using maximizing set and minimizing set. Fuzzy Sets Syst. **105**, 365–375 (1999)
2. Baldwin, J.F., Guild, N.C.F.: Comparison of fuzzy sets on the same decision space. Fuzzy Sets Syst. **2**, 213–231 (1979)
3. Bass S.M., Kwakernaak, H.: Rating and ranking of multiple aspect alternatives using fuzzy sets. Automatica 13, 47–57 (1977)
4. Bortolan, G., Degani, R.: A review of some methods for ranking fuzzy subsets. Fuzzy Sets Syst. **15**, 1–19 (1985)
5. Hipel, K.W., Radford, K.J., Fang, L.: Multiple participant multiple criteria decision-making. IEEE Trans. Syst. Man Cybern. **23**, 1184–1189 (1993)
6. Hwang, C.L., Yoon, K.: Multiple Attribute Decision Making: Methods and Applications. Springer, New York (1981)
7. Jain, R.: Decision making in the presence of fuzzy variables. IEEE Trans. Syst. Man Cybern. **6**, 698–702 (1976)
8. Jain, R.: A procedure for multiple aspect decision making using fuzzy sets. Int. J. Syst. Sci. **8**, 1–7 (1977)
9. Kaufmann, A., Gupta, M.M.: Fuzzy Mathematical Models in Engineering and Management Science. North-Holland, Amsterdam (1988)
10. Klir, G.J., Yuan, B.: Fuzzy Sets and Fuzzy Logic: Theory and Applications. Prentice Hall of India Private Limited, New Delhi (2003)
11. Lad, R.K., Desai, N.G., Christian, R.A., Deshpande, A.W.: Fuzzy modeling for environmental pollution potential ranking of industries. Int. J. Environ. Progr. American Institute of Chemical Engineers (AIChE). **27**(1), 84–90 (2008)
12. Singh, D., Tiong, R.L.K.: A fuzzy decision framework for contractor selection. J. Constr. Eng. Manag. ASCE. **131**(1), 62–70 (2005)
13. Zadeh, L.A.: Fuzzy sets. Inf. Control **8**(3), 338–353 (1965)
14. Zimmerman, H.J.: Fuzzy Set Theory and Its Application. Kluwer Nijhoff, Hingham (1985)

Fuzzy Formal Concept Analysis Approach for Information Retrieval

Cherukuri Aswani Kumar, Subramanian Chandra Mouliswaran, Pandey Amriteya and S.R. Arun

Abstract Recently Formal Concept Analysis (FCA), a mathematical framework based on partial ordering relations has become popular for knowledge representation and reasoning. Further this framework is extended as Fuzzy FCA, Rough FCA, etc. to deal with practical applications. There are investigations in the literature applying FCA for Information Retrieval (IR) applications. The objective of this paper is to apply Fuzzy FCA approach for IR. While adopting Fuzzy FCA, we follow a fast algorithm to generate the fuzzy concepts rather than classical algorithms that are based on residuated methods. Further we follow an approach that retrieves the relevant documents even during absence of exact match of the keywords.

Keywords FCA · Fuzzy FCA · Fuzzy context · Fuzzy concept · Fuzzy concept lattice · Information retrieval

1 Introduction

At present, information retrieval (IR) systems play an important role in various online search applications. Starting from traditional application such as digital library search and web search to recent popular applications such as social search and recommender system, IR systems have a great significance. The interest on IR systems has evolved numerous IR models since 1960 such as Boolean models [1, 2], vector space models [3–6], probabilistic model [7], graph theory model [8] etc. The major challenge of these IR systems is to retrieve the relevant and precise information or documents as per the user needs modeled through keyword search or

C. Aswani Kumar (✉) · S. Chandra Mouliswaran · P. Amriteya · S.R. Arun
School of Information Technology and Engineering, VIT University, Vellore, India
e-mail: cherukuri@acm.org

S. Chandra Mouliswaran
e-mail: scmwaran@gmail.com

V. Ravi et al. (eds.), *Proceedings of the Fifth International Conference on Fuzzy and Neuro Computing (FANCCO - 2015)*, Advances in Intelligent Systems and Computing 415, DOI 10.1007/978-3-319-27212-2_20

query search [6]. The literature has shown the common practice of using the natural language to input the search query and keyword, even though they are vague and uncertain in nature [9]. The fuzzy based IR models are introduced to resolve the issue of mismatch on using uncertain or fuzzy input with concrete or crisp system of IR [10].

Ganter and Willie [11] have proposed Formal Concept Analysis (FCA), a mathematical framework based on lattice theory and it is commonly used for data analysis. This framework produces concept lattice to describe the conceptual structure among the data. There are research interests in adopting FCA for IR applications [12–14]. Priss [15] has introduced a practical IR system called FaIR. Very recently Codocedo and Napoli [3] have provided a detailed review on FCA for IR systems. This survey highlighted several lattice based IR models such as BR-Explorer, CREDO, Fooca, JBrainDead, etc. Kollewe et al. [16] have described that the documents can be related to formal objects, index terms can be related to formal attributes and document-term matrix. This matrix can be transformed to formal context which is the main input to FCA. From this context FCA derives the formal concepts, attribute implications and an ordered hierarchical structure called concept lattice [11, 17]. This same document-term matrix is considered as input to the models such as Vector Space Model (VSM), Latent Semantic Indexing (LSI). Aswani Kumar et al. [18] have compared the performance of VSM, LSI and FCA models for IR application. In an interesting work, Muthukrishnan [19] has developed an IR system based on concept lattices. Exploiting the properties of concept lattice structures, Muthukrishnan [19] was able to process complex queries. However this work was concentrated on FCA with crisp settings. Very recently Kumar et al. [20] have proposed a bi-directional associative memory model that mimics the abilities such as learn, memorize and recall of human brain based on FCA. For practical applications, FCA was extended into fuzzy settings under the framework of Fuzzy FCA. For practical applications, FCA was extended into fuzzy settings under the framework of Fuzzy FCA. Krajci [21] have introduced and defined the one sided fuzzy concept lattices. Butka et al. [22] have devised a proposal for one sided fuzzy formal context based on one sided fuzzy concept lattices. Martin and Majidian [23] have developed an approach for fuzzy concept generation that produces fewer concepts when compared to the existing approaches that are based on residuated methods.

Based on these above observations, in this paper we extend the work of Muthukrishanan [19] in fuzzy environment with the help of Fuzzy FCA algorithm proposed by Martin and Majidian [23]. The rest of this paper is organized as follows. Section 2 provides the brief background of FFCA and the Sect. 2.3 provides the related work on various IR models. We present the proposed FFCA approach for IR in Sect. 3 and the illustration of experiments along with their results in Sect. 4.

2 Background

2.1 *Formal Concept Analysis*

FCA is a framework which explores conceptual dependencies that exists in the given formal context.

Definition 1 (*Formal context*) A formal context K has three set of elements and those elements are defined as $K_C = (P, Q, R)$, Here, the set of objects are called as P, the set of attributes are called as Q and the binary relation between the P and Q is represented as R.

Definition 2 (*Formal concept*) Let the given context $K_C = (P, Q, R)$. Consider the set of elements A ∈ P and the set of elements B ∈ Q. Then, consider another dual sets A′ and B′ such that A′ represents the set of attributes applying to all the objects belonging to B, B′ is the set of objects having all the attributes belonging to A respectively. Therefore,

$$A' = \{q \in Q | pRq \forall p \in P\} \text{ and } B' = \{p \in P | pRq \forall q \in Q\}$$

A formal concept is a concept extent and concept intent pair (A, B) such that $A \subseteq P$, $B \subseteq Q$ and it satisfies the conditions such as $A' = B$ and $B' = A$. For more details on FCA under crisp setting, interested readers can refer [24, 25].

2.2 *Fuzzy Formal Concept Analysis*

FFCA is a mathematical theory which combines fuzzy logic and FCA in order to represent uncertain information in the formal context.

Definition 3 (*Fuzzy Formal Context*) The fuzzy formal context or fuzzy context has three set of elements and those elements are defined as $K_F = (P, Q, R_F = \phi(P, Q))$. Here, the set of objects are called as P, the set of attributes are called as Q and the fuzzy relation between the P and Q is defined as R_F and it has the membership value $\mu(p, q)$ in the range of zero to one i.e. [0,1]. The pair $(p, q) \in R_F$ with the membership value $\mu(p, q)$ can be read as "the object p has the attribute q with the membership value of $\mu(p, q)$.

Definition 4 (*Fuzzy Formal Concept*) Let the given fuzzy context $K_F = (P, Q, R_F = \phi(P, Q))$, with the confidence threshold T, the set of elements A such that $A \subseteq P$ and set of elements B such that $B \subseteq Q$. Consider another dual sets A and B such that $A' = \{q \in Q | \mu(p, q) \geq T \forall p \in A\} B' = \{p \in P | \mu(p, q) \geq T \forall q \in B\}$.

A fuzzy formal concept is defined as a pair $(\phi(A), B)$ on the fuzzy context K_F with confidence threshold T, $A = B$ and $B = A$. Here, A represents the fuzzy extent and B represent the fuzzy intent of the fuzzy concept and every object p belongs to A has the membership value μ_0. It is defined through the membership value $\mu(p, q)$ between the object p and the attribute q in such a way that $\mu_0 = \min_{q \in B} \mu(p, q)$.

Definition 5 (*Super Concept and Sub concept*) Consider the fuzzy concept pairs $(\phi(A1), B1)$ and $(\phi(A2), B2)$ of the fuzzy context $K_F = (P, Q, R_F = \phi(P, Q))$, Here,$(\phi(A1), B1)$ is the super concept of $(\phi(A2), B2)$, denoted as $((\phi(A2), B2) \geq (\phi(A1), B1))$, if and only if $(\phi A2) \supseteq \phi(A1) \Leftrightarrow (B1 \supseteq B2)$. In the similar way, $(\phi(A1), B1)$ is the sub concept of $(\phi(A2), B2)$.

For more details on these definitions readers can refer [26, 27]. As discussed by Martin and Majidian [23], in this paper, we use with crisp set of objects and fuzzy set of attributes and formalize the definition of concept forming operator by considering the fuzzy formal concept as a pair <*A*, *B*> where A is a set of objects with crisp values and B is a set of attributes with fuzzy values such that $A = B$ and $B = A$ where $A^{\uparrow} = \{b \in B | \forall a \in A: \mu_R(a, b) \geq \mu_A(a)$ and $B^{\downarrow} = \left\{ a/\mu_a(a) | \mu_a(a) = \min_{b \in B}(\mu_R(a, b)) \right\}$.

2.3 Related Work

There are several IR models based on algebra. Dominich [5] has provided a detailed analysis on models such VSM, LSI, etc. These models process the given document collection or database as document—term matrix and apply algebraic techniques. Applications of these models can be found in [28]. Interesting applications clustering techniques for IR can be found in [6, 29, 30]. There are several investigations reported in the literature using lattice theory for IR. Very recently Codocedo and Napoli [3] have provided a detailed survey on these investigations. For browsing and query retrieval, lattice based conceptual clustering has been used [31]. Carpineto and Romano [32] discussed the natural language IR system. This system uses the linguistically motivated indexing (LMI) to discover critical semantic aspects of the documents. Concept lattices has been used in concept based retrieval systems [33]. Carpineto and Romano [17] have introduced the Boolean query IR system REFINER which construct and show the part of concept lattice for searched document. Some of the lattice representation techniques [34] are used for query refinement based on Boolean operators which are part of the query. Valverde and Palaez-Moremov [35] differentiate FCA in IR with FCA for IR. Further, they have also discussed the possibility of augmenting IR with ideas of FCA. Instead of Boolean querying system, Godin et al. [36] have introduced the capability of concept lattices and retrieval system. Alam and Napoli [37] have used the concept lattice as a classification of SPARQL answers. Further, several FCA based IR systems [3] such as CREDO, JBrianDead, FooCA, BR-Explorer, Camelis and

CrechainDoare introduced using different factors such as explicit relevance feedback, pivoting and ranking of datasets etc. FCA is also extended to different models such as fuzzy graph, interval valued fuzzy lattice, etc. [38, 39]. Recently, Kumar et al. [20] have proposed formal concept analysis approach to perform some cognitive functions such as learn, memorize and recall the information that is associated with the cue with the support of object-attribute relation. In his dissertation work, Muthukrishnan [19] has developed concept lattice model for IR. The model considers three different approaches for handling the query matching. However the model considers term-document collection as formal context in crisp setting. But the practical applications term-document matrices contain the entries indicating the term strength in a particular document. So to handle such document collections, in this paper we propose FFCA based approach for IR. The proposed approach utilizes the fuzzy concept generation algorithm proposed by Martin and Majidian [23]. In the following section, we illustrate our proposal.

3 Fuzzy FCA for IR

The proposed model considers the document collection as term-document matrix where the entries of the matrix indicate the strength of the term in the document. We do not discuss here the term—document matrix generation from the given document collection and other pre-processing stages. The proposed model considers the normalized term document matrix for the analysis. For adopting fuzzy FCA algorithm, this model transforms this matrix as a formal fuzzy context where the keywords are considered as fuzzy formal objects and documents as crisp attributes and the entries are the fuzzy values which represent the fuzzy relational strength keywords in the documents. Following steps illustrates this.

1. Identify the set of keywords (K), set of documents (D) and the relational strength (R) of individual keyword on different documents. Here, the relational strengths are the fuzzy values.
2. Derive the matrix ($M_{K \times D}$) of identified keywords (K) and documents (D) by assigning the fuzzy relational strength value between documents and keywords.
3. Transform the matrix ($M_{K \times D}$) derived in step 2 as the fuzzy formal context $F_{K,D} = F$ (K, D, $I^{K,D}$) as the mappings the keywords with documents in fuzzy keyword-document context.
4. Obtain the different fuzzy concepts from the fuzzy formal context ($F_{K,D}$) derived in step-3. Here, the fuzzy concepts are the pair fuzzy keywords and crisp documents.
5. Construct the fuzzy concept lattice or keyword-document lattice from the fuzzy formal concepts obtained in step 4. Here, the documents or formal attributes are organized in various levels of the lattice depends upon how the various documents are associated with the different keywords or formal objects.

```
Function: FuzzyConceptsGeneration(<X, Y>, m, M)
Input : M = ordered set of documents 1… n

<X, Y> = a known keyword-document formal concept where X is a set
of objects or keywords with fuzzy value and Y is an ordered subset of M
with crisp values.

m = an integer in the range of 1 to  n+1 such that m ∉ Y
Output : F = a set of fuzzy formal concepts
Local variables : O = fuzzy set of objects or keywords
A = ordered subset of attributes or documents   and
j = integer
P := empty list
IF (Y ≠ M) AND m ≤  n
THEN
        FOR j = m to n DO
                IF j ∉ Y THEN
                        O = X ∩{ j }↓
                        A := O↑
                        IF Y∪Mj = D ∩ Yj THEN
                        F := FuzzyConceptsGeneration (<O, A>,
                        j+1, M)
                        ENDIF
                ENDIF
        ENDFOR
ENDIF

RETURN <X,Y> ⊕P
```

Fig. 1 Fuzzy concept generation algorithm

In the above procedure, to generate the fuzzy concepts, we use the fuzzy concept generation algorithm [23] as shown in Fig. 1. From the generated formal concepts, we formalize the fuzzy concept lattice and the same lattice is queried using the concept lattice query matching algorithm [19] depicted in Fig. 2.

3.1 Fuzzy Concept Generation Algorithm

This fuzzy concept generation algorithm is used to generate the list of fuzzy concepts from the given input fuzzy context such as keyword-document context. Here, we follow the Fuzzy FCA relation incident (I) as described below.

The relation (I) for forming fuzzy FCA could be defined as I: $K \times D \rightarrow [0, 1]$ instead of defining it in the classical crisp form such as I: $K \times D \rightarrow \{0, 1\}$. Here, the context generated $F(K, D, I^{K,D})$ every value can range between [0, 1]. We define the

Variables:
A: Set of input keywords
O: Set of objects or documents
CF: Concept File which contains the concepts in extents or keyword sorted order.
C: The individual concept stored in the concept file (CF) as lattice.
R: Result of the concept.
I: Stores the index of the resultant concept
R_{alt} : Resultant set which contains the alternate result concept

Function:
KeywordQueryProcess( )

1. Start
2. in= 0
3. while (not EOF(CF))
    a. C = next C from CF
    b. if (extent(C) $\supseteq$ A)
        i. result = C
        ii. index = in
    c. if (A $\supseteq$ extent(C))
        i. R_{alt} = C
    d. in = in + 1
4. Return result, R_{alt} and index
5. Stop

Fig. 2 Concept lattice query matching algorithm

fuzzy concept of fuzzy formal context as a pair of K, D where K is the set of keywords fuzzy strength value and D is the crisp set of documents such that, $K^{\uparrow} = D$ and $D^{\downarrow} = K$. Here,

$$K^{\uparrow} = \{d \in D | \forall k \in K: \mu_I(k, d) \geq \mu_K(k)\} \tag{1}$$

$$D^{\downarrow} = \{k/\mu_K(k) | \mu_K(k) = \min_{d \in D}(\mu_I(k, d))\} \tag{2}$$

The concept generation operators (↑ and ↓) described in Eqs. 1 and 2 are used to find the fuzzy extent and crisp intent respectively. The algorithm to generate the fuzzy concept pair such as fuzzy attributes and crisp objects is depicted in Fig. 1. Here, keyword-document context is given as input and the set of fuzzy concepts such as fuzzy keywords and crisp documents pair is received as the output.

These resultant fuzzy concepts are generated and formalized as the fuzzy lattice. Further these fuzzy concepts are stored into a file with its corresponding formalized fuzzy lattice for further processing on IR with the keyword query matching. The algorithm for concept lattice keyword query matching algorithm is described in Sect. 3.2.

3.2 Concept Lattice Query Matching Algorithm

In this query matching algorithm, user submits the set of keywords in which their interest exists as the query. Once, the user input the keywords, those keywords are formalized as the extent for which the user in interested.

Based on this input extent, matching occurs with the various concepts in the hierarchy of resultant fuzzy concept lattice and finds the corresponding intents or document set as the output. Here, the subset rule of concept lattices is used to match the input keyword extents with concepts in the lattice.

For example, if the user input the keywords (K1, K2) and the existing concepts are ((K1), (D1, D2, D3, D4, D5)), ((K1, K2, K3), (D1, D2, D3, D5)) and ((K1, K2, K3, K5), (D1, D2, D5)), then ((K1, K2, K3), (D1, D2, D3, D5)) is retrieved as (K1, K2, K3) is the smallest superset. Similarly, if the user input the keywords (K1, K2, K3, K4, K5) and the existing concepts are ((K1), (D1, D2, D3, D4, D5)), ((K1, K2, K3), (D1, D2, D3, D5)) and ((K1, K2, K3, K5), (D1, D2, D5)), then ((K1, K2, K3, K5), (D1, D2, D5)) is retrieved as (K1, K2, K3, K5) is the largest subset. The intents of these resultant concepts are returned as the resultant document set as the output of the submitted input query. The above process is depicted as the concept lattice query matching algorithm in Fig. 2. By this algorithm, we can obtain the matching fuzzy formal concepts as the result. From these resultant concepts, the intents or the documents set is extracted and given as the result. Further, to get the clarity in understanding of this approach, the flow diagram of the above described concept lattice query matching process is shown in Fig. 3.

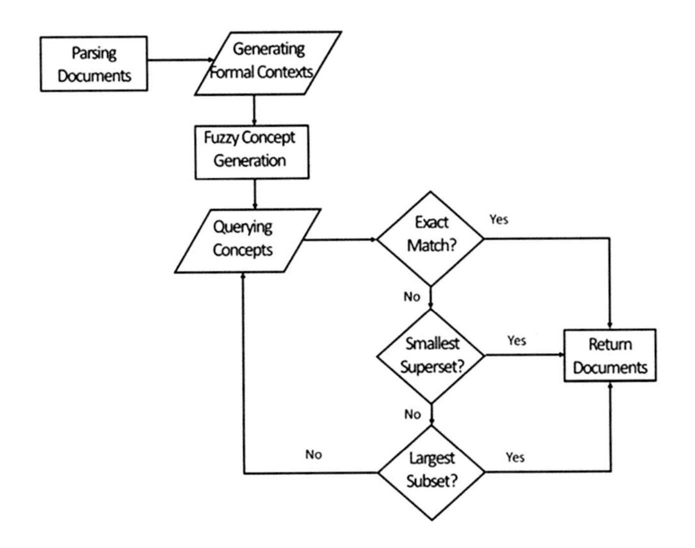

Fig. 3 Flow diagram of concept lattice query matching algorithm

4 Experimental Results

To demonstrate our proposed approach, we consider a sample fuzzy keyword-document relation with eight input keywords and four documents. Consider the entries of this relational matrix are fuzzy values which represent the relational strength of different keywords in various documents. In this table, the rows represent the input keywords from a1 to a8 such as a1, a2, a3, a4, a5, a6, a7 and a8. The columns represent the four input documents such as p, q, r and s. The entries in this table are the fuzzy values from zero to one. The same matrix is formalized as the keyword-document fuzzy formal context as described in Table 1 by considering the keywords as the fuzzy objects and documents as the crisp attributes and their relation as the fuzzy values ranging from zero to one.

Further, we derive the various fuzzy concepts from the fuzzy keyword-document context described in Table 1using the fuzzy concept generation algorithm depicted in Fig. 1.

Various iterative executional steps behind the fuzzy concept generation algorithm are shown below. Here, the values for j, m, X and Y in every iterative step is described.

The initial value of Y is empty and the value of X is all the attributes with membership value 1.

X = [a1:1, a2:1, a3:1, a4:1, a5:1, a6:1, a7:1, a8:1]

Y = Ø

Step 1: initial call to the algorithm is made.

Initial call m = 1, X = [a1:1, a2:1, a3:1, a4:1, a5:1, a6:1, a7:1, a8:1] and Y = Ø

	p	q	r	s
a1	0	1	1	0
a2	0.2	0.6	1	0.2
a3	1	0	0	0
a4	0.3	0.6	0.7	0.3
a5	1	0	0	1
a6	0	1	0	0
a7	0.1	0.5	0.5	0.1
a8	0.5	0.3	0	0.5

j = 1, O = [a2: 0.2, a3: 1, a4: 0.3, a5: 1, a7: 0.1, a8: 0.5], A = [p]

Table 1 Sample fuzzy keyword-document context

	p	q	r	s
a1	0	1	1	0
a2	0.2	0.6	1	0.2
a3	1	0	0	0
a4	0.3	0.6	0.7	0.3
a5	1	0	0	1
a6	0	1	0	0
a7	0.1	0.5	0.5	0.1
a8	0.5	0.3	0	0.5

Step 2: Recursive call takes place in this step.
m = 2, X = [a2: 0.2, a3: 1, a4: 0.3, a5: 1, a7: 0.1, a8: 0.5], Y = [p]

	p	q	r	s
a1	0	1	1	0
a2	0.2	0.6	1	0.2
a3	1	0	0	0
a4	0.3	0.6	0.7	0.3
a5	1	0	0	1
a6	0	1	0	0
a7	0.1	0.5	0.5	0.1
a8	0.5	0.3	0	0.5

j = 2, O = [a2: 0.2, a4: 0.3, a7: 0.1, a8: 0.3], A = [p, q, s]

Step 3: m = 2, X = [a2: 0.2, a3: 1, a5: 1, a4: 0.3, a7: 0.1, a8: 0.5], Y = [p]

	p	q	r	s
a1	0	1	1	0
a2	0.2	0.6	1	0.2
a3	1	0	0	0
a4	0.3	0.6	0.7	0.3
a5	1	0	0	1
a6	0	1	0	0
a7	0.1	0.5	0.5	0.1
a8	0.5	0.3	0	0.5

j = 3, O = [a2: 0.2, a4: 0.3, a7: 0.1], A = [p, q, r, s]

Step 4: m = 2, X = [a2: 0.2, a3: 1, a5: 1, a4: 0.3, a7: 0.1, a8: 0.5], Y = [p]

	p	q	r	s
a1	0	1	1	0
a2	0.2	0.6	1	0.2
a3	1	0	0	0
a4	0.3	0.6	0.7	0.3
a5	1	0	0	1
a6	0	1	0	0
a7	0.1	0.5	0.5	0.1
a8	0.5	0.3	0	0.5

j = 4, O = [a2: 0.2, a5: 1, a4: 0.3, a7: 0.1, a8: 0.5], A = [p, s]

Step 5: m = 1, X = [a1:1, a2:1, a3:1, a4:1, a5:1, a6:1, a7:1, a8:1], Y = Ø

	p	q	r	s
a1	0	1	1	0
a2	0.2	0.6	1	0.2
a3	1	0	0	0
a4	0.3	0.6	0.7	0.3
a5	1	0	0	1
a6	0	1	0	0
a7	0.1	0.5	0.5	0.1
a8	0.5	0.3	0	0.5

j = 2, O = [a1: 1, a2: 0.6, a4: 0.6, a6: 1, a7: 0.5, a8: 0.3], A = [q]

Step 6: m = 3, X = [a1: 1, a2: 0.6, a4: 0.6, a6: 1, a7: 0.5, a8: 0.3], Y = [q]

	p	q	r	s
a1	0	1	1	0
a2	0.2	0.6	1	0.2
a3	1	0	0	0
a4	0.3	0.6	0.7	0.3
a5	1	0	0	1
a6	0	1	0	0
a7	0.1	0.5	0.5	0.1
a8	0.5	0.3	0	0.5

j = 3, O = [a1: 1, a2: 0.6, a4: 0.6, a7: 0.5], A = [q, r]

Step 7: y = 3, X = [a1: 1, a2: 0.6, a4: 0.6, a6: 1, a7: 0.5, a8: 0.3], Y = [q]

	p	q	r	s
a1	0	1	1	0
a2	0.2	0.6	1	0.2
a3	1	0	0	0
a4	0.3	0.6	0.7	0.3
a5	1	0	0	1
a6	0	1	0	0
a7	0.1	0.5	0.5	0.1
a8	0.5	0.3	0	0.5

j = 4, O = [a8: 0.3, a2: 0.2, a4: 0.3, a7: 0.1], A = [p, q, s]

Step 8: y = 1, X = [a1:1, a2:1, a3:1, a4:1, a5:1, a6:1, a7:1, a8:1], Y = Ø

	p	q	r	s
a1	0	1	1	0
a2	0.2	0.6	1	0.2
a3	1	0	0	0
a4	0.3	0.6	0.7	0.3
a5	1	0	0	1
a6	0	1	0	0
a7	0.1	0.5	0.5	0.1
a8	0.5	0.3	0	0.5

j = 3, O = [a1: 1, a2: 1, a4: 0.7, a7: 0.5], A = [r]

Step 9: y = 4, X = [a1: 1, a2: 1, a4: 0.7, a7: 0.5], Y = [r]

	p	q	r	s
a1	0	1	1	0
a2	0.2	0.6	1	0.2
a3	1	0	0	0
a4	0.3	0.6	0.7	0.3
a5	1	0	0	1
a6	0	1	0	0
a7	0.1	0.5	0.5	0.1
a8	0.5	0.3	0	0.5

j = 4, O = [a2: 0.2, a4: 0.3, a7: 0.1], A = [p, q, r, s]

Step 10: Final step.
y = 1, X = [a1:1, a2:1, a3:1, a4:1, a5:1, a6:1, a7:1, a8:1], Y = Ø

	p	q	r	s
a1	0	1	1	0
a2	0.2	0.6	1	0.2
a3	1	0	0	0
a4	0.3	0.6	0.7	0.3
a5	1	0	0	1
a6	0	1	0	0
a7	0.1	0.5	0.5	0.1
a8	0.5	0.3	0	0.5

j = 4, O = [a2: 0.2, a4: 0.3, a5: 1, a7: 0.1, a8: 0.5], A = [p, s]

The generated fuzzy concepts are listed in Table 2. Here, the fuzzy concepts are the pair of values with fuzzy set of keywords and crisp set of documents.

Table 2 Fuzzy concepts of the keyword-document context described in Table 1

S. No.	Extents (Fuzzy valued keywords)	Intents (Crisp valued documents)
C1	{a2: 0.2, a3: 1, a4: 0.3, a5: 1, a7: 0.1, a8: 0.5}	{p}
C2	{a2: 0.2, a4: 0.3, a7: 0.1, a8: 0.3}	{p, q, s}
C3	{a2: 0.2, a4: 0.3, a7: 0.1}	{p, q, r, s}
C4	{a2: 0.2, a4: 0.3, a5: 1, a7: 0.1, a8: 0.5}	{p, s}
C5	{a1: 1, a2: 0.6, a4: 0.6, a6: 1, a7: 0.5, a8: 0.3}	{q}
C6	{a1: 1, a2: 0.6, a4: 0.6, a7: 0.5}	{q, r}
C7	{a1: 1, a2: 1, a4: 0.7, a7: 0.5}	{r}
C8	{a1: 1, a2: 1, a3: 1, a4: 1, a5: 1, a6: 1, a7: 1, a8: 1}	Ø

From this Table-2, we can understand the fuzzy relation between the set of keywords and set of documents. For example, the concept C3 represents the keywords a2, a4 and a7 are related to all the documents p, q, r and s with the relation strength of 0.2, 0.3 and 0.1 respectively. Similarly, the concept C7 represents the keywords a1, a2, a4 and a7 are related to one document r with the relation strength of 1, 1, 0.7 and 0.5 respectively. In this way, we can understand the fuzzy relation strength of set of keywords and set of documents referred as fuzzy concepts of keyword-document context. Totally, we have received eight different concepts. This fuzzy formal analysis helps to classify the various fuzzy concepts and helps to understand how the set of keywords are relevant with the set of documents along with their relational strength value.

Next, we construct the fuzzy concept lattice or keyword-document lattice from the fuzzy formal concepts listed in Table 3. The resultant fuzzy concept lattice is shown in Fig. 4. In this lattice structure, the nodes are representing the fuzzy concepts and these fuzzy concepts are representing the association of set of keywords along with the set of documents with their role strength. The addition and deletion of keywords and documents into this keyword-documents fuzzy formal context or the updates into the relational strength values will bring the necessary updates into the resultant fuzzy concept lattice.

In this lattice structure, we can visualize those fuzzy concepts containing all of the eight keywords in the top level and the four documents in the bottom of the concept. The fuzzy concepts, containing single, two and three documents are placed at various levels depends upon the relevance of different set of keywords with various set of documents. Based on the lattice structure generated in Fig. 4, we query for our searched keywords using the algorithm depicted in Fig. 2. In Table 3, we have listed the various sample the query calls along with their results. In addition, this table shows the methodology followed column which describes how the searched sets of keywords are matched with the fuzzy concept lattice to retrieve the relevant document as result.

Table 3 List of sample query calls along with their results

Searched keywords	Retrieved relevant document	Methodology followed
{a1, a2}	{q, s}	Search for exact match
		Exact match not found
		Search for minimum superset
		Match found
{a2, a4, a7}	{p, q, r, s}	Search for Exact match
		Match found
{a1, a4}	{q}	Search for exact match
		Exact match not found
		Search for minimum superset
		Match found
{a1, a2, a4}	{q, r}	Search for exact match
		Exact match not found
		Search for minimum superset
		Match found
{a7, a8}	{p, q, s}	Search for exact match
		Exact match not found
		Search for minimum superset
		Match found
{a1, a2, a3}	{r}	Search for exact match
		Not found
		Search for minimum superset
		Minimum superset returns null
		Search largest subset
		Match found
{a1, a2, a4, a7}	{q, r}	Search for Exact match
		Match found
{a1, a2, a3, a4, a5}	{p}	Search for exact match
		Not found
		Search for minimum superset
		Minimum superset returns null
		Search largest subset
		Match found
{a1, a2, a3, a4, a5, a6, a7}	{p}	Exact Match found
		Null result returned
		Search for largest subset
		Match found

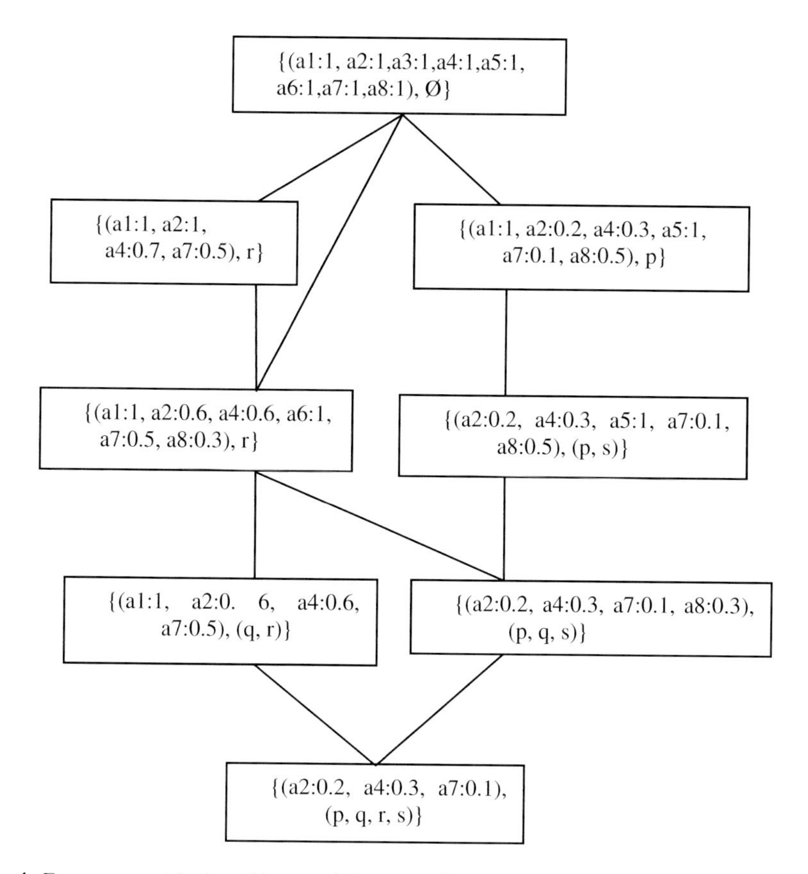

Fig. 4 Fuzzy concept lattice of keyword-document fuzzy concepts

5 Conclusion

In this paper, we have presented the fuzzy formal concept analysis approach for information retrieval. To demonstrate this approach, we have formalized the fuzzy keyword-document context which contains the set of keywords as fuzzy objects and the set of documents as crisp attributes. The entries of this fuzzy context are the fuzzy values representing the relevance of keywords in different documents. The proposed approach adopts a fast fuzzy concept generation algorithm for deriving the concepts and query matching algorithm that exploit the properties of concept lattice structure. Experimental results demonstrated the proposed approach on a toy dataset.

Acknowledgments One of the Authors, Ch. A.K sincerely acknowledges the research grant under cognitive science research initiative schemeSR/CSRI/118/2014 of DST, Govt. of India.

References

1. Waller, W.G., Kraft, D.H.: A mathematical model of a weighted Boolean retrieval system. Inf. Process. Manage. **15**(5), 235–245 (1979)
2. Cerulo, L., Canfora, G.: A taxonomy of information retrieval models and tools. CIT. J. Comput. Inf. Technol. **12**(3), 175–194 (2004)
3. Codocedo, V., Napoli, A.: Formal concept analysis and information retrieval–a survey. In: Formal Concept Analysis, pp. 61–77. Springer International Publishing (2015)
4. Manning, C.D., Raghavan, P., Schütze, H.: Introduction to Information Retrieval, vol. 1, p. 496. Cambridge University Press, Cambridge (2008)
5. Dominich, S.: The Modern Algebra of Information Retrieval, pp. 74–93. Springer, Heidelberg (2008)
6. Baeza-Yates, R., &Ribeiro-Neto, B.: Modern Information Retrieval (vol. 463). ACM press, New York (1999)
7. Wong, S.K.M., Yao, Y.Y.: On modeling information retrieval with probabilistic inference. ACM Trans. Inf. Syst. (TOIS) **13**(1), 38–68 (1995)
8. Flake, G.W., Lawrence, S., Giles, C.L., Coetzee, F.M.: Self-organization and identification of web communities. Computer **35**(3), 66–70 (2002)
9. Subtil, P., Mouaddib, N., Foucaut, O.: A fuzzy information retrieval and management system and its applications. In: Proceedings of the 1996 ACM Symposium on Applied Computing, pp.537–541. ACM (1996)
10. Oussalah, M., Khan, S., Nefti, S.: Personalized information retrieval system in the framework of fuzzy logic. Expert Syst. Appl. **35**(1), 423–433 (2008)
11. Ganter, B., Wille, R.: Formal Concept Analysis: Mathematical Foundations. Springer Science & Business Media (2012)
12. Mooers, C.N.: A Mathematical Theory of Language Symbols in Retrieval. Zator (1958)
13. Kumar, C.A., Srinivas, S.: A note on the effect of term weighting on selecting intrinsic dimensionality of data. Cybern. Inf. Technol. **9**(1), 5–12 (2009)
14. Kumar, C.A.: Fuzzy clustering-based formal concept analysis for association rules mining. Appl. Artif. Intell. **26**(3), 274–301 (2012)
15. Priss, U.: Lattice-based information retrieval. Knowl. Organ. **27**(3), 132–142 (2000)
16. Kollewe, W., Sander, C., Schmiede, R., Wille, R.: TOSCANA als Instrument der bibliothekarischenSacherschließung. Aufbau und ErschließungbegrifflicherDatenbanken, pp. 95–114. BIS-Verlag, Oldenburg (2000)
17. Carpineto, C., Romano, G.: Using concept lattices for text retrieval and mining. In: Formal Concept Analysis, pp. 161–179. Springer, Berlin (2005)
18. Aswani Kumar, C., Radvansky, M., Annapurna, J.: Analysis of a vector space model, latent semantic indexing and formal concept analysis for information retrieval. Cybern. Inf. Technol. **12**(1), 34–48 (2012)
19. Muthukrishnan, A.K.: Information Retrieval Using Concept Lattices (M.S. dissertation, University of Cincinnati) (2006)
20. Kumar, C.A., Ishwarya, M. S., Loo, C.K.: Formal concept analysis approach to cognitive functionalities of bidirectional associative memory. Biol. Inspir. Cogn. Archit. **12**, 20–33 (2015)
21. Krajci, S.: Cluster based efficient generation of fuzzy concepts. Neural Netw. World **13**(5), 521–530 (2003)
22. Butka, P., Pócsová, J., Pócs, J.: A proposal of the information retrieval system based on the generalized one-sided concept lattices. In: Applied Computational Intelligence in Engineering and Information Technology, pp. 59–70. Springer, Berlin, Heidelberg (2012)
23. Martin, T., Majidian, A.: Finding fuzzy concepts for creative knowledge discovery. Int. J. Intell. Syst. **28**(1), 93–114 (2013)
24. Kumar, C.A., Srinivas, S.: Concept lattice reduction using fuzzy K-Means clustering. Expert Syst. Appl. **37**(3), 2696–2704 (2010)

25. Aswani Kumar, C., Srinivas, S.: Mining associations in health care data using formal concept analysis and singular value decomposition. J. Biol. Syst. **18**(04), 787–807 (2010)
26. Formica, A.: Semantic Web search based on rough sets and Fuzzy Formal Concept Analysis. Knowl. Based Syst. **26**, 40–47 (2012)
27. Quan, T.T., Hui, S.C., Cao, T.H.: A Fuzzy FCA-based approach to conceptual clustering for automatic generation of concept hierarchy on uncertainty data. In: CLA, pp. 1–12 (2004)
28. Letsche, T.A., Berry, M.W.: Large-scale information retrieval with latent semantic indexing. Inf. Sci. **100**(1), 105–137 (1997)
29. Baeza-Yates, R., Ribeiro-Neto, B.: Modern Information Retrieval (vol. 463). ACM press, New York (1999)
30. Barbut, M., Monjardet, B.: Ordreet classification algèbre et combinatoirs. Hachette (1970)
31. Carpineto, C., Romano, G.: A lattice conceptual clustering system and its application to browsing retrieval. Mach. Learn. **24**(2), 95–122 (1996)
32. Carpineto, C., Mizzaro, S., Romano, G., Snidero, M.: Mobile information retrieval with search results clustering: prototypes and evaluations. J. Am. Soc. Inform. Sci. Technol. **60**(5), 877–895 (2009)
33. Carpineto, C., Romano, G.: Galois: an order-theoretic approach to conceptual clustering. In: Proceedings of ICML, vol. 293, pp. 33–40 (1993)
34. Spoerri, A.: InfoCrystal: integrating exact and partial matching approaches through visualization. In: RIAO 94: recherched'informationassistée par ordinateur. Conférence, pp. 687–696 (1994)
35. Valverde-Albacete, F.J., Peláez-Moreno, C.: Systems versus methods: an analysis of the affordances of FOrmal concept analysis for information retrieval⋆. In: FCAIR 2012 Formal Concept Analysis Meets Information Retrieval Workshop co-located with the 35th European Conference on Information Retrieval (ECIR 2013), (p. 113). Moscow, Russia (2013)
36. Godin, R., Missaoui, R., April, A.: Experimental comparison of navigation in a Galois lattice with conventional information retrieval methods. Int. J. Man Mach. Stud. **38**(5), 747–767 (1993)
37. Alam, M., & Napoli, A.: Defining views with formal concept analysis for understanding SPARQL query results. In: Proceedings of the Eleventh International Conference on Concept Lattices and Their Applications (2014)
38. Singh, P.K., &Aswani Kumar, Ch.: Bipolar fuzzy graph representation of concept lattice. Inf. Sci. **288**, 437–448 (2014)
39. Singh, P.K., Aswani Kumar, C., Li, J.: Knowledge representation using interval valued fuzzy formal concept lattice. Soft Comput. (2015)

Privacy Preserving Collaborative Clustering Using SOM for Horizontal Data Distribution

Latha Gadepaka and Bapi Raju Surampudi

Abstract In view of present advancements in computing, with the development of distributed environment, many problems have to deal with distributed input data where individual data privacy is the most important issue to be addressed, for the concern of data owner by extending the privacy preserving notion to the original learning algorithms. Privacy Preserving Data Mining has become an active research area in addressing various privacy issues while bringing out solutions for them. There has been lot of progress in developing secure algorithms and models, able to preserve privacy using various data mining techniques like association, classification and clustering, where as importance of privacy preserving techniques applied for learning algorithms related to neural networks for mining problems are still in infancy. Our work in this paper focused on preserving privacy of an individual, using self organizing map (SOM) adopted for collaborative clustering of distributed data between multiple parties. We present Privacy Preserving Collaborative Clustering method using SOM (PPCSOM) for Horizontal Data Distribution, which allows multiple parties perform clustering in a collaborative approach using SOM neural network, without revealing their data directly to each other, in order to preserve privacy of all parties. Our simulation results shows that implementation of PPCSOM method achieves comparable accuracy and assured Privacy than the original non privacy preserving SOM algorithm.

Keywords Privacy preserving · Data partitioning · Clustering · Neural networks · SOM · Collaboration

---

L. Gadepaka (✉) · B.R. Surampudi
School of Computer and Information Sciences, University of Hyderabad, Hyderabad, India
e-mail: latha.gadepaka@gmail.com

B.R. Surampudi
Cognitive Science Lab, International Institute of Information Technology, Hyderabad, India
e-mail: bapiks@yahoo.co.in

V. Ravi et al. (eds.), *Proceedings of the Fifth International Conference on Fuzzy and Neuro Computing (FANCCO - 2015)*, Advances in Intelligent Systems and Computing 415, DOI 10.1007/978-3-319-27212-2_21

1 Introduction

Most data mining techniques are specified to acquire knowledge from highly quantified databases, while most of the information generated is already available for access, and some times distributed among multiple environments under different authority, which makes violating privacy at the time of obtaining data. This situation may lead to some misuse difficulties as there is always a threat of identifying specific information about an individual, resulting in major loss of privacy which is the serious issue to be addressed. For example U.S Department of defence banned all data-mining tasks including research and development deeply concerning growing problem of violating an individual's privacy as per Data-Mining Moratorium Act [19]. Hence the importance of Privacy Preserving Data Mining has been increased to enable these types of situations aiming for privacy protection while extracting private details of an individual [3].

2 Privacy Preserving

Privacy preserving is an important task to solve serious privacy issues. Privacy Preserving Data Mining (PPDM) applied through various data mining techniques provide many privacy preserving methods [1, 19]. Methods like Data Swapping, Data Perturbation, Noise addition, oblivious transfer, data anonymity, and many other methods follow different ways of private computations to produce final results without effecting on privacy of an individual, while exchanging their own information among multiple parties [11, 18]. These methods worked well in assuring privacy. Our main motivation was to use a neural network to preserve privacy while data is distributed among multiple parties to gain combined results same as secure multi party computing [14].

Our work in this paper motivated from [8], where the number of parties cluster their distributed data off line without greatly violating their privacy. They used collaborative filtering [21] scheme applied in SOM neural network [6], and their analysis shows that there is a possibility of collaborative predictions while maintaining privacy of data owner where data is horizontally distributed between various companies. We present a collaborative clustering method using self organizing map (SOM), in Multi Party Distributed Computing environment [4], without any requirement of any party to reveal their own input data to the other directly [2].

3 Data Partitioning

In any distributed computing environment [9, 17], there is always necessity of data partitioning where dataset is partitioned into some specified number of data partitions using many kinds of partitioning methods available. A dataset can be divided

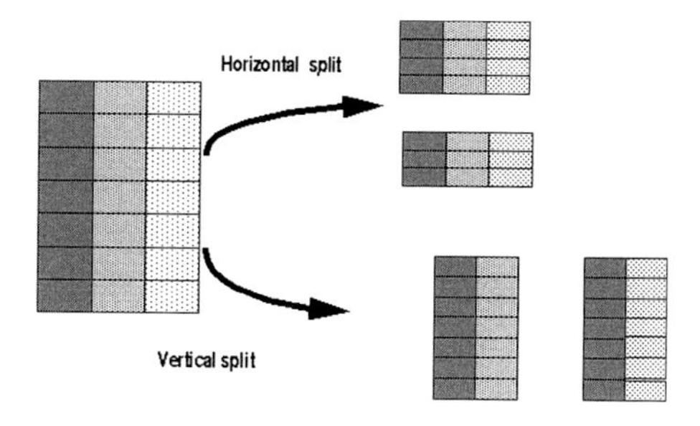

Fig. 1 Horizontal and vertical distribution of data

and distributed to a particular data mining model, in order to gain combined results in small amount of communication time. There are good number of partitioning methods available like horizontal, vertical, and arbitrary partitioning [7] etc., among them we discuss two basic ways of partitioning mostly used by many data mining applications.

If we define a dataset D=(E,I) where E is set of entities for whom information is collected, and I is the set of features, that is collected. Pre assume that there are n number of parties from $P_1, P_2, \ldots P_n$. Given dataset D is partitioned into k number of datasets from $D_1, D_2, \ldots D_k$ and each partition is denoted as $D_1 = (E_1, I_1) \ldots, D_n = (E_n, I_n)$ respectively. Here we present both horizontal and vertical partitioning of any dataset [2, 19].

- **Horizontal Partitioning**: It assumes parties from different locations extract same set of features I_n for dissimilar set of entities E_n, that means a dataset is divided into n partitions where each partition holds set of horizontal records (rows). For example all banks gather similar information of customers, though the purposes of account holders and the bank may be different.
- **Vertical Partitioning**: It assumes that parties from different locations collect dissimilar set of feature I_n for the same set of entities E_n, which means a dataset is divided into n where each partition holds set of vertical records (columns). For example a hospital and an insurance company gather collect medical records of a same person to be used for different purposes (Fig. 1).

4 Clustering Using Self Organizing Map (SOM)

Clustering is a predictive approach where a larger dataset brought into smaller levels of groups with similarity in their features that are specified by the clustering approach. Clustering using Self Organizing Map (SOM) is a self supervised learning method in neural networks domain through which input dataset clustered into possible number of clusters based on the euclidean distance between the neurons of various data tuples [10]. The output layer of SOM appears as a grid map most commonly in a 2-dimensional, rectangular or hexagonal structure map interconnected

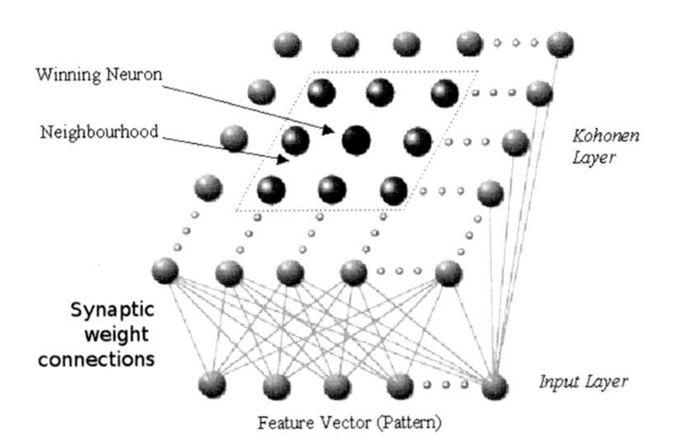

Fig. 2 Kohonens SOM model

with neurons, hence it is known as a Self Organizing Feature Map. SOM is a feed forward neural network introduced by Kohonen with ability of dimensionality reduction, consists one input layer and one output computation layer denoted as kohonen's layer, where all neurons are arranged in rows and columns that are fully connected to input layer source nodes [12].

Interesting aspect of this unsupervised learning model is competitive learning method, through which neurons participate in competition among themselves to be activated at one instance of time and the activated neuron after competing is called as the winning neuron, then the next level of learning is allowed only for the winning neuron. This algorithm is referred as winner-takes-all learning algorithm for which all neurons are forced to organize themselves (self organized) to involve in competition and get activated by winning the competition, and for these reasons the network is called as Self Organized Map [6] (Fig. 2).

For a given network with ***a-b*** configuration, a single input vector is denoted as $[x_1, x_2, \ldots, x_a]$, output vector nodes denoted as $[o_1, o_2 \ldots, o_b]$, and w^o_{ij} denotes the weight vector, connecting j to the output layer node i, where $1 \leq j \leq$ a, $1 \leq i \leq$ b. Clustering performed in SOM in following phases:

1. **Competition**: Competition held among neurons and neuron with largest activation value is decided as the winner neuron. Suppose,

   x input vector $\boldsymbol{x} = \lfloor x_1, x_2, \ldots, x_m \rfloor$ which is (m dimensional).

   w_j weight vector of neuron j connected to input.

   $j = 1, 2, 3\ldots l$ (l-number of output neurons).
   Declaring Winner Neuron: The neuron with the largest inner product $w_j \cdot x$ or neuron with minimum euclidean distance with respect to the input is declared as winner using equation:

$$i(x) = argmin_j \|x_m - w_j\| \tag{1}$$

$$j = 1, 2, 3 \ldots l.$$

i is best-matching or winning neuron for input vector x.

2. **Co-operation**: Spatial location of topological neighbourhood of winner neuron is determined using neighbourhood function $h_{j,i(x)}$ for input vector x which is given as:

$$h_{j,i(x)} = \exp\left(-\frac{d_{j,i}^2}{2\sigma^2}\right) \tag{2}$$

$h_{j,i}$ topological neighbourhood centred on winning neuron i.

$d_{j,i}$ lateral distance between winning neuron i excited neuron j.

$h_{j,i(x)}$ monotonically decreases with the lateral distance $d_{j,i}$.

3. **Weight Updation**: Weights are updated for winning neuron and it's neighbour neurons using equation:

$$w_j(n+1) = w_j(n) + \eta(n)h_{j,i(x)}(n)\left(x_d - w_j\right) \tag{3}$$

x_d is the input pattern for $d = 1, 2, 3 \ldots m$

$h_{j,i(x)}(n)$ is the neighbourhood function.

$w_j(n)$ is a decay term to stop weights going to infinity, each update includes a small degree of decay of previous weight value.

$\eta(n)$ is learning rate parameter, that should decrease gradually with time n. A simple schedule of learning rate is:

$$\eta\,(n) = \eta_0 \exp\left(-\frac{n}{\tau_2}\right) \text{ for } n = 0, 1, 2, \ldots$$

the constant τ_2 determines the slope of the graph of $\eta\,(n)$ against n making τ_2 larger which makes the learning rate $\eta\,(n)$ decrease very slowly.

5 Privacy-Preserving Collaborative Clustering Using SOM (PPCSOM)

In this privacy preserving collaborative clustering SOM (PPCSOM) method, the dataset to be clustered is first horizontally partitioned and distributed between n number of parties to form C number of clusters, with which each data partition is owned and resides at each party, then all parties cluster their own data while collaborating with each other without directly revealing their private information. PPCSOM method is explained in a stepwise process as follows:

1. First the dataset is horizontally partitioned between n parties, then parties(data owners) decide c number of clusters, and determine the sequence of active parties to be followed in complete clustering process.
2. The first active party(say p_1) is represented as Initial Party(IP) which first assigns w_j vectors for all c clusters and declare required constant parameters.
3. Now IP(p_1) select a random user among all users it holds, then start clustering by deciding winner using Eq. 1, and updates weights using Eq. 3, then increases step s by one.
4. p_1 repeats step 2 and 3 unless all it's users are allocated to a cluster, then sends the final w_j vectors and increased s value to next party(say p_2), which is presently declared as an active party in the sequence.
5. Now the present active party p_2 repeats step 2 and 3 same as p_1, until all users held at p_2 are allocated to a cluster, then it sends new s value and updated w_j vectors to the next party.
6. Each party do the same as previous parties does till all their users are allocated to clusters, and an epoch is completed when all parties allocate their users to a cluster, then last party updates w_j vectors and send to the IP(p_1).
7. Setps 3–6 are continued until no noticeable change in the future map, which means all parties completed assigned their data to clusters.

6 Working Process of PPCSOM

Dataset in this method is horizontally partitioned and distributed between multiple parties. For example if data set with say 1000 records is horizontally partitioned between two parties say p_1 and p_2 each party consists of equal number of records (500 at each party). Now we apply clustering on this data resided at each party, using PPCSOM method [4, 8]. First the Initial party p_1 chooses w_j weight vectors and update them n number of times and sends the updated w_j vector to next active party p_2, this is done after assigning data at that party p_1 to a cluster. Now party p_2 do the same as party p_1 did, then sends updated weight vector w_j to next party say p_n and finally party p_n do the same and sends updated weights to p_1. This entire process will be continued until each data attribute which has been horizontally partitioned is assigned to any one of the clusters (Fig. 3).

Here all parties assign their own data to any cluster based on the winning neuron decided by the minimum euclidean distance between neighbouring neurons of SOM at kohonen layer [12] using Eq. 1, then weight vector w_j is updated using Eq. 3. All parties decide winning neurons and updates their weight vectors at their private active state to assign their own data to the various clusters, and sends updated w_j vector to the next party, hence next party cannot learn the true values of an individ-

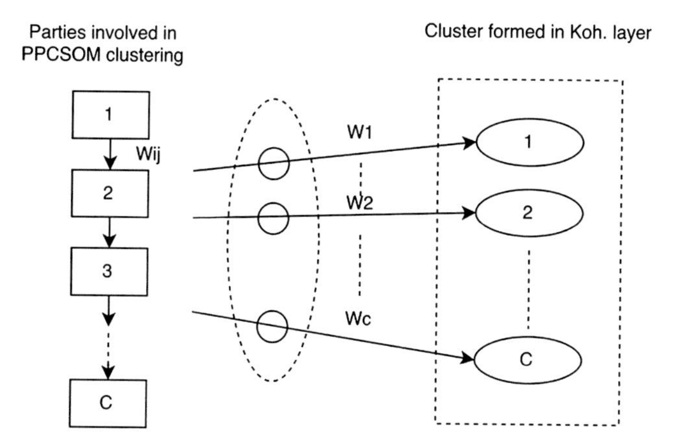

Fig. 3 Process of privacy preserving collaborative clustering using SOM

ual user held by previous party. There will not be a serious problem though a next party knows number of users held by a preceding party, because next active party can't get information like, specifically rated items by a specific user and the distance values between users and w_j vectors.

This is the main privacy preserving way which is also applicable for all parties involved in this collaborative clustering process. Ultimately in PPCSOM method all parties exchange only updated w_j vectors from their own locations without colluding with the previous or next parties in the sequence. Always an updated weight vector w_j consists changed weight values while sending to other party, hence parties cannot learn private details of users held by any other previous party, hence PPCSOM process assures preserving privacy while exchanging information without revealing private information to any party to determine their clusters.

7 Results Analysis

Our experimental results are derived through built-in SOM tool in matlab, the results shows how performance accuracy differs due to collaboration and privacy preservation among multiple number of parties [11]. Experiments held using 4 different datasets adopted from UCI Machine learning repository. We expected our method to perform collaborative clustering among number of parties that are priorly specified before starting the entire process of PPCSOM clustering. Priorly specified number of parties from 1 to 5 as basic level then we increased number of parties to be collaborated to form clusters with their own data residing at their own locations. We also pre specified how many number of clusters to be formed at the end of this process. Each dataset is horizontally partitioned between specified number of collaborating

Table 1 Description of datasets

Dataset	No of instances	No of attributes	No of classes
Iris	150	4	3
Glass	214	10	7
Wine	278	13	3
Pima Indians	768	8	2

Table 2 Average time and error rate comparison of PPCSOM with SOM

Dataset	SOM-Error	SOM-AvgTime	PPCSOM-Error	PPCSOM-AvgTime
Iris	22.18	4.23	23.92	6.23
Glass	27.36	17.33	29.01	23.39
Wine	21.12	19.56	23.12	25.25
Pima Indians	29.36	67.23	33.56	89.23

parties, then PPCSOM method is applied to get combined results from every party involved in clustering.

Datasets (Iris, Glass, Wine, and Pima Indians) used in our experiments from UCI-ML repository, given in Table 1 with their description, Table 2 presents performance ratings in the form of average time(in seconds) and error rate(in percentage) of our PPCSOM and non privacy preserving SOM algorithm. When parties(data owners) decide to collaborate with each other in forming clusters, they can achieve better results with an assured privacy using this method. Average time taken to complete the process for PPCSOM and non PPCSOM are almost similar, the similarity appears for error rate also.

Individual process time has been increased when number of parties and partitions are more at the time of producing collaborative clusters, but this increase rate is considerable,as this difference is appeared as per size of dataset and number of records held at each party. All the experiments shown good results in achieving considerable accuracy which has been measured in the form of Average time taken to complete the process. Privacy preserving collaborative clustering using SOM method is compared with the existing non privacy preserving SOM algorithm, as our focus was mainly on preserving privacy while parties collaborating with each other in forming clusters using PPCSOM model, hence the parties can preserve privacy of their data with better performance (Fig. 4).

7.1 Performance Analysis

We compare our approach with non privacy preserving SOM Algorithm, to evaluate the clustering accuracy of PPCSOM while horizontally distributing data among

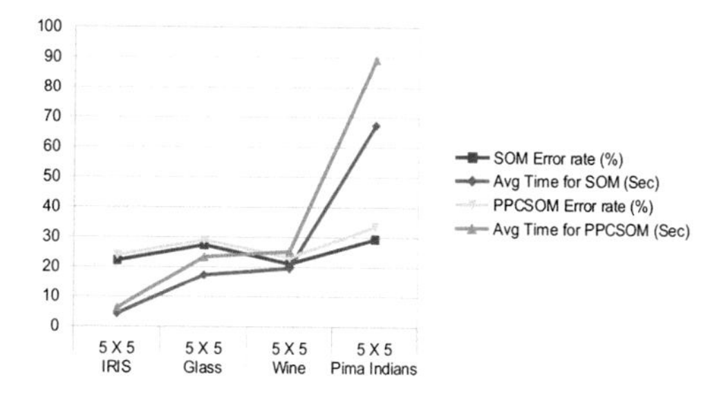

Fig. 4 Error rate and average time comparison graph for both SOM and PPCSOM

multiple parties distributed with various equal sized partitions among them. As our method is aimed for privacy preserving in collaborative model, there is much possibility of increase in communication and computational costs, that are inevitable as all these performance measures conflict with each other [5]. To compare our results with non privacy preserving algorithm we used the performance measures as average time taken and error rate observed for each dataset for different number of parties and partitions.

Experimental results emphasize that it is possible to achieve privacy while undertaking collaborative computations among multiple parties using SOM clustering that we refer as collaborative SOM clustering. Results are most similar when compared with non privacy preserving SOM which doesn't involve with collaborative approach [8, 16]. We could come up with acceptable performance for off line communication, later if applied online the results may show unexpected communication and computation costs still privacy is preserved. Hence we suggest to conduct only off line computations at initial stage of verifying performance of the method to achieve better accuracy online. Overall PPCSOM process shown acceptable communication cost due to collaboration between multiple parties, when compared with non privacy preserving SOM Algorithm.

7.2 *Privacy Analysis*

Privacy is defined by Lindell and Pinkas [14] in terms of distributed data mining, no individual should learn any information about a user other than its output. Hence the same is adopted for our method, where the parties can't learn the exact values of their private information belongs to any other party. To analyse privacy of our proposed method, we identified some possible attacks on privacy may be caused by the parties involved in this collaborative clustering process, which can be harmful to the confidentiality of other parties. We came up with some solutions for such type of attacks also shown how we assure better privacy compared with non privacy preserving. The main target while implementing this work has been achieved promising better privacy in distributive collaborative clustering environment with less communication cost, within acceptable error rate and average time.

Our work of this paper preserves privacy of data owners. Parties involved are able to cluster without exchanging private information, as they exchange only updated weight vectors and s values in entire process. We assure that collaborative clustering in SOM does not lead to any privacy and legal issues [13, 20]. Parties send their own data to any cluster based on winning neuron decided by the minimum euclidean distance between neighbouring neurons of SOM at kohonen layer. Weights are updated at each party in their own locations. This is a direct approach where parties doesn't directly send their information to each other, the only information is exchanged or revealed is the updated weight vector w_j which can't effect ones individual information of any party. There is no data loss or privacy violation observed due to distribution of data while clustering [15]. with this means PPCSOM method is able to assure better privacy.

8 Conclusions

We mainly focused and addressed the privacy issue in distributed environment through a neural network called SOM. Verified the performance and accuracy of method we proposed in this paper. Presented Privacy preserving collaborative SOM clustering method applied through SOM neural network while data is distributed in number of horizontal partitions between multiple parties, where parties share decided no of data partitions and collaboratively construct clusters. The overall goal of our work is to securely construct clusters using SOM without violating privacy and with an assured less communication overhead at the time of distributing and collaborating between parties. Our proposed method could acquire better accuracy with an assured privacy with less communication overhead though most of the privacy issues make accuracy losses. Hence we conclude that our PPCSOM method working well for horizontal partitioning of data among multiple parties to form clusters with assured privacy and better performance.

9 Future Scope

The same work can be applied for vertical and arbitrary ways of data partitioning. Collaborations in PPCSOM can be enhanced for better online performance. In Future we undertake investigations for vertical partitioning adopted for cryptographic approaches to secretly share the information among huge number of parties as the current use of information in general is rapidly growing. We have plan to build a secure neural network learning model which can be applied for other data mining techniques available. We may undertake work to modify the present recommender system and come up with a new secured collaborative system.

References

1. Agrawal, R., Srikant, R.: Privacy-preserving data mining. SIGMOD Rec. **29**(2), 439–450 (2000). http://doi.acm.org/10.1145/335191.335438
2. Ceri, S., Negri, M., Pelagatti, G.: Horizontal data partitioning in database design. In: Proceedings of the 1982 ACM SIGMOD International Conference on Management of Data, pp. 128–136. SIGMOD '82, ACM, New York, NY, USA (1982). http://doi.acm.org/10.1145/582353.582376
3. Clifton, C., Kantarcioglu, M., Vaidya, J., Lin, X., Zhu, M.: Tools for privacy preserving distributed data mining. ACM SIGKDD Explor. **4**(2), 28–34 (2002). http://www.cs.purdue.edu/homes/clifton/DistDM/kddexp.pdf
4. Gorgônio, F.L., Costa, J.A.F.: Parallel self-organizing maps with application in clustering distributed data. In: IEEE International Joint Conference on Neural Networks, 2008. IJCNN 2008. (IEEE World Congress on Computational Intelligence), pp. 3276–3283. IEEE (2008)
5. Han, S., Ng, W.-K.: Privacy-preserving self-organizing map. In: Song, I.-Y., Eder, J., Nguyen, T.M. (eds.) DaWaK 2007. LNCS, vol. 4654, pp. 428–437. Springer, Heidelberg (2007)
6. Haykin, S.: Neural Networks: A Comprehensive Foundation, 2nd edn. Prentice Hall PTR, Upper Saddle River, NJ, USA (1998)
7. Jagannathan, G., Wright, R.N.: Privacy-preserving distributed k-means clustering over arbitrarily partitioned data. In: Proceedings of the Eleventh ACM SIGKDD International Conference on Knowledge Discovery in Data Mining, pp. 593–599. KDD '05, ACM, New York, NY, USA (2005). http://doi.acm.org/10.1145/1081870.1081942
8. Kaleli, C., Polat, H.: Privacy-preserving som-based recommendations on horizontally distributed data. Knowl. Based Syst. **33**, 124–135 (2012)
9. Kantarcioglu, M., Clifton, C.: Privacy-preserving distributed mining of association rules on horizontally partitioned data. In: ACM SIGMOD Workshop on Research Issues on DMKD'02 (June 2002). http://citeseer.nj.nec.com/kantarcioglu02privacypreserving.html
10. Kantarcioglu, M., Clifton, C.: Privacy-preserving distributed mining of association rules on horizontally partitioned data. IEEE Trans. Knowl. Data Eng. **16**, 1026–1037 (2004). http://www.cs.purdue.edu/homes/jsvaidya/pub-papers/kdd02.pdf
11. Kargupta, H., Das, K., Liu, K.: Multi-party, privacy-preserving distributed data mining using a game theoretic framework. In: Kok, J.N., Koronacki, J., Lopez de Mantaras, R., Matwin, S., Mladenič, D., Skowron, A. (eds.) PKDD 2007. LNCS (LNAI), vol. 4702, pp. 523–531. Springer, Heidelberg (2007)
12. Kohonen, T., Oja, E., Simula, O., Visa, A., Kangas, J.: Engineering applications of the self-organizing map. Proc. IEEE **84**(10), 1358–1384 (1996)
13. Lin, X., Clifton, C., Zhu, M.: Privacy preserving clustering with distributed EM mixture modeling. Knowledge and Information Systems (to appear)
14. Lindell, Y., Pinkas, B.: Secure multiparty computation for privacy-preserving data mining. J. Priv. Confidentiality **1**(1), 5 (2009)
15. Merugu, S., Ghosh, J.: Privacy-preserving distributed clustering using generative models. In: The Third IEEE International Conference on Data Mining (ICDM'03). Melbourne, FL (2003). http://citeseer.ist.psu.edu/705042.html
16. Pedrycz, W., Vukovich, G.: Clustering in the framework of collaborative agents. In: Proceedings of the 2002 IEEE International Conference on Fuzzy Systems, 2002. FUZZ-IEEE'02, vol. 1, pp. 134–138. IEEE (2002)
17. Sanil, A., Karr, A., Lin, X., Reiter, J.: Privacy preserving regression modelling via distributed computation. In: 10th ACM SIGKDD International Conference on Knowledge Discovery and Data Mining. Seattle, WA (2004). http://www.niss.org/dgii/TR/secure-coefficients.pdf
18. Strehl, A., Ghosh, J.: Cluster ensembles—a knowledge reuse framework for combining multiple partitions. J. Mach. Learn. Res. **3**, 583–617 (2003)
19. Vaidya, J., Clifton, C.W., Zhu, Y.M.: Privacy preserving data mining, vol. 19. Springer Science & Business Media, Berlin (2006)

20. Yang, Y., Tan, W., Li, T., Ruan, D.: Consensus clustering based on constrained self-organizing map and improved cop-kmeans ensemble in intelligent decision support systems. Know. Based Syst. **32**, 101–115 (2012). http://dx.doi.org/10.1016/j.knosys.2011.08.011
21. Zhang, S., Ford, J., Makedon, F.: A privacy-preserving collaborative filtering scheme with two-way communication. In: Proceedings of the 7th ACM Conference on Electronic Commerce, pp. 316–323. EC '06, ACM (2006). http://doi.acm.org/10.1145/1134707.1134742

A Hybrid Approach to Classification of Categorical Data Based on Information-Theoretic Context Selection

Madhavi Alamuri, Bapi Raju Surampudi and Atul Negi

Abstract Clustering or classification of data described by categorical attributes is a challenging task in data mining. This is because it is difficult to define a measure between pairs of values of a categorical attributes. The difficulty arises due to lack of ordering information between various pairs of categorical attributes. In this paper we introduce a Hybrid Approach which combines set based context selection with distance computation using KL divergence method. In the literature context based approaches have been introduced recently. Current approaches look at categorical attributes individually, however our approach proposes a novel scheme inspired from information theory. We consider the interdependence redundancy measure to select the significant attributes for context selection. The proposed approach gives encouraging results for low dimensional benchmark UCI datasets with k-nearest neighbor classifier based on the proposed measure. On these datasets the proposed measure performed well in comparison to other distance measures while using various classifiers such as SVM, Naive Bayes and C4.5.

Keywords Categorical data · Similarity · Context · Classification

1 Introduction

Similarity or distance between two objects plays a significant role in many data mining and machine learning tasks like classification, clustering and outlier detection. In general distance computation is a built-in step for these learning algorithms and different distance measures can be conveniently used. However the effectiveness of

M. Alamuri (✉) · B.R. Surampudi · A. Negi
School of Computer and Information Sciences, University of Hyderabad, Hyderabad, India
e-mail: madhavi_alamuri@yahoo.com

B.R. Surampudi
Cognitive Science Lab, International Institute of Information Technology, Hyderabad, India

V. Ravi et al. (eds.), *Proceedings of the Fifth International Conference on Fuzzy and Neuro Computing (FANCCO - 2015)*, Advances in Intelligent Systems and Computing 415, DOI 10.1007/978-3-319-27212-2_22

the proposed distance measure/metric usually has significant influence on the tasks like classification and clustering.

For numerical data sets, the distance computation is well understood and most commonly used measures such as minkowski distance, mahalanobis distance can be applied. However, measuring similarity or distance is not straight forward for categorical data sets as there is no explicit ordering of categorical values and it is very difficult to determine how much one symbol differs from another. By categorical data we mean the values that are nominal e.g. $color = \{black, blue, green\}$ or ordinal e.g. $size = \{small, large, verylarge\}$.

In this paper we study the similarity/distance measures for categorical data objects and propose a hybrid approach, based on Set based Context Selection (SBCS) and distance computation on the context using KL divergence method. In our Hybrid Approach (HA) context for each attribute refers to the subset of attributes which gives some contextual interpretation over the attribute set.

The proposed approach can well quantify the distance in supervised learning environment. In this paper we also focus on a data driven methods for selecting a good context for a given attribute. We provide information theoretic approaches for context selection of each and every attribute. The underlying assumption in our approach is that if the similarity function has high value for the given context, then the objects represented by the given context description are similar. Recently, increasing attention is being paid to find the grouping structure/classification of categorical data.

The rest of the paper is organized as follows: In Sect. 2 we discuss the state of the art in categorical similarity/dissimilarity measures. In Sect. 3 we present the theoretical details of Hybrid Approach. In Sect. 4 we present the technical details and in Sect. 5 we present the results of set of experiments on low dimensional UCI benchmark datasets.

2 Background

According to Michalski [16], the conventional measures of similarity are "*Context-free*" i.e. the distance between any two data objects X and Y is a function of these points only, and does not depend on the relationship of these points to other data points.

$$Similarity(X, Y) = f(X, Y)$$

Context-free similarity measures may be inadequate in some clustering/classification problems. Hence recent approaches for finding similarity measures include "*Context-sensitive*" methods, where

$$Similarity(X, Y) = f(X, Y, Context)$$

Here the similarity between X and Y depends not only on X and Y, but also on the relationship of X and Y to other data points, represented by Context.

In a user driven approach context refers to the subset of attributes of interest and is application dependent.

Pearson's [17] Chi-square statistic was often used in the late 1800s to test the independence between categorical variables in a contingency table. Sneath and Sokal [19] were among the first to put all the similarity measures together. The conventional methods of similarity measures used to binarize the attribute values where bit 1 indicates the presence and bit 0 indicates absence of a possible attribute value. After obtaining binary vectors, binary similarity measures are used to apply on them. The major drawback is transforming objects into a binary vector may leave many minute insights in to the dataset. The most Commonly used similarity measures for categorical data is overlap, where similar values are assigned a distance of 1, and dissimilar values are assigned a distance of 0. The drawback with this measure is, it does not distinguish the different values taken by an attribute.

In general the similarity measures for categorical data are categorized based on how they utilize the context of the given attributes. There are several supervised and unsupervised measures from the literature are existing to find the similarity between categorical feature values.

- The supervised similarity measures do consider the class attribute information. The supervised measures are further divided into learning, non-learning approaches. The learning approaches can be IVDM [21], WVDM [21], Learned Dissimilarity measure [22]. The non-learning approaches can be VDM [20], MVDM [7].
- The unsupervised similarity measures do not consider the class attribute information. These measures are further classified into Probabilistic, Information theoretic, and Frequency based approaches.
  - Probabilistic approaches: Goodall [9], Smirnov [18], Anderberg [2].
  - Information theoretic approaches: Lin [15], Lin1 [4], Burnaby [6].
  - Frequency based approaches: OF [11], IOF [11].

Boriah et al. [4] classified the similarity measures based on the storage mechanism of similarity values in the similarity matrix, parameters used to propose the measure and based on the weight of the frequency of the attribute values.

Jierui Xie [22] proposed a learning algorithm, which learns a dissiimilarity measure by mapping each categorical value into random numbers. This learning algorithm is guided by the classification error for effective classification of the data points.

The taxonomy of various distance/similarity measures for categorical data is explored in [1].

Apart from these categories of similarity measures, recent attention has been paid to context-based approaches. Dino et al. [10] present a context based approach to compute distance between pair of values of a categorical data. They proposed a distance learning framework for context selection and validated in a hierarchical clustering algorithm.

Zeinab et al. [12] proposed a novel approach for distance learning based on the context information. This method is also used to compute dissimilarity between probability distributions. Both these approaches use the context information for dissimilarity computation.

We extend the current context based approaches by introducing a Hybrid Approach which utilizes information-theoretic measures. The proposed approach is explored in Sect. 3.

3 Proposed Hybrid Approach

In this section we present a Hybrid Approach for computing distance between any pair of values of a categorical attribute. We also introduce "Set Based Context Selection Algorithm" (SBCS) for the effective selection of the context followed by calculating the distance by using KL divergence [13, 14] method. Our Hybrid Approach is formulated based on the following two steps.

1. Set Based Context Selection: This method selects a subset of correlated features based on a given attribute. i.e., selection of a meta attribute set of a given attribute which is relevant in terms of information theoretic measure is calculated in this step.
2. Distance Computation: Computation of the distance measure between pair of values of an attribute using the meta attributes set defined in the previous step. KL divergence method is applied on the context to measure the difference between the probability distributions.

The essential premise in formulating this algorithm is with an open minded, fairness in the importance of the attributes, it excludes weightage or bias towards a certain set of attributes, unless it is explicitly defined in the context.

The notations used in this algorithm are as follows: Consider the set $F = \{A_1, A_2, \ldots A_m\}$ of m categorical attributes and let the set of instances set $D = \{X_1, X_2 \ldots X_n\}$. We denote by a lower case letter $a_i \epsilon A_i$, a specific value of an attribute A_i.

3.1 Set Based Context Selection

The selection of a good context is not trivial, when data are high dimensional. In the SBCS, we use mutual information normalized with joint entropy to get the context for each and every attribute.

The Set Based Context Selection Algorithm which we propose considers a score for each feature independently of others. A useful and relevant set of features may not only be individually relevant but also may not be redundant with respect to each other. The selecting criterion of a context is based on the relevance index which quantifies whether a particular feature can be included in a context set or not.

In the following sub section we introduce some basic concepts of information theoretic measures followed by how they are utilized to tackle the problem of context selection.

3.2 Entropy and Mutual Information

The *entropy* of a random variable [8], is the fundamental unit of information which quantifies the amount of uncertainty present in the distribution of the random variable.

The entropy of a random variable A_i is defined as,

$$H(A_i) = -\sum_{k \epsilon A_i} p(a_k^i) log_2 p(a_k^i) \quad (1)$$

where $p(a_k^i)$ is the probability of value of a_k of attribute A_i.

The entropy of a random variable can be conditioned on other variables. The conditional entropy of A_i given A_j is,

$$H(A_i|A_j) = -\sum_{k \epsilon A_j} p(a_k^j) \sum_{l \epsilon A_i} p(a_l^i|a_k^j) log_2 p(a_l^i|a_k^j) \quad (2)$$

where $p(a_l^i|a_k^j)$ is the probability that $A_i = a_l$ after observing the value $A_j = a_k$. This can be interpreted as the amount of uncertainty present in A_i after observing the variable A_j.

The amount of Information shared between A_i and A_j, which is also called as mutual information is defined by,

$$I(A_i; A_j) = H(A_i) - H(A_i|A_j) \quad (3)$$

This is the difference between two entropies which can be interpreted as the amount of uncertainty in A_i which is removed by knowing A_j.

The mutual information between two attributes also measures the average reduction in uncertainty with another attribute. A smaller value of mutual information indicates lesser dependence and a larger value of mutual information indicates greater dependence.

The main drawback of using this measure is that the mutual information value increases with the number of distinct values that can be chosen by each attribute. To overcome this problem Au et al. [3] proposed interdependence redundancy measure where mutual information is normalized with joint entropy, which is defined as,

$$IDR(A_i, A_j) = \frac{I(A_i; A_j)}{H(A_i, A_j)} \quad (4)$$

where the joint entropy $H(A_i, A_j)$ is calculated as,

$$H(A_i, A_j) = -\sum_{k \epsilon A_i} \sum_{l \epsilon A_j} p(a_k^i, a_l^j) log_2 p(a_k^i, a_l^j) \tag{5}$$

According to [3] the interdependence measure evaluates the degree of dependency between two attributes. Unlike mutual information, where the number of possible values which an attribute can take effect, the interdependency measure has no effect on the number of distinct values taken by an attribute. Hence IDR measure is considered as more ideal index to rank the attributes in terms of dependency. $IDR(A_i, A_j) = 1$ means that the attributes A_i and A_j are dependent on each other while $IDR(A_i, A_j) = 0$ indicates that the attributes are statistically independent. When the value of IDR lies between 0 and 1 the attributes are partially independent. By using this IDR measure, we can maintain a $m \times m$ matrix IDR to store the dependency degree of pair of attributes.

For each and every attribute A_i we find all the attributes that have interdependency with it and store them in a context set. In order to not to unnecessarily increase the size of the context set we introduce a threshold t to include the significant attributes in the context set.

$$context(A_i) = \left\{A_k | IDR(A_i; A_k > t, A_i, A_k \epsilon F)\right\} \tag{6}$$

It is being conjectured that for lower values of threshold t we may obtain higher values of classification accuracy. However for larger values of the threshold the classification accuracy may drop.

The relationship between these quantities is explored in [5] and can be observed from the Fig. 1.

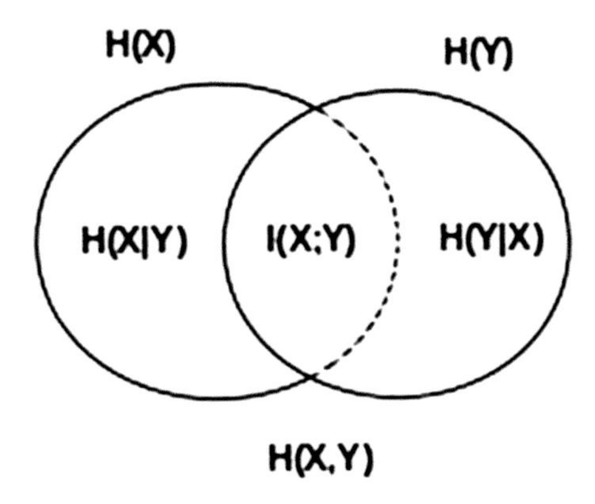

Fig. 1 Relationship of various information-theoretic measures

To get the context of a given attribute we make use of these information-theoretic measures by adding/removing the relevant features in the context set.

3.3 Distance Computation

The distance between pair of values (x_j, y_j) of a categorical attribute A_j is formulated using KL divergence method as,

$$d_{A_j}(x_j, y_j) = \sum_{A_i \epsilon context(A_j)} \sum_{v_i \epsilon A_i} p(v_i|x_j) log \frac{p(v_i|x_j)}{p(v_i|y_j)} + p(v_i|y_j) log \frac{p(v_i|y_j)}{p(v_i|x_j)} \quad (7)$$

The distance defined above depends on the meta attribute set associated to each attribute, where meta attribute set is derived from the SBCS algorithm. The dissimilarity between two probability distributions using KL divergence is symmetric, and hence the distance between pair of values of an attribute is also symmetric.

Let X and Y be two instances of the dataset then the above calculated pairwise distance between attribute values is embedded in the total distance as,

$$D(X, Y) = \sum_{j=1}^{m} d_{A_j}(x_j, y_j) \quad (8)$$

4 Hybrid Approach Implementation

In this section we introduce the algorithmic details for the implementation of (i) Set Based Context Selection and (ii) Distance Computation of Hybrid Approach.

In Algorithm 1 we propose interdependence redundancy based context selection for each attribute in the feature set. Initially the selected set S is empty and the unselected set US is the set of all features from the feature set. At line 6 this algorithm selects an attribute to be included in the context set based on the threshold t and at line 8, it deselects the attribute from the unselected set US. When the context is chosen for a given attribute Distance Computation function computes the distance matrix between each pair of values of the attribute.

In Algorithm 2 distance measure is computed between pair of values of $A_i \epsilon F$ using context set derived from the first step of the Hybrid Approach.

The total distance between two objects is then calculated by using Eq. 8 defined in Sect. 3.3.

Algorithm 1 Hybrid Algorithm

1: **procedure** HYBRID APPROACH(D, F)
2: selectedSet S = { };
3: unselectedSet US = $\{A_1, A_2...A_m\}$;
4: For each pair of attributes $(A_i, A_j), (i,j)\epsilon\{1, 2, ..., F\})$ calculate $IDR(A_i, A_j)$ according to Eq. 4
5: **for** `all` $A_i \epsilon F$ **do**
6: a. Find the feature A_k from the US such that $IDR(i, k) > t$
7: `b. insert(`A_k`,S)`
8: `c. remove(`A_k`,US)`
9: `d. contextSet(`A_i`) = S`
10: **end for**
11: $DistanceMatrix_{A_i} = calculateDistance(A_i, contextSet(A_i))$;
12: **return** $DistanceMatrix_{A_i}$
13: **end procedure**

Algorithm 2 Distance Computation

1: **procedure** CALCULATEDISTANCE(A_j, $Context(A_j)$)
2: **for** `all` $x_j, y_j \epsilon A_j$ **do**
3: **if** $x_j \neq y_j$ **then**
4:

$$d_{A_j}(x_j, y_j) = \sum_{A_i \epsilon context(A_j)} \sum_{v_i \epsilon A_i} p(v_i|x_j) log \frac{p(v_i|x_j)}{p(v_i|y_j)} + p(v_i|y_j) log \frac{p(v_i|y_j)}{p(v_i|x_j)}$$

5: **else**
6: $d_{A_j}(x_j, y_j) = 0$
7: **end for**
8: **return** d_{A_j}
9: **end procedure**

5 Experimental Results

To evaluate the proposed Hybrid Approach, we compare our approach with other base-line similarity measures explored in Sect. 2 and with the other classifiers. We present results on 5 benchmark categorical datasets, which are taken from the UCI machine learning repository (Table 1).

5.1 Results Description

We compare our Hybrid Approach with the 5 similarity measures Overlap, Lin, Gambaryan, OF, IOF described in Sect. 2. We evaluate the classification accuracy of the nearest neighbor classifier (k = 7) with five fold cross validation. The last row of the Tables 2 and 3 gives the average performance over all the datasets. In summary

Table 1 Dataset description

Dataset	Size	Dimension	Attributes and symbols	No.of classes
Mushroom	8124	22	Various sizes from 2 to 12, e.g. $cap - surface = \{fibrous, grooves, scaly, smooth\}$	2
Tic-tac-toe	958	9	Each attribute takes on $\{x, o, b\}$	2
Balance scale	625	4	Each attribute takes on $\{1, 2, 3, 4, 5\}$	3
Car evaluation	1728	6	Each attribute takes different values e.g. $buying = \{vhigh, high, med, low\}$	4
Hayes-Roth	160	5	Each attribute takes on different values e.g. $hobby = \{1, 2, 3\}$	3

Table 2 Performance comparison with various similarity measures with knn (k = 7)

Dataset	Overlap	Lin	Gambaryan	OF	IOF	HA
Mushroom	**100**	98.75	53.00	98.9	99.95	97.23
Tic-tac-toe	82.35	**97.30**	80.45	72.49	95.13	92.34
Balance Scale	72.31	72.21	72.32	73.59	72.34	**85.12**
car Evaluation	88.34	**93.10**	83.20	92.43	87.13	90.46
Hayes-Roth	68.50	71.00	67.52	60.00	65.50	**83.55**
Average	82.3	86.472	71.298	79.482	84.01	**89.74**

Table 3 Performance comparison with various classifiers

Dataset	SVM	NB	C4.5	HA
Mushroom	**100**	96.5	**100**	98
Tic-tac-toe	77	75.34	84.12	**92**
Balance scale	95.5	**96.3**	73	88.21
Car evaluation	88.2	92.1	**96.51**	95.35
Hayes-Roth	64	68.5	71	**80**
Average	84.94	85.748	84.926	**90.712**

the proposed Hybrid Approach achieves best rank in two datasets namely Balance Scale and Hayes Roth and stands best on average classifier accuracy.

We also compare our Hybrid Approach with the algorithms implemented in Weka 3.6.10 including SVM, C4.5 and Naive Bayes (NB). Our method uses the set based context selection with KL divergence as a distance measure whereas the other methods use the Euclidean distance with simple matching between categorical objects. As shown in Table 3, the Hybrid Approach performs better in Tic-Tac-Toe and Hayes-Roth and stands best in average performance with 5 datasets.

6 Conclusions

In this paper we propose a hybrid approach to measure similarity between categorical attribute values. This algorithm uses *Set Based Context Selection* method inspired from information-theoretic measures. We tested our approach on five benchmark datasets from UCI machine learning repository. The proposed approach gives promising results for low-dimensional datasets.

The proposed approach gives superior results for datasets with dimensionality approximately below 10. Computation of context selection is very expensive for high dimensional datasets. This is a limitation of the proposed approach. We are exploring generalization of dimensionality reduction techniques for categorical attributes so that the proposed approach can be combined with these methods in future. We also investigate the impact of the threshold parameter t on the proposed distance measure for large datasets in future.

References

1. Alamuri, M., Surampudi, B.R., Negi, A.: A survey of distance/similarity measures for categorical data. In: 2014 International Joint Conference on Neural Networks (IJCNN), pp. 1907–1914. IEEE (2014)
2. Anderberg, M.R.: Cluster analysis for applications. Technical report, DTIC Document (1973)
3. Au, W.H., Chan, K.C., Wong, A.K., Wang, Y.: Attribute clustering for grouping, selection, and classification of gene expression data. IEEE/ACM Trans. Comput. Biol. Bioinf. (TCBB) **2**(2), 83–101 (2005)
4. Boriah, S., Chandola, V., Kumar, V.: Similarity measures for categorical data: a comparative evaluation. In: Proceedings SIAM International Conference on Data Mining. SIAM, Atlanta (2008)
5. Brown, G., Pocock, A., Zhao, M.J., Luján, M.: Conditional likelihood maximisation: a unifying framework for information theoretic feature selection. J. Mach. Learn. Res. **13**(1), 27–66 (2012)
6. Burnaby, T.P.: On a method for character weighting a similarity coefficient, employing the concept of information. J. Int. Assoc. Math. Geol. **2**(1), 25–38 (1970)
7. Cost, S., Salzberg, S.: A weighted nearest neighbor algorithm for learning with symbolic features. Mach. Learn. **10**(1), 57–78 (1993)
8. Cover, T.M., Thomas, J.A.: Elements of Information Theory. John Wiley & Sons, New York (1991)
9. Goodall, D.W.: A new similarity index based on probability. Biometrics **22**(4), 882–907 (1966)
10. Ienco, D., Pensa, R.G., Meo, R.: From context to distance: learning dissimilarity for categorical data clustering. ACM Trans. Knowl. Discov. Data (TKDD) **6**(1), 1 (2012)
11. Jones, K.S.: A statistical interpretation of term specificity and its application in retrieval. J. Documentation **28**(1), 11–21 (1972)

12. Khorshidpour, Z., Hashemi, S., Hamzeh, A.: Distance learning for categorical attribute based on context information. In: 2010 2nd International Conference on Software Technology and Engineering (ICSTE), vol. 2, pp. V2–296. IEEE (2010)
13. Kullback, S.: Information Theory and Statistics. John Wiley & Sons, New York (1959)
14. Kullback, S., Leibler, R.A.: On information and sufficiency. Ann. Math. Stat. **22**, 79–86 (1951)
15. Lin, D.: An information-theoretic definition of similarity. Proc. ICML **98**, 296–304 (1998)
16. Michalski, R.S.: Knowledge acquisition through conceptual clustering: a theoretical framework and algorithm for partitioning data into conjunctive concepts. Int. J. Policy Anal. Inf. Syst. **4**(3), 219–244 (1980)
17. Pearson, K.: On the general theory of multiple contingency with special reference to partial contingency. Biometrika **11**(3), 145–158 (1916)
18. Smirnov, E.S.: On exact methods in systematics. Syst. Biol. **17**(1), 1–13 (1968)
19. Sneath, P.H.A., Sokal, R.: Numerical taxonomy. The principles and practice of numerical classification. Freeman, San Francisco (1973)
20. Stanfill, C., Waltz, D.: Toward memory-based reasoning. Commun. ACM **29**(12), 1213–1228 (1986)
21. Wilson, D.R., Martinez, T.R.: Improved heterogeneous distance functions. J. Artif. Intell. Res. **6**, 1–34 (1997)
22. Xie, J., Szymanski, B., Zaki, M.J.: Learning dissimilarities for categorical symbols. J. Mach. Learn. Res. Proc. Track **10**, 97–106 (2010)

Robust Stabilization of Uncertain T-S Fuzzy Systems: An Integral Error Dependent Sliding Surface Design Approach

Nilanjan Mukherjee, T.K. Bhattacharya and Goshaidas Ray

Abstract In this paper, an unique control law for a class of systems having bounded uncertainty, represented by T-S fuzzy models using sliding mode technique has been discussed. The proposed theory explores the possibility of incorporating an integral error while designing sliding surface that reduces the settling time and the reaching time as well. Subsequently, the conditions necessary for robustly stabilizing the closed-loop system has been derived by obtaining the solution for a given linear matrix inequality(LMI). Finally, the simulation results for an inverted cart-pendulum system subject to bounded external disturbances have been presented to justify the usefulness and uniqueness of the approach.

Keywords Uncertain systems · Sliding mode controllers · Robust control · Linear matrix inequality · Integral error · Takagi-Sugeno (T-S) fuzzy model · Fuzzy stability

1 Introduction

Fuzzy logic and fuzzy models have been commonly used in the control of complex non-linear plants and appears to be one of the most useful and effective control scheme because of its simplicity. T-S fuzzy system finds its use in most of the real

N. Mukherjee (✉) · T.K. Bhattacharya · G. Ray
Department of Electrical Engineering, Indian Institute of Technology, Kharagpur, India
e-mail: nmnilu6@gmail.com

T.K. Bhattacharya
e-mail: tapas@ee.iitkgp.ernet.in

G. Ray
e-mail: gray@ee.iitkgp.ernet.in

V. Ravi et al. (eds.), *Proceedings of the Fifth International Conference on Fuzzy and Neuro Computing (FANCCO - 2015)*, Advances in Intelligent Systems and Computing 415, DOI 10.1007/978-3-319-27212-2_23

world plants. Linear models that are obtained after linearizing the non-linear plant have been found to represent very closely to the original non-linear plant. Therefore, one can control the nonlinear system by clustering all the linear models, for achieving an effective control of the non-linear plant. Recently, sliding mode control (SMC) have also gained tremendous momentum and its major advantages are directed to strong robustness, good transient performance and fast response. Infact, it is regarded as an effective control method for robustly stabilizing an uncertain system and therefore, it's use is found in a wide variety of practical applications as in [1–3]. So, it would be quite desirable to develop a systematic SMC approach for non-linear systems since most of the real world plants are non-linear in nature. Therefore, it would be quite helpful if one can able to combine these two concepts i.e. the SMC technique which is designed based T-S fuzzy models will be effective in controlling the non-linear plants. For large interconnected plants, the SMC is designed for each subsystems and then a sliding mode controller is obtained by integrating each of the local sliding mode controllers using fuzzy reasoning approach in [4] which is valid for the overall system. Recently, there has been an improvement in the development of a new sliding mode control technique in LMI framework for a non-linear system having a time delay where the sliding surface is based on both the system states and the control input and subsequently, the conditions needed to asymptotically stabilize the system was established in [5]. Results presented therein have considered only the pendulum dynamics to show the effectiveness of the method. However, this method suffers from two distinct disadvantages having high settling time and the reaching time of the system states. In this paper, the present authors proposed a design method to reduce the settling time as well as the reaching time by introducing an integral error term along with the system states and the control input while designing the sliding surface. The present method has considered both the pendulum and the cart dynamics while designing the proposed controller and obtaining the LMI which is used to derive conditions necessary for the system to satisfy in order to achieve stability.

This paper is arranged as follows: Part 2 considers the statement of the problem and mathematical modeling. Part 3 deals with the control scheme considering the integral error term while designing the sliding surface which is used to robustly stabilize the non-linear system with time delay and having bounded uncertainty. Part 4 presents the simulation results for the inverted cart pendulum system to illustrate the usefulness of the discussed approach. Part 5 presents the conclusions.

Notations: The notations used here are quite simple. The notation "*" are used to indicate symmetrical terms in a matrix, "T" is used to denote transpose of a vector or a matrix. I_{nxn} and 0_{mxn} are used to indicate *nxn* identity matrix and *mxn* zero matrix respectively. Finally, $L_2[0, \mathrm{T}]$ are used to indicate the vector valued functions, i.e., ω: $[0\,\mathrm{T}]$ maps to $\Re^p \in L_2[0, \mathrm{T}]$.

2 Problem Formulation

A non-linear plant with an incorporated time delay and represented T-S fuzzy model has been considered to justify robust stability. A dynamic sliding mode controller has been designed where the sliding surface depends on the system states, control input and an additional integral error term in order to reduce the settling time and reaching time. This, in turn, needs to solve a linear matrix inequality constraint which is used to establish the conditions needed by the system to satisfy to justify stability and also to obtain the design parameters of the sliding surface matrices. The design technique have been applied to a inverted cart pendulum system to present the usefulness and uniqueness of the given method.

2.1 System Modeling

A T-S fuzzy time delay model of an uncertain non-linear plant with r rules can be described by the following:

Plant Rule: **If** $x_1(t)$ is M_1^l and. . . . .$x_g(t)$ is M_g^1 **Then**

$$\begin{aligned} \dot{x}(t) &= (A_l + \Delta A_l)x(t) + (E_l + \Delta E_l)x(t-\tau) + (B_l + \Delta B_l)u(t) + (H_l + \Delta H_l)w(t) \\ y(t) &= C_l x(t) + C_{ld}u(t) + D_l x(t-\tau), x(t) = \varphi(t), l \in L = \{1,2,3\ldots\ldots\}\ t \in [-\tau, 0] \end{aligned} \tag{1}$$

where, $\mathfrak{R}^l$ denotes the lth fuzzy inference rule, M_i^l are the fuzzy sets with Membership Functions $\mu_i^l, x(t) \in \mathfrak{R}^n$ are the state vectors, $u(t) \in \mathfrak{R}^m$ are the input vector, $w(t) \in \mathfrak{R}^p$ is the external disturbance, $A_l \in \mathfrak{R}^{nxn}, E_l \in \mathfrak{R}^{nxn}, B_l \in \mathfrak{R}^{nxm}$ and $H_l \in \mathfrak{R}^{nxp}$ are the constant system matrices, τ is the constant time delay.

The norm-bounded parameter uncertainties are given as ε_{Al}, ε_{El} and ε_{Hl} such that

$$\|\Delta A_l,\ \Delta B_l\| \le \varepsilon_{Al}, \|\Delta E_l\| \le \varepsilon_{El}, \Delta H_l \Delta H_l^T \le \varepsilon_h^2 \mathrm{I}_n.$$

Considering a delayed input signal $u(t-\tau)$ in the output equation, the resultant fuzzy system inferred from plant (1) can be written as follows:

$$\begin{aligned} \dot{x}(t) = \sum_{l=1}^{r} \omega_l(\theta(t)) &[(A_l + \Delta A_l)x(t) + (E_l +_l)x(t-\tau) \\ &+ (B_l + \Delta B_l)u(t) + H_l w(t)] \\ y(t) = \overline{\mathrm{C}}_l &\left[x^T(t), u^T(t)\right]^T + \overline{\mathrm{D}}_l\left[x^T(t-\tau), u^T(t-\tau)\right]^T \end{aligned} \tag{2}$$

where,

$$\overline{\mathrm{C}}_l = \sum_{l=1}^{r} \omega_l(\theta(t))\,\overline{\mathrm{C}}_l,\ \overline{\mathrm{D}}_l = \sum_{l=1}^{r} \omega_l(\theta(t))\overline{\mathrm{D}}_l,\ \overline{\mathrm{C}}_l = [C_l, C_{ld}],$$
$$\overline{\mathrm{D}}_l = [D_l,\ 0_{nXm}].$$

$$\overline{\mathrm{H}}_l = \sum_{l=1}^{r} \omega_l(\theta(t))(H_l + \Delta H_l).$$

It is given that $\omega_l(\theta(t)) = \frac{\prod_{i=1}^{g} \mu_i^l(\theta(t))}{\sum_{l=1}^{r} \prod_{i=1}^{g} \mu_i^l(\theta(t))}$ which is regarded as the normalized weight of each **IF-THEN** rules satisfying $\omega_l(\theta(t)) \geq 0$ and $\sum_{l=1}^{r} \omega_l(\theta(t)) = 1$.

The aim of the designed controller described in this paper is to robustly stabilize the non linear uncertain systems, and also to justify the reaching condition of the system states in a given time and also to achieve the asymptotic stability even when external disturbances are present. The results of this paper resolves two important issues substantially as compared with the results in [5–11] i.e., (i) The proposed method reduces the settling time as compared with the existing SMC. (ii) Design method of the SMC law exhibits less reaching time even in presence of external disturbances with pre-specified performance index γ, i.e., $\int_{t_0}^{\infty} \|y(t)\|^2 dt < \gamma^2 \int_{t_0}^{\infty} \|w(t)\|^2 dt$ to be satisfied for all non-zero $w(t)$, where, $t_0 = t_r + \tau$ and t_r is the reaching time to the sliding surface.

3 Proposed Design of the Sliding Surface

For system defined in (1) or (2), a new sliding surface is has been considered which is d- ependent on the system states $x(t)$, control input $u(t)$ and the integral of the error $e(t)$ as follows:

$$s(t) = S_x x(t) + S_u u(t) + S_e \int_0^t e(t)\,dt = S\overline{x}(t) + S_e \int_0^t e(t)dt = 0 \tag{3}$$

where, $S_x \in \Re^{mxn}, S_u \in \Re^{mxm}, S_e \in \Re^{mxn}$, $S = [S_x, S_u]$ and the new state vector $\overline{x}(t) = [x_1(t), x_2(t) \ldots\ldots\ldots x_n(t), u_1(t), u_2(t) \ldots\ldots .u_m(t)]^T$. S_u is assumed be non-singular.

At the sliding surface, the control strategy is described as: $u(t) = -S_u^{-1} S_x x(t) - S_u^{-1} S_e \int_0^t e(t)\,dt$ and the corresponding derivative is obtained as follows:

$$\dot{u}(t) = -S_u^{-1}S_x\dot{x}(t) - S_u^{-1}S_e e(t).$$

Putting the value of $\dot{x}(t)$ in the above equation we get,

$$\dot{u}(t) = -\sum_{l=1}^{r}\omega_l(\theta(t))[S_{ux}(A_l x(t) + E_l x(t-\tau) + B_l u(t) - (\alpha + \varsigma(t))S_u^{-1} \\ sign(s(t))] - S_{ue}e(t) \quad (4)$$

where,

$$\varsigma(t) = \sum_{l=1}^{r}\omega_l(\theta(t))\|S_x\|\left[\varepsilon_{Al}\|\bar{x}(t)\| + \varepsilon_{El}\|\bar{x}(t-\tau)\| + \left(\|H_l\| + \varepsilon_h^2\right)\rho(t)\right]$$

and $S_{ux} = S_u^{-1}S_x$, $S_{ue} = S_u^{-1}S_e$, α is a positive constant and $\rho(t)$ is the uniform upper bound of $w(t)$, i.e., $\|w(t)\| \leq \rho(t)$.

Since,

$$R_1 = [I_n,\ 0_{nxm}]^T,\ R_2 = [0_{mxn},\ I_m]^T,\ \overline{A_l} = [A_l,\ B_l] \\ \Delta\overline{A_l} = [\Delta A_l,\ _l],\ \overline{E_l} = [E_l,\ 0_{nxm}],\ \Delta\overline{E_l} = [\Delta E_l,\ 0_{nxm}].$$

Therefore, overall system can be described in compact form as follows:

$$\dot{\bar{x}}(t) = \sum_{l=1}^{r}\omega_l(\theta(t))\{[(R_1 - R_2S_{ux})\bar{A}_l + R_1\Delta\bar{A}_l]\bar{x}(t) + [(R_1 - R_2S_{ux})\bar{E}_l + R_1\Delta\bar{E}_l]\bar{x}(t-\tau) \\ + R_1\overline{\mathrm{H}}_l w(t)\} - R_2(\alpha + \varsigma(\mathrm{t}))S_u^{-1}sign\ (s(t)) - R_2S_{ue}e(t) \\ y(t) = \overline{\mathrm{C}}_l\bar{x}(t) + \overline{\mathrm{D}}_l\bar{x}(t-\tau),\ \bar{x}(t) = \left[\Phi^T(t),\ 0\right]^{\mathrm{T}}. \quad (5)$$

Theorem 1 *Considering the system described in* (5), *the system states will hit the sliding surface in a given time, provided the derivative of the Lyapunov Function, i.e.,* $2s^T(\mathrm{t})\dot{s}(t) < 0$ *where* $s(t)$ *is the sliding surface.*

Proof Considering a Lyapunov Function $S(t) = s^T(\mathrm{t})s(t)$ for $\mathrm{t} > 0$, the derivative of the sliding surface along the trajectories of the system (5) can be expressed as:

$$\dot{S}(\mathrm{t}) = 2s^T(\mathrm{t})\dot{s}(t) \\ = 2s^T(t)[S_x x(t) + S_u u(t)] + 2s^T(t)S_e e(t) \\ = 2s^T(t)S\dot{\bar{x}}(t) + 2s^T(t)S_e e(t) \\ = 2\sum_{1=1}^{\mathrm{r}}\omega_1(\theta)s^T(\mathrm{t})S\{\left[(R_1 - R_2S_{ux})\overline{A_l} + R_1\Delta\overline{A_l}\right]\bar{x}(t) + \left[(R_1 - R_2S_{ux})E_l + R_1\Delta\overline{E_l}\right] \\ \bar{x}(t-\tau) + R_1\overline{\mathrm{H}}_l w(t) - R_2(\alpha + \varsigma(t))S_u^{-1}sign\,(s(t)) - R_2S_{ue}e(t)\} + 2s^T(t)S_e e(t) \quad (6)$$

Since,

$$
\begin{aligned}
S(R_1 - R_2 S_{ux}) = 0,\ \varsigma(t) = &\sum_{l=1}^{r} \omega_l(\theta(t)) \|S_x\| [\varepsilon_{Al} \|\bar{x}(t)\| + \varepsilon_{El} \|\bar{x}(t-\tau)\| \\
&+ \left(\|H_l\| + \varepsilon_h^2\right)\rho(t)] \\
\dot{S}(t) = 2 &\sum_{l=1}^{r} \omega_l(\theta) s^T(t) \{ S_x \left[\Delta\overline{A}_l \bar{x}(t) + \Delta E_l \bar{x}(t-\tau) + \overline{\mathrm{H}}_l w(t)\right] - 2\varsigma(t)\|s(t)\| \\
&- 2\alpha \|s(t)\| - S_e e(t) \} + 2s^T(t) S_e e(t). \\
&\leq -2\alpha \|s(t)\| \\
= &-2\alpha \sqrt{S(t)}.
\end{aligned}
\tag{7}
$$

This implies that the system states hit the sliding surface within a given time. □

Remark The sliding surface matrix parameters S_x, S_u and S_e can be obtained easily based on LMI framework and detailed design methodology and the conditions which are needed by the system to satisfy so that it justifies the stabilizing property of the designed controller has been established in the following sections.

3.1 Design and Stability Analysis of the T-S Fuzzy Model Based Sliding Mode Control System

When the system trajectories reach the sliding surface, the augmented system dynamics are represented as:

$$
\begin{aligned}
\dot{\bar{x}}(t) = &\sum_{l=1}^{r} \omega_l(\theta(t)) \{ \left[(R_1 - R_2 S_{ux})\overline{A}_l + R_1 \Delta\overline{A}_l\right] \bar{x}(t) + \left[(R_1 - R_2 S_{ux}) E_l + R_1 \Delta\overline{E}_l\right] \\
&\bar{x}(t-\tau) + R_1 \overline{\mathrm{H}}_l w(t) \} - R_2 S_{ue} e(t). \\
y(t) = &\overline{\mathrm{C}}_l x(t) + \overline{\mathrm{D}}_l \bar{x}(t-\tau),\ \bar{x}(t) = \left[\Phi^T(t),\ 0\right]^{T.}
\end{aligned}
\tag{8}
$$

Theorem 2 *Given a constant* $\gamma > 0$, *the sliding mode controller derived from* (4) *will be able to stabilize asymptotically closed-loop system* (8) *if there exists two positive definite matrices* P, $Q \in R^{(m+n)x\ (m+n)}$, *two sets of matrices* $W_{l1}, W_{l2} \in R^{mx(m+n)}$ *and positive constants* $\delta_l, l \in L$ *such that the following LMI is satisfied.*

$$\begin{bmatrix} \Xi_l & * & * & * & * & * \\ P\bar{E}_l^T R_1^T + W_{l2}^T R_2^T & -Q & * & * & * & * \\ -PS_{ue}^T R_2^T & 0 & -\delta_l I & * & * & * \\ \bar{C}_l P & \bar{D}_l P & 0 & -I & * & * \\ -\varepsilon_{Al} P & 0 & 0 & 0 & -\delta_l I & * \\ 0 & -\varepsilon_{El} P & 0 & 0 & 0 & -\delta_l I \end{bmatrix} < 0 \tag{9}$$

where,

$$\Xi_l = R_1 \bar{A}_l P + R_2 W_{1l} + P\bar{A}_l^T R_1^T + W_{1l}^T R_2^T + Q + \in_l R_1 R_1^T + \frac{2}{\gamma^2} R_1 \left(H_l H_l^T + \varepsilon_h^2 I_n\right) R_1^T \tag{10}$$

Proof

Case 1:

Considering the case when $w(t) = 0$, let us consider a Lyapunov-Krasovskii functional candidate, $V(t) = \bar{x}^T(t) X \bar{x}(t) + \int_{t-\tau}^{t} \bar{x}^T(\varphi) Y \bar{x}(\varphi) d\varphi$ for all $t \geq t_r + \tau$, where, $X = P^{-1}$, $Y = P^{-1} Q P^{-1}$ and t_r is the hitting time of the system states.

Since, $S = R_2^T X$, so $R_2^T X \bar{x}(t) = 0,$.

Now, defining a new vector $\xi(t) = [\bar{x}^T(t),\ \bar{x}^T(t-\tau),\ e^T(t)]^T$ one can rewrite Thus, for the augmented system (8), the derivative $V(t)$ is expressed as follows:

$$\begin{aligned}
\dot{V}(t) &= 2\bar{x}^T(t) X \dot{\bar{x}}(t) + \bar{x}^T(t) Y \bar{x}(t) - \bar{x}^T(t-\tau) Y \bar{x}(t-\tau) \\
&= 2\sum_{l=1}^{r} \omega_l(\theta(t))\, \bar{x}^T(t) X \left\{R_1(\bar{A}_l + \Delta\bar{A}_l)\bar{x}(t) + R_1(\bar{E}_l + \Delta\bar{E}_l)\bar{x}(t-\tau)\right\} \\
&\quad - 2\sum_{l=1}^{r} \omega_l(\theta(t))\, \bar{x}^T(t) X R_2 \left\{S_{ux}\bar{A}_l \bar{x}(t) + S_{ux}\bar{E}_l \bar{x}(t-\tau) - S_{ue} e(t)\right\} \\
&\quad - \bar{x}^T(t-\tau) Y \bar{x}(t-\tau) + \bar{x}^T(t) Y \bar{x}(t) \\
&= 2\sum_{l=1}^{r} \omega_l(\theta(t))\, \bar{x}^T(t) X \left\{R_1(\bar{A}_l + \Delta\bar{A}_l)\bar{x}(t) + R_1(\bar{E}_l + \Delta\bar{E}_l)\bar{x}(t-\tau)\right\} \\
&\quad + \bar{x}^T(t) Y \bar{x}(t) - 2\sum_{l=1}^{r} \omega_l(\theta(t))\, \bar{x}^T(t) X R_2 \{K_{1l}\bar{x}(t) + K_{2l}\bar{x}(t-\tau) - S_{ue} e(t)\} \\
&\quad - \bar{x}^T(t-\tau) Y \bar{x}(t-\tau)
\end{aligned} \tag{11}$$

where, $K_{1l} = S_{ux}\bar{A}_l$ and $K_{2l} = S_{ux}\bar{E}_l$ are the matrices to be determined.

the perturbed terms in (11) associated with ΔA_l, ΔE_l as the inequality

$$2\bar{x}^T(t)XR_1(\Delta\bar{A}_l\bar{x}(t)+\Delta\bar{E}_l\bar{x}(t-\tau)\leq\bar{x}^T(t)(\delta_l XR_1R_1^TX+\delta_l^{-1}\varepsilon_{El}^2\bar{x}^T(t-\tau)\\ \bar{x}(t-\tau) \tag{12}$$

Using (12), (11) can be rewritten in the form of matrix inequality as follows:

$$\dot{V}(t)\leq\sum_{l=1}^{r}\omega_l(\theta(t))\xi^T(t)\begin{bmatrix}\gamma_l & * & * \\ (R_1\bar{E}_l+R_2K_{2l})^TX & \delta_l^{-1}\varepsilon_{El}^2I-Y & 0 \\ -S_{ue}^TR_2^TX & 0 & 0\end{bmatrix}\xi(t)$$

where,

$$\gamma_l=X(R_1\bar{A}_l+R_2K_{1l})+(R_1\bar{A}_l+R_2K_{1l})^TX+\delta_lXR_1R_1^TX\\ +\delta_l^{-1}\varepsilon_{Al}^2I+Y \tag{13}$$

It may be noted that $\dot{V}(t)<0$ if

$$\begin{bmatrix}\gamma_l & * & * \\ (R_1\bar{E}_l+R_2K_{2l})^TX & \delta_l^{-1}\varepsilon_{El}^2I-Y & 0 \\ -S_{ue}^TR_2^TX & 0 & 0\end{bmatrix}<0 \tag{14}$$

Equation (14) is not an LMI, in order to make LMI form the following seque- of operations are performed. Multiplying diag(P,P,P) from both sides of (14), we get the following LMI.

$$\begin{bmatrix}\Lambda_l & * & * \\ P(R_1\bar{E}_l+R_2K_{2l})^T & \delta_l^{-1}\varepsilon_{El}^2PP-Q & 0 \\ -PS_{ue}^TR_2^T & 0 & 0\end{bmatrix}<0 \tag{15}$$

where,

$$\Lambda_l=(R_1\bar{A}_l+R_2K_{1l})P+P(R_1\bar{A}_l+R_2K_{1l})^T+\delta_lR_1R_1^T\\ +\delta_l^{-1}\varepsilon_{Al}^2PP+Q$$

Using Schur-Complement lemma in (15) and putting $W_{1l}=K_{1l}P$ and W_{2l} $K_{2l}P$, we have

$$\begin{bmatrix}\Theta_l & * & * & * & * \\ P\bar{E}_l^TR_1^T+W_{12}^TR_2^T & -Q & * & * & * \\ -PS_{ue}^TR_2^T & 0 & -\delta_lI & * & * \\ \varepsilon_{Al}P & 0 & 0 & -\delta_lI & * \\ 0 & \varepsilon_{El}P & 0 & 0 & -\delta_lI\end{bmatrix}<0 \tag{16}$$

where, $\Theta_l = P\bar{A}_l^T R_1^T + R_1\bar{A}_l P + W_{l1}^T R_2^T + R_2 W_{l1} + \delta_l R_1 R_1^T + Q$.
If (16) holds then the system (8) with $w(t)$ is asymptotically stable.

Case 2:
Now we consider $w(t) \neq 0$. Here the objective is to design a SMC in the presence of external disturbances and also to ensure the performance inequality $\int_{t_r}^{\infty} \|y(t)\|^2 dt < \gamma^2 \int_{t_r}^{\infty} \|\omega(t)\|^2 dt$ for all non zero $w(t) \in L_2[0.\infty)$ under the initial conditions $x(\theta) = 0$ for $\theta \in [t_r, t_r + \tau]$. Considering an augmented vector $\vartheta(t) = [\bar{x}^T(t), \bar{x}^T(t-\tau), e^T(t), w^T(t)]^T$. Combining (13) with the performance inequality constraint described above to regulate the effect of bounded external disturbance at the system output, the resulting inequality is obtained.

$$\dot{V}(t) + y^T(t)y(t) - \gamma^2 w^T(t)w(t) \leq \sum_{l=1}^{r} \omega_l(\theta(t))\vartheta^T(t) \times \begin{bmatrix} \gamma_l & * & * & * \\ (R_1\bar{E}_l + R_2 K_{2l})^T X & \delta_l^{-1}\varepsilon_{El}^2 I - Y & * & * \\ -S_{ue}^T R_2^T X & 0 & 0 & * \\ 0 & 0 & 0 & * \end{bmatrix} \vartheta(t) + \vartheta^T(t) \begin{bmatrix} \overline{C}_l^T \overline{C}_l & * & * & * \\ \overline{D}_l^T \overline{C}_l & \overline{D}_l^T \overline{D}_l & * & * \\ 0 & 0 & 0 & * \\ \overline{H}_l^T R_1^T X & 0 & 0 & -\gamma^2 I \end{bmatrix} \vartheta(t) \tag{17}$$

where, γ_l has been defined in (13).
Now, (17) can be expressed in the LMI form as follows:

$$\begin{bmatrix} \gamma_l & * & * & * \\ (R_1\bar{E}_l + R_2 K_{2l})^T X & \delta_l^{-1}\varepsilon_{El}^2 I - Y & * & * \\ -S_{ue}^T R_2^T X & 0 & 0 & * \\ 0 & 0 & 0 & * \end{bmatrix} + \begin{bmatrix} \overline{C}_l^T \overline{C}_l & * & * & * \\ \overline{D}_l^T \overline{C}_l & \overline{D}_l^T \overline{D}_l & * & * \\ 0 & 0 & 0 & * \\ \overline{H}_l^T R_1^T X & 0 & 0 & -\gamma^2 I \end{bmatrix} < 0 \tag{18}$$

Adopting the procedure as before such as multiplying (18) with $diag(P,P,P,P)$ with $P = X^{-1}$ and then applying Schur-Complement lemma, we get,

$$\begin{bmatrix} \Lambda_l + \frac{1}{\gamma^2} R_1 \overline{\mathrm{H}}_l \overline{\mathrm{H}}_l^T R_1^T & * & * & * \\ P\left(R_1 \overline{E}_l + R_2 K_{2l}\right)^T & \varepsilon_l^{-1} \in_{El}^2 PP - Q & * & * \\ -PS_{ue}^T R_2^T & 0 & 0 & * \\ \overline{\mathrm{C}}_l P & \overline{\mathrm{D}}_l P & 0 & -I \end{bmatrix} < 0 \tag{19}$$

where, Λ_l has been defined in (15).

Using Young's Inequality, $Z^T U + U^T Z \leq \varepsilon_1 Z^T Z + \varepsilon_1^{-1} U^T U$, where, ε_1 is a any scalar and Z and U are vectors, one can rewrite $R_1 \overline{\mathrm{H}}_l \overline{\mathrm{H}}_l^T R_1^T \leq 2R_1 \left(\overline{\mathrm{H}}_l \overline{\mathrm{H}}_l^T + \varepsilon_h^2 I_n\right) R_1^T$ and putting the values of $\overline{\mathrm{C}}_l$, D_l and $\overline{\mathrm{H}}_l$ in (19), the following LMI is obtained.

$$\begin{bmatrix} \Lambda_l + \frac{2}{\gamma^2} R_1 (H_l H_l^T + \varepsilon_h^2 I_n) R_1^T & * & * & * \\ P(R_1 \overline{E}_l + R_2 K_{2l})^T & \varepsilon_l^{-1} \in_{El}^2 PP - Q & * & * \\ -PS_{ue}^T R_2^T & 0 & 0 & * \\ \overline{C}_l P & \overline{D}_l P & 0 & -I \end{bmatrix} < 0 \tag{20}$$

Using Schur-Complement lemma, (20) can be used to derive LMI (9). □

Remarks The Convex Optimization problem to obtain the minimum H_∞ performance index can be stated as follows:

Minimize $-\frac{1}{\gamma^2}$ subject to the LMI (9) with the conditions P, Q, W_{l1}, $W_{l2} > 0$

4 Simulation Results

Let us consider an inverted cart pendulum system which is an unstable nonlinear physical system. Our task is to stabilize the inverted pendulum at the upward position (within an angle range of ±4°) at a desired cart position by applying the sliding mode control strategy that has been developed using an additional integral error term in the sliding surface design in contrast to [5] where the results presented have considered only the pendulum dynamics.

The dynamical equations of the cart pendulum system are described as follows:

$$\dot{x}_1(t) = x_2(t) + w(t), \quad \dot{x}_2(t) = f_1(x(t), u(t)), \quad \dot{x}_3(t) = x_4(t) + w(t)$$
$$\dot{x}_4(t) = f_2(x(t), u(t))$$

where, $f_1(x,u) = g\ \sin\ (x_1) - a\,m\,l\,x_2^2 \sin\ (2x_1)/2 - a\ \cos(x_1)u$

$$f_2(x,u) = \frac{-mag\ \sin x_1 \cos x_1 + \frac{4mla}{3} x_2^2 \sin x_1 + \frac{4au}{3}}{\frac{4}{3} - ma\cos^2 x_1}$$

where, x_1 = pendulum angle from the vertical (radian), x_2 = pendulum angular velocity (radians/sec) x_3 = horizontal cart displacement (m), x_4 = horizontal cart velocity (m/sec), g = 9.8 m/s^2 is the gravity constant, m = pendulum mass, M = cart mass, 2l = length of the pendulum and w is the disturbance, a = 1/(m + M) = 0.1

Here, M = 8 kg, m = 2 kg, 2l = 1 m has been taken.

By linearizing the plant model around the following operating points are considered $[\pm 0.3,\ 0]$

The discrepancy between the between the original nonlinear plant model and linearized plant model is considered as the norm-bounded uncertainties. The plant model obtained is as follows:

Plant Rule $\Re^l$: **If** $x_1(t)$ is M_1^l and $x_2(t)$ is M_2^l and $x_3(t)$ is M_3^l and $x_4(t)$ is M_4^l **Then**,

$$\begin{aligned} \dot{x}(t) &= (A_l + \Delta A_l)x(t) + (B_l + \Delta B_l)u(t) + (H_l + \Delta H_l)w(t) \\ y(t) &= C_l x(t) + C_{ld}\, u(t),\ l \in L = \{1, 2, 3, 4, 5, ,,,,,21\} \end{aligned}$$

The membership functions of the 4 states which are symmetric about the origin are shown in Fig. 1a–d where M^1 denotes Small, M^2 denotes Zero and M^3 denotes Big in terms of linguistic variables.

Some of the system matrices corresponding to different operating points are given as follows:

$$A_1 = \begin{bmatrix} 0 & 1 & 0 & 0 \\ 17.29 & 0 & 0 & 0 \\ 0 & 0 & 1 & 0 \\ -1.729 & 0 & 0 & 0 \end{bmatrix},\ A_2 = \begin{bmatrix} 0 & 1 & 0 & 0 \\ 13.29 & 0 & 0 & 0 \\ 0 & 1 & 0 & 0 \\ -1.329 & 0 & 0 & 0 \end{bmatrix}$$

$$A_5 = \begin{bmatrix} 0 & 1 & 0 & 0 \\ 12.45 & 0 & 0 & 0 \\ 0 & 1 & 0 & 0 \\ -1.245 & 0 & 0 & 0 \end{bmatrix},\ A_{13} = \begin{bmatrix} 0 & 1 & 0 & 0 \\ 11.68 & 0 & 0 & 0 \\ 0 & 1 & 0 & 0 \\ -1.168 & 0 & 0 & 0 \end{bmatrix}$$

$B_1 = [0\ \ -0.17\ \ 0\ \ 0.17]^T$, $B_2 = [0\ \ -0.13\ \ 0\ \ 0.13]^T$, $B_5 = [0\ \ -0.10\ \ 0\ \ 0.10]^T$
$B_{13} = [0\ \ -0.6\ \ 0\ \ 0.6]^T$, $C_1 = C_2 = C_5 = C_{13} = [0.5\ \ 1]$, $C_{1d} = 0.05$,
$C_{2d} = 0.05$, $C_{5d} = 0.01$, $C_{13d} = 0.0004$, $H_1 = H_2 = H_5 = H_{13} = [1\ \ 0\ \ 1\ \ 0]^T$

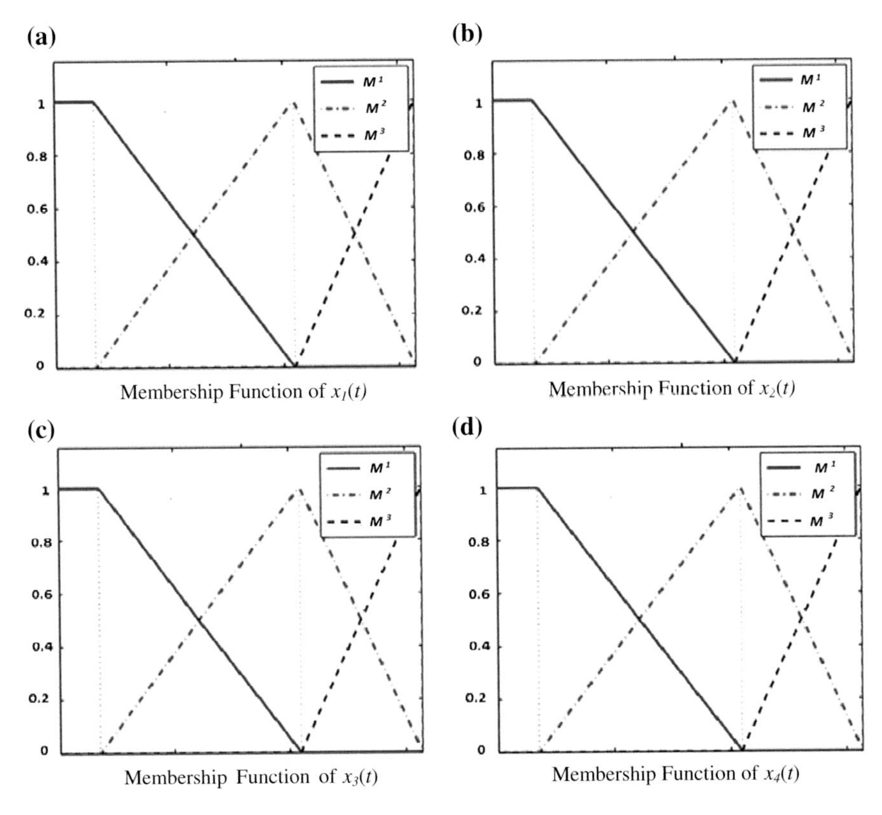

Fig. 1 **a** Membership function of $x_1(t)$. **b** Membership function of $x_2(t)$. **c** Membership function of $x_3(t)$. **d** Membership function of $x_4(t)$

The designed a sliding mode controller will be able to stabilize the closed-loop system with H_∞ performance index $\gamma > 0$ as minimum as possible. The performance index is defined as the ratio of the system output signal energy to the disturbance signal energy. So, lower the performance index γ, system is less sensitive to external disturbances. Here, ε_{Al} is taken as 0.03 and $\varepsilon_{\mathrm{El}}$ is taken as 0.003. The simulation results have been carried out with initial states $[x_1, x_2, x_3, x_4] = [0.16, 0, 0.10, 0]^{\mathrm{T}}$ and the desired position of the cart is chosen as $x_{3d} = 0.2$ m. The sliding mode controller is found to achieve stability for the overall system with $\alpha = 2.5$.

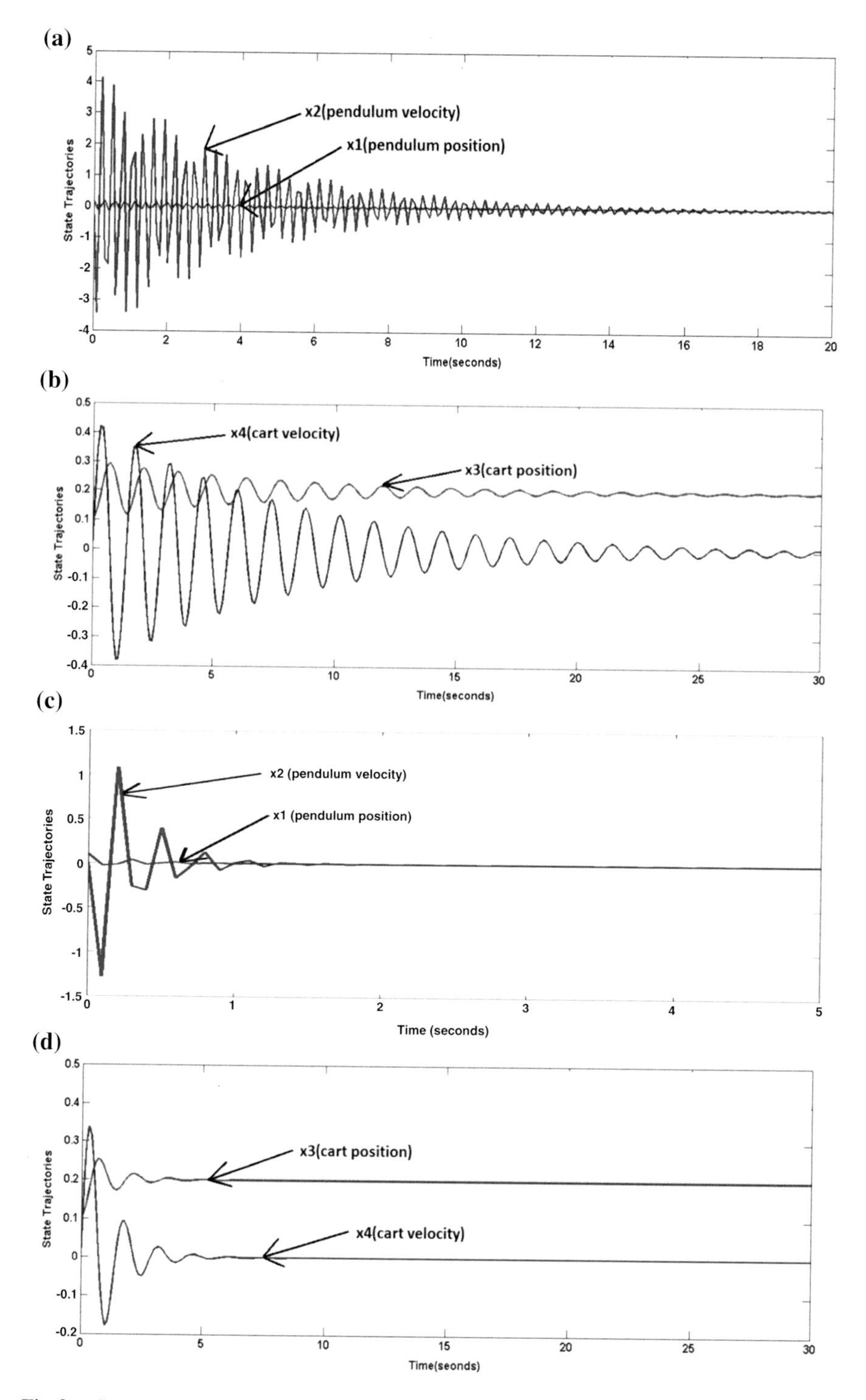

Fig. 2 **a** System state of the pendulum without integral error. **b** System state of the cart without integral error. **c** System state of the pendulum with integral error. **d** System state of the cart with integral error

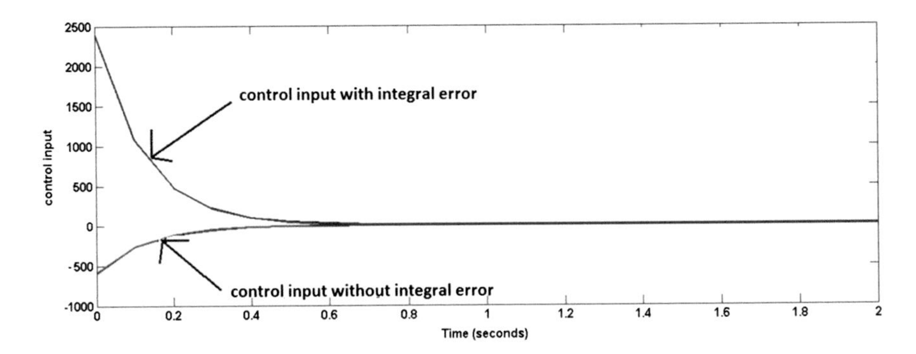

Fig. 3 Control input

Solving (9) using LMI tool-box, the corresponding sliding surface matrix S and the relevant positive definite matrices P and Q are as follows:

$$S = \begin{bmatrix} -0.9281 & -0.8334 & 0.6534 & 2.1437 & 0.0040 \end{bmatrix}$$

$$P = \begin{bmatrix} 0.0546 & -0.0426 & 3.8224 & 0.4223 & 4.2810 \\ -0.0426 & 0.592 & 2.4804 & 0.1234 & 6.1526 \\ 3.8224 & 2.4804 & 3.2489 & 0.1568 & 1.5102 \\ 0.4223 & 0.1239 & 0.1568 & 2.1456 & 8.9608 \\ 4.2810 & 6.1526 & 15102 & 8.9608 & 1.6787 \end{bmatrix}$$

$$Q = \begin{bmatrix} 0.0302 & -0.0219 & 0.7699 & 0.6123 & 1.5671 \\ -0.0219 & 0.0316 & -0.2521 & 0.0212 & -3.4661 \\ 0.7699 & -0.2521 & 1.2349 & 0.1564 & 5.6923 \\ 0.6123 & 0.0212 & 0.1564 & 9.2568 & 8.4562 \\ 1.5671 & -3.4661 & 5.6923 & 8.4562 & 303.45 \end{bmatrix}$$

The state trajectories of the plant with and without the integral error term introduced while designing the sliding surface are given in the Fig. 2a–d. The corresp- nding control input response are shown in Fig. 3. The ratio of the system output energy to the effect of the bounded disturbance energy, i.e., the performance index γ is plotted in Fig. 5, where the disturbance is taken as $e^{-0.02*sin(3.14*t)}$. The sliding surface are depicted in Fig. 4.

5 Discussion of Results

From the state trajectories, it is observed that the integral error approach reduces the settling time to less than 7 s approximately while the settling time takes more than 25 s in case of non-integral error approach. From the sliding surface plot in Fig. 4, one can see that the sliding surface has been reached to zero value much faster when

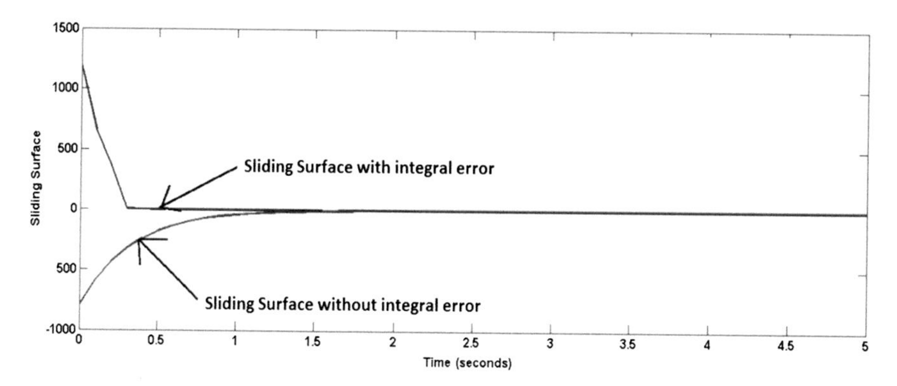

Fig. 4 Sliding surface

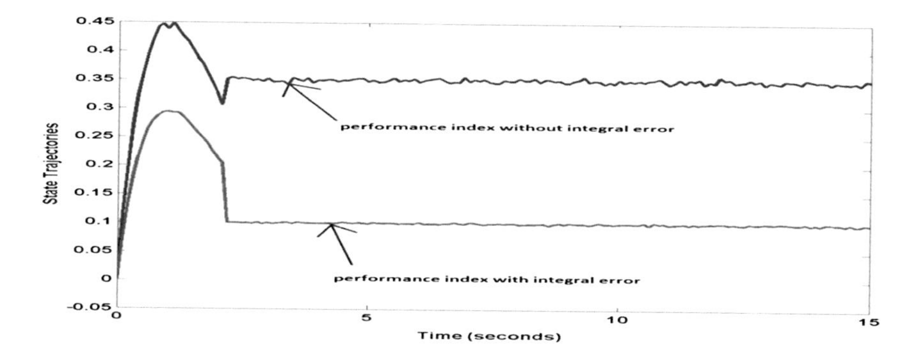

Fig. 5 Performance index $\gamma = \int_0^t y^T(\varphi)y(\varphi)\,d\varphi / \int_0^t w^T(\varphi)w(\varphi)$

the integral error is included in the design procedure. Figure 5 indicates that the integral error approach improves the performance index γ to 0.3064 in presence of bounded disturbances which means that the method is able to achieve sufficient robustness. In case, where the integral error is not included, the performance index γ achieves a value of 0.786. Further, it may be noted that the settling time and the reaching time have been reduced to significantly lower values by sacrificing the control effort.

6 Conclusion

Here, a dynamic sliding mode controller has been designed in LMI framework by introducing an integral error term while designing the sliding surface. The sliding surface design parameters have been computed by means of a set of LMIs that has been obtained to derive the conditions necessary for achieving the stability of the system. Moreover, the settling time of the system to reach the desired cart and pendulum position have been also reduced considerably which is an essential improvement over the existing control scheme. However, to reduce the control effort, an anti wind-up control strategy will be considered along with the present scheme for future work. Simulation results justify the usefulness and uniqueness of the designed controller over the existing SMC controller scheme as presented in [5].

References

1. Zhang, J., Shi, P., Xia, Y.: Robust adaptive sliding-mode control for fuzzy systems with mismatched uncertainties. IEEE Trans. Fuzzy Syst. **18**(4), 700–711 (2010)
2. Feng, G.: A survey on analysis and design of model-based fuzzy control systems. IEEE Trans. Fuzzy Syst. **14**(5), 676–697 (2006)
3. Xi, Z., Feng, G., Hesketh, T.: Piecewise sliding-mode control for T-S fuzzy systems. IEEE Trans. Fuzzy Syst. **19**(4), 707–716 (2011)
4. Yu, X., Man, Z., Wu, B.: Design of fuzzy sliding-mode control systems. Fuzzy Sets Syst. **95** (3), 295–306 (1998)
5. Qing, Gao, Feng, G., Xi, Z., Wang, Y., Qiu, J.: Robust H-infinity Control of T-S fuzzy time-delay systems via a new sliding-mode control scheme. IEEE Trans. Fuzzy Syst. **22**(2), 459–465 (2014)
6. Choi, H.H.: Robust stabilization of uncertain fuzzy-time-delay systems using sliding-mode-control approach. IEEE Trans. Fuzzy Syst. **18**(5), 979–984 (2010)
7. Feng, G.: Analysis and Synthesis of Fuzzy Control Systems: A Model-Based Approach. CRC, Boca Raton, FL, USA (2010)
8. Lin, C., Wang, Q., Lee, T.H.: Stabilization of uncertain fuzzy time- delay systems via variable structure control approach. IEEE Trans. Fuzzy Syst. **13**(6), 787–798 (2005)
9. Seuret, A., Edwards, C., Spurgeon, S.K., Fridman, K.E.: Static output feedback sliding mode control design via an artificial stabilizing delay. IEEE Trans. Autom. Control **54**(2), 256–265 (2009)
10. Zheng, F., Wang, Q., Lee, T.H.: Output tracking control of MIMO fuzzy nonlinear systems using variable structure control approach. IEEE Trans. Fuzzy Syst. **10**(6), 686–697 (2002)
11. Ho, D.W.C., Niu, Y.: Robust fuzzy design for nonlinear uncertain stochastic systems via sliding-mode control. IEEE Trans. Fuzzy Syst. **15**(3), 350–358 (2007)

Extreme Learning Machine for Eukaryotic and Prokaryotic Promoter Prediction

Praveen Kumar Vesapogu and Bapi Raju Surampudi

Abstract Promoters are DNA sequences containing regulatory elements required to guide and modulate the transcription initiation of the gene. Predicting promoter sequences in genomic sequences is a significant task in genome annotation and understanding transcriptional regulation. In the past decade many methods with many feature extraction schemes have been proposed for the prediction of eukaryotic and prokaryotic promoters. Still there is great need for more accurate and faster methods. In this paper we employed extreme learning machine algorithm (ELM), for promoter prediction in DNA sequences of *H. sapiens*, *D. melanogaster*, *A. thaliana*, *C. elegans* and *E. coli*. We extracted dinucleotide and CpG island features, and achieved accuracy above 90 % for all the five species. Performance is compared with the feed forward back propagation algorithm (BP) and support vector machines (SVM) and the results establish the viability of the presented approach.

Keywords Promoter · Dinucleotide · CpG-island · Extreme learning machine

1 Introduction

Predicting promoter regions in DNA sequences is an important problem in bioinformatics. Promoters are regions in genomic sequences having elements which regulate and control the transcription initiation process of a gene. Transcription is the process by which the information in a strand of DNA is copied into a new molecule of messenger RNA (mRNA). It is significant to use computational techniques for promoter prediction to discover genes that are missed by gene predictiors and devising

P.K. Vesapogu (✉) · B.R. Surampudi
School of Computer and Information Sciences, University of Hyderabad,
Hyderabad 500046, Telangana, India
e-mail: praveencs@uohyd.ac.in

B.R. Surampudi
Cognitive Science Lab, International Institute of Information Technology,
Hyderbad 500032, Telangana, India
e-mail: raju.bapi@iiit.ac.in

V. Ravi et al. (eds.), *Proceedings of the Fifth International Conference on Fuzzy and Neuro Computing (FANCCO - 2015)*, Advances in Intelligent Systems and Computing 415, DOI 10.1007/978-3-319-27212-2_24

experiments [1]. Even though several promoter recognition methods have been proposed, still there is a necessity for accurate and expeditious promoter prediction techniques.

In the past decades many algorithms have been proposed for promoter prediction. Rani and Bapi [29] used n-grams and neural networks for promoter recognition. Burden et al. [9], Pederson et al. [26], Reese [34] used neural networks. Gordan et al. [15], Anwar et al. [4] applied support vector machines (SVM). Ohler et al. [23] and Ohler [25] used hidden Markov model (HMM). Lawrance and Reilly [20] used Expectation Maximization (EM) algorithm. Ben-Gal et al. [7] proposed Bayesian Network model for promoter recognition. Li and Lin [21], Akan and Deloukas [2] used position weight matrix (PWM). Gordan et al. [15] extracted signal features using matching function between sequences and used support vector machine for classification. Anuj et al. [3] extracted dinucleotide feature from the sequences around binding sites and classified the sequences using feed forward neural network. Rani and Bapi [29] extracted n-grams as features and fed to multi layer feed forward neural network and applied to whole genome promoter prediction in *E. coli* and *D. melanogaster*.

For promoter recognition problem, to extract the most discriminative features from promoter and non-promoter sequences is an important problem [44]. Zeng et al. [44] extracted context, structure and signal features and proposed a hierarchical classification system for entire human genome promoter recognition. Rani et al. [33], Anuj et al. [3] used dinucleotide features. Ponger and Mouchiroud [28] extracted CpG-island features for recognizing promoter regions in human and mouse genome sequences. In the literature many authors have extracted different types of features from promoter and non-pormoter DNA sequences, and used different classification models for promoter recognition problem. Performances of the existing methods, still are not fulfilling expectations. In this paper we extracted 16 dinucleotide and two CpG island signal features, total 18 features are fed to the Extreme learning machine (ELM) algorithm. Dinucleotide features nor CpG island features alone didn't give satisfying results, we used combined features, and achieved accuracy above 90 % for all the species we mentioned.

2 Materials and Methods

2.1 *Datasets and Performance Measures*

Eukaryotic promoter sequences are taken from EPDnew [12, 13]. EPDnew is a recently introduced collection of automatically compiled, promoter lists complementing the old experimentally characterized eukaryotic POL II promoters EPD [27]. We used promoter sequences of length 251 [−200, +50] bp around Transcription Start Sites (TSS) from EPDnew. For negative eukaryotic promoter dataset, we have taken randomly extracted exon and intron sequences of length 251 bp from Advances in the EID [39].

We took both protein coding regions and protein non-coding sequences for negative dataset. Rani and Bapi [29, 33], Zang et al. [43, 44] and Lin et al. [22] have used more than one dataset for negative promoter dataset. Rani et al. [29, 33] used both coding regions and non-coding regions for negative dataset. Zang et al. [44] have taken coding sequences, introns and untranslated region ('UTR) sequences for negative dataset. Lin et al. [22] used coding sequences and introns for negative promoter dataset. For negative promoter dataset considering sequences other than promoter regions i.e., from both coding regions and intergenic regions seems biologically meaningful and realistic than individually.

For *H. sapiens* 23,300 promoter sequences from EPDnew [12], 23,125 coding sequences and 23,100 intron sequences are randomly extracted from Advances in the EID [39]. For *D. Melanogaster* 15,000 promoter sequences, 14,990 coding sequences and 14,900 introns are randomly extracted. We used 10,000 promoter sequences, 9,990 coding sequences and 9,990 introns sequences for *A. thaliana*. For *C. elegans* 7,000 promoter sequences, 6,980 coding sequences and 6,900 introns sequences were used.

Prokaryotic promoter sequences and non-promoter sequences are taken from gordan et al. [15]. For positive *E. coli* dataset 669 promoter sequences of length 80 [−60, +20] bp around Transcription Start Site (TSS). For negative *E. coli* dataset 660 sequences of coding and 660 sequences of non-coding of length 80 bp were taken.

The performance of the algorithms is measured in terms of sensitivity, specificity and accuracy [10]. The formulae are: Sensitivity, Sn = TP/(TP+ FN), specificity, Sp = TN/(TN+FP), and accuracy Ac = (TP+TN)/(TP+TN+FP+FN), where TP stands for the true positives, which is the number of promoter sequences that are correctly identified as promoter. TN is the true negatives, i.e., the number of non-promoter sequences that are correctly identified as non-promoters. FN is the false negatives, i.e., the number of wrongly identified non-promoter sequences. FP represents the false positives, which is the number of incorrectly identified promoter sequences. S_n is the percentage of promoter sequences that are correctly identified as promoters. S_p is the percentage of non-promoter sequences that are correctly identified as non-promoters. The average of S_n and S_p is accuracy Ac.

2.2 Extreme Learning Machine (ELM)

Extreme learning machine (ELM) is a simple learning method for single layer feed-forward network proposed by Huang et al. [18, 19]. We applied Extreme learning machine algorithm in this paper for promoter prediction. Three step Extreme learning algorithm is given below.

Algorithm 1 Extreme learning machine (ELM) Algorithm

Given a training set $\mathbb{N} = \{(x_i, t_i)|x_i \epsilon R^n, t_i \epsilon R^m, i = 1, ..., N\}$, activation function $f(x)$, and the number of hidden nodes S.

1. Assign randomly input weight vector w_i and hidden node bias $b_1, i = 1, ..., S$.
2. Compute the hidden layer output matrix H.
3. Compute the output weight β : $\beta = H^{\dagger}T$.
   where x_i is an $n \times 1$ input vector and t_i is an $m \times 1$ target vector. sigmoidal activation function $f(x) = \frac{1}{1+e^{-x}}$. $H^{\dagger}$ is the Moore-Penrose generalized inverse of hidden layer output matrix H.
   Target vector $T = \{t_1, t_2, ..., t_N\}^T$.

3 Feature Extraction

3.1 Dinucleotide Features

Dinucleotide features are extracted from genomic sequences by calculating the occurance frequencies of dinucleotides [33]. For the DNA nucleotides A, T, G, C, the set of 16 feasible pairs such as GG, GA, GC, GT, AA. etc. compose all the dinucleotides. let r_i be 16-dimensional feature vector and is given as

$$r_i = \frac{f_i}{|L| - 1}, \quad 1 \leq i \leq 16 \tag{1}$$

In Eq. (1), f_i indicate the occurance frequency of the i-th feature and $|L|$ indicate the length of DNA sequence, and a_i denotes the proportional occurance frequency of i-th feature. A 16-dimensional feature vector $(r_i, r_2, ..., r_{16})$ is used to represent each promoter and non-promoter sequence.

3.2 CpG-Island Features

The CpG-island is a DNA region which is rich in cytosine (C) and guanine (G) nucleotide pairs and longer than 200 bp [43]. Like CGGGCG, CpG islands are n-mers having abundant C and G nucleotides [44]. In this paper we extracted two CpG-island features, one is the overall frequency of G and C nucleotides GC_p and another is the ratio of expected to observed CG dinucleotides [11, 28].

$$GC_p = pe(G) + pe(C) \tag{2}$$

$$CpG_{\frac{o}{e}} = \frac{pe(CG)}{pe(C)pe(G)} \tag{3}$$

where $pe(CG), pe(C)$ and $pe(G)$ are percentages of CG, C and G in a DNA sequence respectively. Roughly 70 % of annotated gene promoters are associated with a CpG-island, indicating that this is the most common promoter feature in the vertebrate genome [35].

4 Results and Analysis

In this section we have applied Extreme learning machine (ELM) [18, 19] for promoter prediction using combined dinucleotide and CpG-island features and the performance of the ELM is compared with the feedforward backpropagation algorithm and support vector machine (SVM). For all the experiments in this paper we used 5-fold cross-validation. All the experiments are done in MATLAB 7.10.0 environment running in Intel core i5, 3.10 GHz processor with 8GB RAM. ELM code is provided by Huang et al. [14]. For feedforward neural network we used the faster Levenberg-Marquardt backpropagation algorithm. We used compiled MATLAB code of SVM packages: LIBSVM [41]. Linear kernel function is used in SVM and sigmoidal function is used in ELM.

For all the experiments in this paper we used three datasets as mentioned in Sect. 2.1. Table 1 shows the results of ELM with two and three datasets. In two, datasets we used promoter sequences for positive dataset and for negative dataset we used only one dataset i.e., protein coding sequences. The performance of ELM with two datasets and threee datasets is almost equal, with two datasets the performance is slightly better than threee datasets. When a classification model is used to predict promoters in entire genome we feel, it is more bioligically realistic to train and test classification model on more than one negative dataset.

The performance of BP, SVM and ELM are compared on four eukaryotic organisms, *H.sapiens*, *D.melanogaster*, *A.thaliana* and *C.elegans*, and one prokaryotic

Table 1 Performance evaluation of Extreme learning machine (ELM) on two and three datasets

Species	No. of datsets	*Sn*	*Sp*	Ac
H. sapiens	Three datasets	0.99	0.99	0.99
	Two datasets	0.99	0.99	0.99
D. melanogaster	Three datasets	0.93	0.94	0.93
	Two datasets	0.93	0.91	0.92
A. thaliana	Three datasets	0.95	0.93	0.95
	Two datasets	0.99	0.99	0.99
C. elegans	Three datasets	0.99	0.98	0.99
	Two datasets	0.99	0.99	0.99
E. coli	Three datasets	0.95	0.98	0.96
	Two datasets	0.99	0.99	0.99

Table 2 Comparison of BP and ELM in terms of accuracy, training and testing time

Species	BP					ELM				
	Sn	*Sp*	Ac	Tr.time (s)	Te.time (s)	*Sn*	*Sp*	Ac	Tr.time (s)	Te.time (s)
H. sapiens	0.99	0.99	0.99	110.43	0.31	0.98	0.98	0.98	2.46	0.05
D. melanogaster	0.99	0.99	0.99	195.39	0.31	0.99	0.99	0.99	2.41	0.09
A. thaliana	0.90	0.95	0.94	71.52	0.28	0.95	0.93	0.95	2.48	0.062
C. elegans	0.98	0.99	0.99	90.39	0.31	0.99	0.98	0.99	1.73	0.049
E. coli	0.99	0.99	0.99	7.73	0.043	0.95	0.98	0.96	0.025	0.1e-4

Table 3 Comparison of SVM and ELM in terms of Accuracy, Training and Testing time

Species	SVM					ELM				
	Sn	*Sp*	Ac	Tr.time (s)	Te.time (s)	*Sn*	*Sp*	Ac	Tr.time (s)	Te.time (s)
H. sapiens	0.99	0.99	0.99	38.11	2.43	0.99	0.99	0.99	5.72	0.16
D. melanogaster	0.78	0.85	0.83	48.81	6.82	0.93	0.94	0.93	3.62	0.10
A. thaliana	0.89	0.93	0.92	13.06	1.79	0.95	0.93	0.95	2.48	0.062
C. elegans	0.82	0.85	0.84	10.07	1.39	0.99	0.98	0.99	1.73	0.049
E. coli	0.87	0.93	0.91	0.062	0.018	0.95	0.98	0.96	0.025	0.1e-4

organism *E.coli*. Table 2 shows the results of BP and ELM, the accuracy of BP algorithm for *E.coli* and *H.sapiens* is slightly better than ELM. The accuracy of ELM for *D.melanogaster* is slightly better than BP algorithm, for remaining organisms the accuracies are same. We assigned 200 hidden nodes for ELM for all organisms except *E.coli*, for E.coli we assigned 50 hidden nodes.We assigned 20 hidden nodes for BP algorithm, for all the experiments. For *D.melanogaster*, BP algorithm took 195.39 s for training and 0.31 s for testing, where as ELM algorithm took 2.41 s of training time and 0.09 s of testing time. ELM algorithm is 81 times faster than BP algorithm in training time for *D.melanogaster*. ELM algorithm is faster than BP algorithm for all organisms in training and testing time. In Table 2 because of the out of memory problem when running BP algorithm on *H.sapiens* and *D.melanogaster* we used only 10,000 sequences of each promoter, protein coding and non-coding sequences. We can observe the differences in the results of ELM in Tables 2 and 3 , for these two species. The accuracy of ELM for *D.melanogaster* is decreased, when dataset size is increased and for *H.sapiens* accuracy is slightly increased, when dataset size is increased. For these two species we can also see from Tables 2 and 3, the training and testing time of ELM is relative to the size of the dataset.

Table 3 shows the results of SVM algorithm and ELM algorithm on five different species. From the Table 3 we can see that the ELM algorithm performed better than SVM in terms of accuracy, training and testing time. The accuracy of SVM algorithm on *H.sapiens* is equal to ELM algorithm, for all remaining four organims accuracy of ELM is better than SVM. For *H.sapiens* SVM took 38.11 s of training time and 2.43 s, ELM took 5.72 s for training and 0.16 s for testing. ELM is 6 times and 15 times faster than SVM in training and testing respectively. For all the species ELM algorithm is faster than SVM algorithm in testing and training for promoter prediction problem.

5 Conclusion

We applied Extreme learning machine (ELM) algorithm for promoter prediction problem on eukaryotic and prokaryotic DNA sequences. Dinucleotide and CpG island features are extracted from promoter and non-promoter sequences taken from bench mark datasets. Sensitivity, specificity and accuracy of the ELM algorithm is above 90 % for all the five species which we used in this paper. Performance of the ELM algorithm is compared with the conventional feedforward backpropagation algorithm and support vetor machines (SVM) algorithm. ELM performed equally well when compared to BP algorithm in terms of accuracy. ELM algorithm performed extremely better than BP algorithm in terms of training and testing time. ELM is also superior than SVM algorithm in terms of accuracy, training time and testing time for promoter recognition problem.

References

1. Abeel, T., Saeys, Y., Rouze, P., van de Peer, Y.: ProSOM: core promoter prediction based on unsupervised clustering of DNA physical profiles. Bioinformatics **24**, i24–i31 (2008a)
2. Akan, P., Deloukas, P.: DNA sequence and structural properties as predictors of human and mouse promoters. Gene **410**, 165–176 (2008)
3. Anuj, K., Pramod, K., Rani, T.S., Bhavani, S.D., Bapi, R.S.: Identification of promoter region in a DNA sequence using EM algorithm and neural networks. In: Proceedings of the 1st Indian International Conference on Artificial Intelligence, pp. 676–684. IICAI, Hyderabad (2003)
4. Anwar, F., Baker, S.M., Jabid, T., Mehedi, H.M., Shoyaib, M., Khan, H., Walshe, R.: Pol II promoter prediction using characteristic4-mer motifs: a machine learning approach. BMC Bioinf. **9**, 414 (2008)
5. Bajic, V.B., Seah, S.H., Chong, A., Zhang, G., Koh, J.L., Brusic, V.: Dragon promoter finder: recognition of vertebrate RNA polymerase II promoters. Bioinformatics **18**, 198–199 (2002)
6. Bajic, V.B., Choudhary, V.: Content analysis of the core promoter region of human genes. Silico Biol. **4**, 109–125 (2004)
7. Ben-Gal, I., Shani1, A., Gohr, A., Grau, J., Arviv, S., Shmilovici, A., Posch, S., Grosse, I.: Identification of transcription factor binding sites with variable-order Bayesian networks. Bioinformatics. **21**, 2657–2666 (2005)
8. Bhavani, S.D., Rani, T.S., Bapi, R.S.: Feature selection using correlation fractal dimension: issues and applications in binary classification problems. Appl. Soft Compt. **8**(1), 555–563 (2008)
9. Burden, S., Lin, Y.X., Zhang, R.: Improving promoter prediction for the NNPP2.2 algorithm: a case study using E. Coli DNA sequences. Bioinformatics **21**, 601–607 (2005)
10. Burset, M., Guigo, R.: Evaluation of gene structure prediction programs. Genomics **34**, 353–367 (1996)
11. Davuluri, R.V., Grosse, I., Zhang, M.Q.: Computational identification of promoters and first exons in the human genome. Nat. Genet. **29**, 412–417 (2001)
12. Dreos, R., Ambrosini, G., Perier, R.C., Bucher, P.: EPD and EPDnew, high-quality promoter resources in the next-generation sequencing era. Nucleic Acids Res. **41**, D157–D164 (2013)
13. Dreos, R., Ambrosini, G., Perier, R.C., Bucher, P.: The eukaryotic promoter database: expansion of EPDnew and new promoter analysis tools. Nucleic Acids Res. **43**, D92–D96 (2015)
14. ELM Source Codes. http://www.ntu.edu.sg/home/egbhuang/ (2015)
15. Gordon, L., Chervonenkis, A.Y., Gammerman, A.J., Shahmuradov, L.A., Solovyev, V.V.: Sequence alignment kernel for recognition of promoter regions. Bioinformatics **19**, 1964–1971 (2003)
16. Gordon, J.J., Towsey, M.W., Hogan, J.M., Mathews, S.A., Timms, P.: Improved prediction of bacterial transcription start sites. Bioinformatics **22**, 142–148 (2006)
17. Hannenhalli, S., Levu, S.: Promoter prediction in the human genome. Bioinformatics **17**, S90–S96 (2001)
18. Huang, G.-B., Zhu, Q.-Y., Siew, C.-K.: Extreme learning machine: a new learning scheme of feedforward neural networks. In: Proceedings of International Joint Conference on Neural Networks, pp. 985–990. IEEE, Hungary (2004)
19. Huang, G.-B., Zhu, Q.-Y., Siew, C.-K.: Extreme learning machine: theory and applications. J. Neurocomputing **70**, 489–501 (2006)
20. Lawrence, C.E., Reilly, A.A.: An expectation maximization (EM) algorithm for the identification and characterization of common sites in unaligned biopolymer sequences. Proteins Struct. Funct. Genet. **7**, 41–51 (1990)
21. Li, Q.Z., Lin, H.: The recognition and prediction of r70 promoters in Escherichia coli K12. J. Theor. Biol. **242**, 135–141 (2006)
22. Lin, H., Li, Q.-Z.: Eukaryotic and prokaryotic prediction using hybrid approach. Theory Biosci. **130**, 91–100 (2011)
23. Ohler, U., Liao, G.C., Niemann, H., Rubin, G.M.: Computational analysis of core promoters in the Drosophila genome. Genome Biol. **3**, RESEARCH0087 (2002)

24. Ohler, U., Harbeck, S., Niemann, H., Noth, E., Reese, M.G.: Interpolated Markov chains for eukaryotic promoter recognition. Bioinformatics **15**, 363–369 (1999)
25. Ohler, U.: Identification of core promoter modules in Drosophila and their application in accurate transcription start site prediction. Nucleic Acids Res. **34**, 5943–5950 (2006)
26. Pedersen, A.G., Baldi, P., Brunak, S., Chauvin, Y.: Characterization of prokaryotic and eukaryotic promoters using Hidden Markov models. Proc. Int. Conf. Intell. Syst. Mol. Biol. **4**, 182–191 (1996)
27. Perier, R.C., Praj, V., Junier, T., Bonnard, C., Bucher, P.: The eukaryotic promoter database (EPD). Nucleic Acids Res. **28**, 302–303 (2000)
28. Ponger, L., Mouchiroud, D.: CpGProD: identifying CpG islands associated with transcription start sites in large genomic mammalian sequences. Bioinf. Appl. Note **18**(4), 631–633 (2002)
29. Rani, T.S., Bapi, R.S.: Analysis of n-Gram based promoter recognition methods and application to whole genome promoter prediction. Silico Biol. **9**(1–2), s1–s16 (2009)
30. Rani, T.S., Bapi, R.S.: Cascaded Multi-level promoter recognition of E. coli using dinucleotide features. In: International Conference on Information Technology, pp. 83–88. ICIT, Bhubaneswar (2008)
31. Rani, T.S., Bapi, R.S.: E.coli promoter recognition through wavelets. In: International Conference on Bioinformatics and Computational Biology, pp. 256–262. BIOCOMP, Las Vegas, Nevada (2008)
32. Rani, T.S., Bhavani, S.D., Bapi, R.S.: Promoter Recognition using dinucleotide features : a case study for E. coli. In: International Conference on Information Technology, pp. 7–10. ICIT, Bhubaneswar (2006)
33. Rani, T.S., Bhavani, S.D., Bapi, R.S.: Analysis of E.coli promoter recognition problem in dinucleotide feature space. Bioinformatics **23**(5), 582–588 (2007)
34. Reese, M.G.: Application of a time-delay neural network to promoter annotation in the Drosophila melanogaster genome. Comput. Chem. **26**, 51–56 (2001)
35. Saxonov, S., Berg, P., Brutlag, D.L.: A genome-wide analysis of CpG dinucleotides in the human genome distinguishes two distinct classes of promoters. Proc. Natl. Acad. Sci. **103**, 1412–1417 (2006)
36. Saxonov, S., Daizadeh, I., Fedorov, A., Gilbert, W.: EID: The Exon-Intron database an exhaustive database of protein-coding intron-containing genes. Nucleic Acids Res. **28**, 185–190 (2000)
37. Schmid, C.D., Perier, R., Praz, V., Bucher, P.: EPD in its twentieth year: towards complete promoter coverage of selected model organisms. Nucleic Acids Res. **34**, D82–D85 (2006)
38. Shen, Y., Jiang, Y., Liu, W., Liu, Y.: Multi-class AdaBoost ELM. In: Proceedings in Adaptation, Learning and Optimization, pp. 179–188. Springer (2015)
39. Shepelev, V., Fedorov, A.: Advances in the exonintron database (EID). Brief. Bioinf. **7**, 178–185 (2006)
40. Singh, R.P., Dabas, N., Chaudhary, V., Nagendra,: Online sequential extreme learning machine for watermarking. In: Proceedings in Adaptation, Learning and Optimization, pp. 115–124. Springer, (2015)
41. SVM Source Codes. http://www.csie.ntu.edu.tw/cjlin/libsvm/ (2015)
42. Takai, D., Jones, P.A.: Comprehensive analysis of CpG Islands in human chromosomes 21 and 22. Nat. Genet. **99**, 3740–3745 (2002)
43. Zeng, J., Zhu, S., Yan, H.: Towards accurate human promoter recognition: a review of currently used sequence features and classification methods. Brief. Bioinf. **10**(5), 498–508 (2009)
44. Zeng, J., Zhao, X.-Y., Cao, X.-Q., Hong, Y.: SCS: signal, context, and structure features for genome-wide human promoter recognition. IEEE/ACM Trans. Comput. Biol. Bioinf. **7**(3), 550–562 (2010)
45. Zhang, H., Yin, Y., Zhang, S., Sun, C.: An improved ELM algorithm based on PCA technique. In: Proceedings in Adaptation, Learning and Optimization, pp. 95–104. Springer (2015)
46. Zhang, R., Huang, G.-B., Sundararajan, N., Saratchandran, P.: Multicategory classification using an extreme learning machine for microarray gene expression cancer diagnosis. IEEE/ACM Trans. Comput. Biol. Bioinf. **4**(3), 485–495 (2007)

A Hybrid Approach to the Maximum Clique Problem in the Domain of Information Management

Demidovskij Alexander, Babkin Eduard and Tatiana Babkina

Abstract In this paper we observe the opportunity to offer new methods of solving NP-hard problems which frequently arise in the domain of information management, including design of database structures and big data processing. In our research we are focusing on the Maximum Clique Problem (MCP) and propose a new approach to solving that problem. The approach combines the artificial neuro-network paradigm and genetic programming. For boosting the convergence of the Hopfield Neural Network (HNN) we propose the genetic algorithm as the selection mechanism for terms of energy function. As a result, we demonstrate the proposed approach on experimental graphs and formulate two hypotheses for further research.

Keywords Information management · Maximum clique problem · Hopfield neural network · Genetic algorithm

1 Introduction

This paper is dedicated to the study of one of classical NP-hard problems, proposed by Karp in 1972, which is the Maximum Clique Problem (MCP) [5, 6]. Special emphasis should be laid on the variety of practical implications of the MCP to solving significant real-life problems in the domain of information management. In that domain not only the problem of big data processing is actual, but also the problem of engineering the structures and algorithms for comparison of large amounts of data. In other words, the process of revealing common elements and general regularity in

---

D. Alexander · B. Eduard (✉) · T. Babkina
Higher School of Economics, Bolshaya Pecherskaya 25, Nizhny Novgorod, Russia
e-mail: eababkin@hse.ru

D. Alexander
e-mail: ademidovskiy@hse.ru

T. Babkina
e-mail: taty-bab@yandex.ru
URL:http://www.hse.ru

V. Ravi et al. (eds.), *Proceedings of the Fifth International Conference on Fuzzy and Neuro Computing (FANCCO - 2015)*, Advances in Intelligent Systems and Computing 415, DOI 10.1007/978-3-319-27212-2_25

various objects of the environment becomes the serious task for modern researchers. It is obvious that these computations are going to be more complex as data amount is rocketing.

To begin with, we propose to consider the opportunity of application of the MCP in issues connected with databases and, in particular, their design. There is no doubt that any information obtained should be stored somewhere at least for reasons of being available for further investigations. Due to the fact that amounts of information are rocketing the problem of the proper design of the database appears to be the crucial one in modern circumstances. On the other hand, besides the availability, the other problem arises the problem of efficiency of the database structure e.g. from the glance of convenient retrieving data or getting rid of extra dependencies and duplication. These reasons make the problem of designing databases structure of the great importance. Artifacts gained after experiments can vary in their content and at the same time they can have very similar features. For instance, when analyzing activities of sportsmen we know that all of them have the frequency of heart beat, frequency of their breathing etc. However, for swimmers time would mean something quite different from time of the runner. This brings to the idea of creating such relational databases that would recognize such similarities and differences of the given data and store it according to these features.

In this field fundamental principles of MCP application were proposed by multiple researchers [2, 3, 13]. In particular, Beeri et al. [2] proposed the idea of using maximum cliques to reveal common properties of the given data and this idea was applied to the special class of database schemes, called acyclic. In their later research Beeri et al. [3] developed some techniques of manipulating existing dependencies and again, after casting the database scheme to the split-free normal form, it can be referred to as the graph, where the clique denotes the combination of closely related features to be joined for storing appropriate data. The amazing breakthrough is that such a method can be already applied to relational (cyclic or acyclic) databases.

In order to emphasize the relevance of our research topic, in addition to the problems mentioned above we should draw attention to plenty of other tasks, which can be easily interpreted in terms of the MCP, where entities denote vertices and edges represent interactions between these entities. To be concise, such tasks are: financial analysis, diseases detection and classification, data mining, expert finding and so on. This problem has various practical implications like template recognition, computer vision, comparison of biological structures, producing of complex medicaments, revealing mutations in human genome etc.

From the very beginning, multiple attempts have been made to solve the MCP problem by different methods. Some of them will be observed further in this paper as, for instance, the modifications of the Branch-and-Bound algorithm [15, 18], which uses special heuristics of graph colouring. The most important drawback of these studies is the loss of universality because of focusing on several special types of graphs even though these approaches over-perform more traditional and universal solutions.

The general goal of our work is to elaborate the neuro-network approach to solving the MCP in the arbitrary graph. After correct formulation of mathematical statement of the task neural networks are considered to be a universal tool, which has a great potential from the point of application of parallel computing technologies. In present time the neural network approach is often considered as the most efficient and winning one both in case of speed and precision of results [11, 17, 19–21]. Despite of that, a neural network is able to recognize even noisy data and provide correct results. Moreover, it is essential to mention that neural networks are in most cases resistant to failure. In particular, one of the most valuable features of the Hopfield Neural Networks (HNN) [10] is a relatively simple parallelization, which reveals the additional potential of the usage of the HNN.

In the current paper we propose the novel approach to solving the MCP. Lets refer to this approach hereinafter to as the hybrid approach. It is based on two basic milestones and combines the flexibility and ease of parallelization of the HNN and the adoption ability of genetic algorithms. The proposed method provides the higher speed of a neural network convergence and as a result it is expected to significantly decrease time of solving the MCP on each instance of known graphs from the large DIMACS[1] (Center for Discrete Mathematics and Theoretical Computer Science) dataset. Therefore, our novel technique to solving the MCP should be considered as the efficient and cost-effective tool.

This paper is organized as follows. In Sect. 2 the MCP is formulated and we observe approaches which were already applied to maximum clique search, especially we focus on the structure of the HNN. Furthermore (in Sect. 3), we explain the novel design of the HNN for this problem, which is based on the usage of the genetic algorithm as the selection mechanism for coefficients of the energy function, then we propose the general technique of solving the MCP on the sample graph, and, finally, we describe the structure of the distributed framework which we use as the experimental platform for validation of the proposed in this paper approach. In Sect. 4 we demonstrate results of experiments on a low-dimensional graph from the Second DIMACS dataset. The final analysis and conclusions are given in Sect. 5.

2 Background of the Study

First of all, several terms and definitions should be given for further formulating the MCP.

Definition 1 (*Graph*) A graph G consists of the set of vertices V and the collection (not only the set) of not ordered pairs of vertices, being referred to as edges. The graph is denoted by G = (V, E).

Definition 2 (*Complete graph*) A complete graph K_n is the graph, which consists of n vertices and for every pair of vertices exists the edge between them.

Definition 3 (*Subgraph*) A graph H = (W, F) is called a subgraph of G = (V, E) if W is a subset of V, F is a subset of E.

[1] http://dimacs.rutgers.edu/Challenges/.

Definition 4 (*Clique*) The clique in graph G is the complete subgraph of graph G.

Definition 5 (*Adjacency matrix*) An adjacency matrix of graph is the matrix $A_{nn} = [a_{ij}]$, where the non diagonal element a_{ij} denotes the existence of the edge between vertices i and j, the diagonal element a_{ii} denotes the existence of cycles in the vertice i. The adjacency matrix of graph is symmetric: $a_{ij} = a_{ji}$ for each i and j. The adjacency matrix in the simple graph is binary and every diagonal element is equal to 0. For the complete graph K_n the diagonal element is equal to 1.

Definition 6 (*Maximum clique*) A maximum clique is the clique which size (the number of vertices) is maximum.

Also let us introduce here some definitions required for basic understanding of the proposed approach in the case of the genetic algorithm as the selection tool for the coefficients of the terms of the energy function:

Definition 7 (*Chromosome*) Cromosome is the bit string which represents the individual. It represents the genetic information.

Definition 8 (*Individual*) Individual is the entity, which stores the chromosome.

Definition 9 (*Population*) Population is number of individuals.

Definition 10 (*Crossover*) Crossover is the operation of creation of new individual by mixing chromosomes of two individuals in the population.

Definition 11 (*Mutation*) Mutation is an operation of arbitrary changing the chromosome of the single individual in the population.

Having defined these definitions it is now possible to represent the mathematical formulation of the problem. The MCP is all about the search of the maximum clique in the graph. Lets denote the size of the clique for graph G as w(G). Due to the fact that we are looking for the maximum clique we have to maximize the number of vertices in the clique. Mathematically it can be posed as follows (1):

$$\begin{aligned} &max(\sum_{i=0}^{n} x_i), \\ &where\ x_i + x_j \leq 1\ \forall (i,j) \subseteq E \\ &x_i \subseteq \{0,1\}\ i = 1,..,n \\ &x_i = \begin{cases} 0, \text{if vertice } i \text{ is in clique} \\ 1, \text{otherwise} \end{cases} \end{aligned} \tag{1}$$

Various attempts have been made to elaborate an efficient approach and apply it to the MCP: enumeration algorithms, genetic algorithms, tabu-search, the branch-and-bound algorithm etc. [5, 6, 18] In one of the recent studies ([15]) the MCP was applied to the analysis of large real-life networks for retrieving overlapping communities. More importantly, not only was the community analysis approach proposed, but also a rather competitive advanced Branch-and-Bound algorithm, where the pruning routine was improved to make the algorithm be applicable to massive graphs. The results were reported not only for standard DIMACS dataset, but also for real-world graphs like those from the Stanford Large Network Dataset Collection. Later, in this paper we will use these results as benchmark ones.

The special emphasis should be laid on the neuro-network paradigm due to the opportunity of parallelization and learning (self-learning). As the MCP is the rather complex problem it is especially crucial for the approach to be able to adapt to the input data and produce qualitative results. First attempts to solve this problem using neural networks are practically not comparable with modern studies due to the fact that researches did not include experimental results in papers and reports. Thus, the earliest approach, which already included results of experiments, was the research of Lin and Lee [14], which used 0–1 formulation, as the basis for the heuristic that they applied to the MCP. The second important contribution to the investigation of this problem was made by Grossman [8], who proposed the discrete HNN with an additional threshold parameter, which defined the degree of stability of the neural network. For manipulation of that parameter it was proposed to use simulation annealing. However, the reported results of experiments on DIMACS graph were over-performed by more modern approaches.

The special attention should be paid to the approach, proposed by Jagota et al. [11]. The new contemporary trend, proposed by these scientists, had immediate consequences for the general direction of research. The idea is to decrease dynamics stochastically as much as possible and combine it with mean field annealing imitation. The core problem with this idea is lack of performance on small graphs when comparing to simpler methods, but on high dimension graphs the algorithm outperformed traditional approaches. More importantly, the subject inspired a lot of substantive debate among professionals not only due to performance, but because of non-optimal final solution as well. Jagota' s research has called into question the widespread trend to develop approaches to find the maximum clique, while it is even more important to solve the generalized problem: finding the clique of a certain size. The annealing imitation idea receives strong support as it provides an opportunity to solve the general problem. Due to this fact the idea can be found in numerous publications dedicated to the hierarchical cluster division in pattern recognition operating with the methods of the graph theory. Stochastic networks slightly overperform other similar approaches in case of producing more precise results when applying simulation annealing in the HNN for escaping from a local minimum. The similar approach is used in Boltzmann machines [9]. However the process of modelling these networks is the complex task even for low-dimensional problems.

Therefore, in the framework of current study we apply the deterministic HNN [10, 11, 19]. In 1982 Hopfield proved that such a network can be used for approximation of optimization problems because the network converges to the state which denotes the minimum of energy function if following conditions are satisfied: the weight matrix W should be symmetric ($w_{ij} = w_{ji}$) and the weight of self-connection of every neuron should be equal to 0: $w_{ii} = 0$. The procedure of the neuron update can be observed as the search of the minimum in the solution space of the energy function.

3 The Novel Design of the HNN Dynamics for Solving the MCP

Due to the fact that the dynamics of the HNN is completely described by the energy function, it is important to adopt the general energy function according to the specific conditions of the MCP. The energy function consists of several summands which are usually referred to as terms of the energy function. Traditionally, when being applied to solving various problems, the HNN is working according to statically chosen and once assigned coefficients of these terms. There is no doubt that these coefficients define the dynamics of the HNN, which in other words means that they have an impact on the speed of convergence of the neural network. That is why it is important to choose their values according to each instance of problem to which the HNN is applied in order to obtain best results. This brings to the lack of flexibility of the proposed approaches and tools. In this paper we propose an innovative technique of applying the genetic paradigm as the selection tool for coefficients of the energy function. In other words we introduce a flexible mechanism of adopting the HNN to every graph, in which we try to find the maximum clique. In the rest part of this section we will examine the proposed genetic approach to solving the MCP.

To begin with, we have to formulate the energy function which fully describes the dynamics of the HNN. The example of formulating the energy function can be found in [19], however, we propose the novel energy function. To ensure that we are able to solve the MCP with the HNN we need the energy function reflect the the MCP. We have already formulated the MCP earlier (1) and we use it to build the respective energy function (2):

$$-1/2 * A * \sum_{i=1}^{N} \sum_{j \neq i}^{N} d_{ij} x_i x_j + B * \sum_{i=1}^{N} x_i, \tag{2}$$

where the output of x_j neuron i :

$$x_j = \begin{cases} 1, & \text{if neuron } j \text{ is in maximum clique} \\ 0, & \text{otherwise} \end{cases} \tag{3}$$

$$d_{ij} = \begin{cases} 1, & \text{if vertices } i \text{ and } j \text{ are connected with an edge} \\ 0, & \text{otherwize} \end{cases} \tag{4}$$

A, B—constants. However, we should modify the energy function to the canonical energy function form of the HNN. For this reason we have modified the weight assignment rule proposed in [20] as follows:

$$w_{ij} = d_{ij}(1 - \delta ij) \quad (5)$$

$$h_i = B \quad (6)$$

$$\delta ij = \begin{cases} 1, & \textit{if i is equal to j} \\ 0, & \textit{otherwise} \end{cases} \quad (7)$$

Now the energy function has the canonical form: (3):

$$-1/2 * A * \sum_{i=1}^{N} \sum_{j \neq i}^{N} w_{ij} x_i x_j + \sum_{i=1}^{N} h_i x_i,$$

as for the activation function—we use the classical sigmoid function.

As we can can see in (3) it has two terms. So, we need to find optimal values of these two coefficients. In order to solve the problem of finding coefficients it is exceedingly important to map it in terms of the genetic paradigm. Therefore, we denote the set of coefficients as the chromosome of the size equals two. Next, we should define the size of the population, the maximum number of generations etc. Moreover, we should define if the crossover and mutation mechanism are used, ranges of probabilities for entities to be involved in crossover or mutation. Finally, the finishing condition should be defined as the number of generations, where the minimum suitability remains stable. Having defined these parameters, we can proceed to the exact algorithm of finding optimal coefficients.

Firstly, we initialize the list of entities, according to the defined earlier rules and generate the initial population. Secondly, we estimate the suitability of the initial population and get that one with the highest estimation. After that we pass parameters obtained to the HNN, which cycles until it does not converge. If it does not converge, then we define the suitability of the corresponding chromosome as the worst one. After that, if there is no optimal solution in initial population, the probabilities of entities to be involved in crossover and mutation are then estimated and, according to these probabilities, we prepare and perform the crossover and mutation algorithms. In particular, when applying crossover it is essential to make it only for those entities, whose probabilities are from the corresponding range, discussed earlier. Then, chromosomes, which passed this filtering, are translated to the Gray code, grouped in pairs, each one split into two halves and, finally, they are joined in such a way, that two halves of one chromosome can not organize the new entity. After that, chromosomes are again translated back, checked for the uniqueness and the suitability of every entity is estimated.

The second step is the mutation procedure. Here we use the random selection mechanism for getting entities for mutation, then we apply the transformation of chromosomes to the Gray code and randomly change bits of the chromosome code.

Again, after finishing the procedure we check the uniqueness of the generated chromosomes and calculate their suitability. Finally, all entities are combined, sorted by their suitability and then the selection procedure is performed. As the stop condition we observe the number of cycles of the genetic algorithm while which the minimum suitability among all entities in population remains stable.

To provide deeper comprehension of this specific design of the HNN we introduce the simple graph which is represented in Fig. 1. When speaking in terms of database design, as we discussed it earlier in the Sect. 1, nodes can represent features that we try to store in the database and the clique then represents that these features, which respective nodes are in the clique, are close to each other and should be joined to store appropriate data.

It is obvious that the maximum clique exists in this graph and it consists of vertices $\{2, 3, 5\}$. As it was already mentioned above, while constructing the neural network we map every edge in the neuron, where the weight between neurons denotes the following: the weight of connection is stronger if and only if these two neurons will be in the final solution. In terms of the HNN this means that the energy function decreases when comparing with its current state.

Firstly, we compose the adjacency matrix of this graph (Table 1):

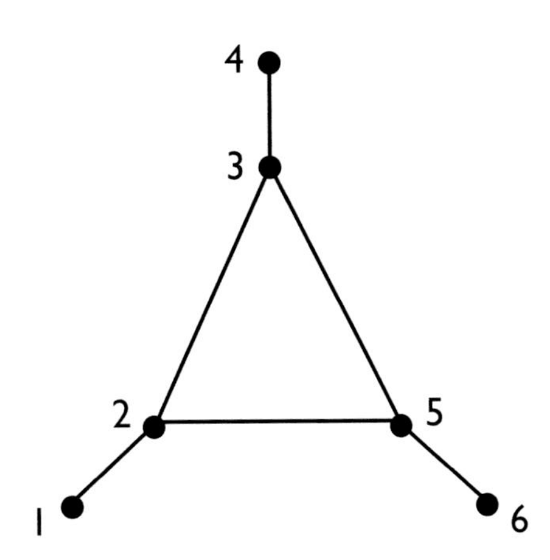

Fig. 1 A sample graph (the maximum clique is {2,3,5})

Table 1 Adjacency matrix of the graph represented in Fig. 1

0	1	0	0	0	0
1	0	1	0	1	0
0	1	0	1	1	0
0	0	1	0	0	0
0	1	1	0	0	1
0	0	0	0	1	0

Next, we make mapping from the 6-vertices graph to the HNN with one layer (Fig. 2).

After the HNN converges we obtain the values at the axons of neurons, which correspond to the output of the neural network. Then we apply the filter to this output and get neurons which output value is extremely close to 1. Having mapped neurons back to the graph structure, we get the resulting set of vertices which stands for the maximum clique. For further investigation and study of the neuro-network approach features in this paper we use the distributed framework described by Karpunina [1]. Besides the HNN the genetic algorithm was also implemented in this framework, which we use in each iteration as the selection mechanism for terms of the energy function. It was implemented in the C++ programming language in the Object-Oriented paradigm. The general principle of the framework functioning is represented in Fig. 3:

In this distributed framework the MPI technology (Message Passing Interface) [4] was used as the mechanism for mass parallelization, which enables us to make computations easily distributed on several clusters, while not losing any computation power and being very reliable. This framework, due to the flexible entity of the HNN, can qualitatively solve the problem if it is formulated correctly. More importantly, we can expect the boost up to 11 times [4] in computing the final solution when comparing with the more traditional approach.

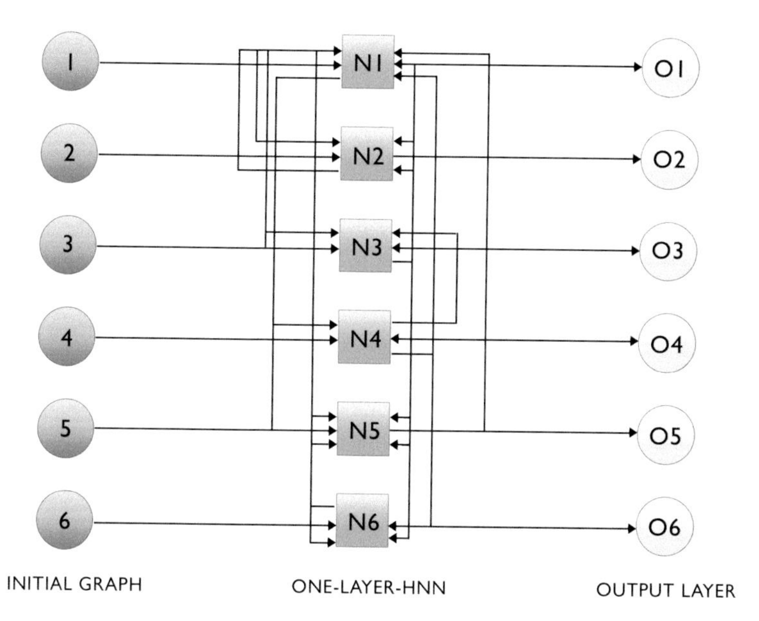

Fig. 2 General technique of solving the MCP on the example of small graph from Fig. 1

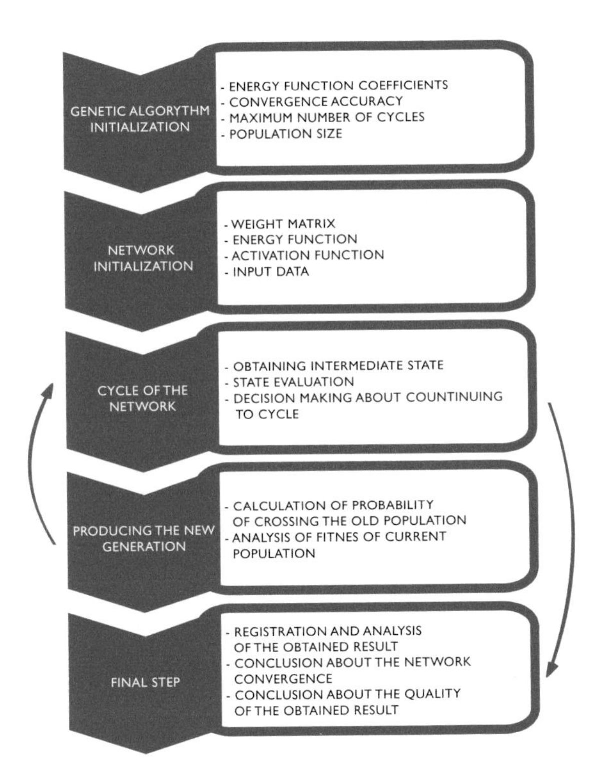

Fig. 3 The structure of the distributed framework

4 Experiments

In this section we introduce results of the preliminary analysis of the proposed approach. The data consists of the general information about the analysed graph and the values of the size of the clique and time spent when applying our innovative technique.

These experiments we carried out on the Intel i5 vPro processor, 4GB CPU, host OS—Windows OS.

Results obtained are demonstrated in Table 4, where:
|V|—number of vertices,
|E|—number of edges,
|W|—actual size of the clique,
$|W_{HNN\&GA}|$—size of the clique obtained by the proposed approach,
$t_{HNN\&GA}$—average time spent on obtaining the result using the proposed approach,
$|W_{BnB[13]}|$—size of the clique obtained by the Branch-and-Bound algorithm [18]
$t_{BnB[13]}$—average time spent on obtaining the result using the Branch-and-Bound algorithm [18]

As we can see from the table we have used some graph instances from the DIMACS benchmark dataset. It contains graphs of various density, size and struc-

Table 2 Results obtained on the several graphs from the DIMACS dataset

Name	$\|V\|$	$\|E\|$	$\|W\|$	$\|W_{BnB[13]}\|$	$t_{BnB[13]}$	$\|W_{HNN\&GA}\|$	$t_{HNN\&GA}$
C125.9	125	6963	34	34	no information	34	0.015
C250.9	250	27984	44	44	no information	44	0.02
C500.9	500	112332	57	57	no information	56	0.1
brock200_2	200	9876	13	13	0.016	13	0.05
brock200_4	200	13089	17	17	0.32	16	0.4

ture. Most of these graphs are divided into groups often referred to as families of graphs, which represent real-life structures and systems. Although his dataset is not the unique one and there are multiple sets of graphs, but historically DIMACS dataset is used in most of papers and, therefore, we also reference to its graphs for benchmarking our proposed solution (Table 2).

From the table we can see that the obtained results are optimal in most of cases, which proves the validity of the proposed approach. However, there is a slight drawback in the size of the maximum clique obtained when comparing with the traditional Branch-and-Bound algorithm [18]. This brings us to the conclusion that our algorithm is rather precise because it produces near-optimal results. Therefore, the usage of the rather computationally complicated state transition mechanism such as the genetic algorithm is a rather efficient technique. Moreover, it is expected to provide much better results on the graphs of the large density and dimension.

However, efficiency of the algorithm is determined not only by the precision of results but also by the cost effectiveness, which means that the efficient approach is a trade-off between the quality and speed of finding the solution. More importantly the proposed approach should be compared to Branch-and-Bound approaches in terms of costs and, in particular, in terms of computational time. To maintain the consistency of the results reported we included information about computational time for one of existing and well-known approaches observed before ([18]). As for the proposed approach we should state that we have got quite similar results, which ensures that the novel design of the dynamics of the HNN provides us with the competitive solution. However, there is the drawback in the minor overhead which stays for the genetic algorithm needs. In general, there is the opportunity of optimizing its primitives to overcome this obstacle in time losses.

5 Conclusions and Directions for Further Research

To sum up, in the framework of this study the broad investigation of the MCP has been conducted, which revealed the actuality of this problem for the modern methods of information management (for example, design of complex data base structures).

The special emphasis should be laid on the neuro-network approach and generic formulations of the problem to allow solving this problem on the arbitrary graph. As we have stated it at the beginning of this paper we refer to the novel proposed algorithm as to the *hybrid* one. The nature of this algorithm combines two paradigms: neural-networks and genetic algorithms, where each one serves the exact aim: basic calculations were performed by the HNN and the selection mechanism of coefficients of the energy function was implemented with the help of the genetic algorithm. The former provides ease of implementation and parallelization while the second considerably cuts the number of cycles of the HNN and improves the flexibility of solving the MCP with the HNN.

In our particular research we offer a special selection mechanism for the coefficients of the energy function of the Hopfield Neural Network (HNN), which provides the flexibility and universalism of the proposed solution. The novelty of our contribution is the efficient combination of the genetic paradigm and the dynamics of the HNN to solving the MCP. To prove the validity of the proposed approach we have implemented it on the basis of the distributed neuro-network framework. Finally, having used the software prototype we compared the quality of results with modifications of another widely used approach - the Branch-and-Bound algorithm [15, 18]. Achieved results form a generic formal framework which may be specialized for particular engineering problems of information management. Specification and analysis of such particular cases are included to our research agenda.

Another valuable result of this paper is that we have identified key hypotheses that can help to build a more powerful and competitive approach to the MCP. They are the following:

1. the application of advanced heuristics can provide much better results. This hypothesis is based on the fact that the HNN is very likely to locate the local optimum instead of the global one, therefore the application of new heuristics can produce better solutions;
2. the application of the distributed implementation of the proposed approach is likely to provide significant speedups in solving the MCP;

Formulated above hypotheses can be considered as potential improvements to the proposed approach and further experiments can prove its validity. These hypotheses are expected to become crucial directions of the further research. In particular, it would be logical to use the advanced distributed Tabu-Machine [1] that is also implemented in the observed above framework. As we have seen in [15] massive parallelization can considerably boost the algorithm and cut computation time.

Finally, we are convinced, that it is important to lay special emphasis on the application of the MCP to the analysis of social networks. Due to the growing interest, inherent complexity and noise in the analysis of social networks we consider that the MCP approach can be efficiently applied to this task and it should be evaluated as the considerably essential tool for such analysis.

More importantly, as millions of people are using Web as the collaboration and work platform, it reveals great opportunities for collecting data about peoples preferences and interests. According to various parameters people can be divided into different groups and subgroups, which are often referred to as *user communities*. Such a division opens new perspectives for business because it enables precise targeting and providing customer-oriented products and services. Such an attempt to reveal user communities was proposed in [16]. The core idea proposed in this research is that it is possible to build a weighted graph according to features specific for users. After applying the threshold filter we get the unweighted graph, where all maximal cliques are found. Not only shows it how people can be grouped, but also helps to obtain some behavioral patterns. From the point of the social network analysis [12] such revealed communities can describe the society analyzed in terms of speed of information exchange, probability of appearance of conflicts and so on. For instance, all these conclusions can be made on the basis of the level of overlapping of these communities. One of considerable examples of analyzing real-life problems is [7], where authors analyzed the social network Ipernity.com as the directed labeled graph.

For the problem of social networks analysis, we propose to imagine that the graph we have analyzed (1) is the small network, for instance, it is the project group. As we have already discussed, we can model such a network as the undirected graph and try to find the maximum clique. Having obtained this clique we can easily detect the core part of the project team, understand the way interactions are performed there and, as a result, it can help the manager or the product owner to better organize the work of this small community. This small and rather artificial example should be considered as the example of application of the MCP to the real-world cases.

References

1. Babkin, E., Karpunina, M.: Towards application of neural networks for optimal synthesis of distributed database systems. In: Proceedings of the 12th IEEE International Conference on Electronics, Circuits, and Systems, Satellite Workshop Modeling, Computation and Services (2005)
2. Beeri, C., Fagin, R., Maier, D., Yannakakis, M.: On the desirability of acyclic database schemes. J. ACM (JACM) **30**, 479–513 (1983)
3. Beeri, C., Kifer, M.: An integrated approach to logical design of relational database schemes. ACM Trans. Database Syst. (TODS) **11**, 134–158 (1986)
4. Dahl, G., McAvinney, A., Newhall, T., et al.: Parallelizing neural network training for cluster systems. In: Proceedings of the IASTED International Conference on Parallel and Distributed Computing and Networks, pp. 220–225. ACTA Press (2008)
5. Du, D.-Z., Pardalos, P.M.: Handbook of Combinatorial Optimization, Second Edn. Springer, Berlin (2013)
6. Du, D.-Z., Pardalos, P.M.: Handbook of Combinatorial Optimization. Springer, New York (1999)
7. Erétéo, G., Buffa, M., Gandon, F., Corby, O.: Analysis of a Real Online Social Network Using Semantic Web Frameworks. Springer, Berlin (2009)
8. Grossman, T.: Applying the INN model to the maximum clique problem. Cliques Color. Satisf. **26**, 125–146 (1996)
9. Hinton, G.E., Sejnowski, T.J.: Learning and relearning in Boltzmann machines. MIT Press, Cambridge, Mass. **1**, 282–317 (1986)
10. Hopfield, J.J.: Neural networks and physical systems with emergent collective computational abilities. Proc. Natl. Acad. Sci. **79**, 2554–2558 (1982)

11. Jagota, A.: Approximating maximum clique with a Hopfield network. IEEE Trans. Neural Netw. **6**, 724–735 (1995)
12. Jamali, M., Abolhassani, H.: Different aspects of social network analysis. In: IEEE/WIC/ACM International Conference on Web Intelligence, pp. 66–72. Springer (2006)
13. Kwok, Y.-K., Karlapalem, K., Ahmad, I., Pun, N.M.: Design and evaluation of data allocation algorithms for distributed multimedia database systems. IEEE J. Sel. Areas Commun. **14**, 1332–1348 (1996)
14. Lin, F.: A parallel computation network for the maximum clique problem. In: ISCAS'93, 1993 IEEE International Symposium on Circuits and Systems, pp. 2549–2552 (1993)
15. Pattabiraman, B., Patwary, Md., Ali, M., Gebremedhin, A.H., Liao, W.-K., Choudhary, A.: Fast Algorithms for the Maximum Clique Problem on Massive Graphs with Applications to Overlapping Community Detection (2014)
16. Pierrakos, D., Paliouras, G., Papatheodorou, C., Karkaletsis, V., Dikaiakos, M.D.: Web community directories: a new approach to web personalization. In: Berendt, B., Hotho, A., Mladenič, D., van Someren, M., Spiliopoulou, M., Stumme, G. (eds.) EWMF 2003. LNCS (LNAI), vol. 3209, pp. 113–129. Springer, Heidelberg (2004)
17. Smith, K.A.: Neural networks for combinatorial optimization: a review of more than a decade of research. INFORMS J. Comput. **11**, 15–34 (1999)
18. Tomita, E., Seki, T.: An efficient branch-and-bound algorithm for finding a maximum clique. Discrete Mathematics and Theoretical Computer Science, pp. 278–289. Springer, Berlin (2003)
19. Wang, R.L., Tang, Z., Cao, Q.P.: An efficient approximation algorithm for finding a maximum clique using Hopfield network learning. Neural Comput. **15**, 1605–1619 (2003)
20. Wang, Y.-F., Cui, G.-Z., Zhang, X.-C., Huang, B.-Y.: An improved hopfield neural network algorithm for computational codeword design. J. Comput. Theoret. Nanosci. **4**, 1257–1262 (2007)
21. Yang, G., Yang, N., Yi, J., Tang, Z.: An improved competitive Hopfield network with inhibitive competitive activation mechanism for maximum clique problem. Neural Comput. **130**, 28–35 (2014)

An Integrated Network Behavior and Policy Based Data Exfiltration Detection Framework

R. Rajamenakshi and G. Padmavathi

Abstract There is a growing concern of exfiltration of sensitive data over the network, with the attackers employing variety of new techniques over wired, wireless networks, distributed platforms and handheld devices. This poses greater challenge for researchers to devise effective detection and mitigation techniques to thwart these attacks. This paper presents an integrated behavior and policy based data-exfiltration detection framework for detecting data exfiltration in the network environment. Firstly, we extend the existing taxonomy for data-exfiltration by including distributed platforms and handheld devices. Secondly we propose an integrated behavior and policy based data-exfiltration detection framework for detecting data exfiltration in the network environment using multiple inputs pertaining to hosts, network and known vulnerabilities. Finally, we present our analysis results that brings out the efficiency of our framework.

Keywords Data exfiltration detection framework • Behavior based model • Policy based model • Data leakage

1 Introduction

Recent day cyber-attacks are well-crafted and pose a challenge to both the research and the user community. Incidents related to system exploitation and sensitive information leakage are on the rise. It is possible to exfiltrate sensitive data by using

R. Rajamenakshi (✉) · G. Padmavathi
Avinashilingam Institute for Home Science and Higher Education for Women, Coimbatore, India
e-mail: menakshi@cdac.in

G. Padmavathi
e-mail: ganapathi.padmavathi@gmail.com

V. Ravi et al. (eds.), *Proceedings of the Fifth International Conference on Fuzzy and Neuro Computing (FANCCO - 2015)*, Advances in Intelligent Systems and Computing 415, DOI 10.1007/978-3-319-27212-2_26

a compromised host that a malicious role within the network. Wilson et al. [1] presents that 20 % of data exfiltration attacks in 2010 are due to malicious insider activities. Kwang [2] discusses briefly the short and long term impact of the cyber threats. The network anomalies are results of (1) malicious activities such as port scanning, DDoS, prefix hijacking etc. and (2) from misconfigurations such as link failures, routing problems.

Most of the organizations take proactive measures to guard their network and hosts by deploying security solutions that would prevent the network being compromised by attackers. Perimeter security solutions like IDS, IPS, firewalls and UTMs are capable of handling the attacks that emanating from outside the network and guard the network from the outside world. Host based security solutions such as anti-virus and anti-malware solutions are designed to protect the hosts from virus and malware and these solutions can only detect the known attacks. It is necessary to note that without the knowledge of the internal network, policies, data storage and access mechanisms, it is impossible to exfiltrate sensitive data. Modern day attacks are capable of taking control of the network by installing root kits, botnets without leaving any traces.

Most of these exfiltration (1) can emanate from a compromised host in a given network or (2) can be done by an insider for many reasons. Key concerns of these types are (1) Majority of the attacks goes unnoticed until it is detected by someone at the time of audit. (2) It is difficult to measure the loss and the impact is huge. (3) From the business perspective, this leads to long term impact and loss of reputation and business. (4) With the increase in the number of Advanced Persistent Threats (APTs) the entire network may be eventually be comprised to the attackers and it may become susceptible to launch other attacks.

These types of attacks pose the following concerns: (1) Majority of the attacks goes unnoticed until it is detected by someone at the time of audit. (2) It is difficult to measure the loss and the impact is huge. (3) From the business perspective, this leads to long term impact and loss of reputation and business. (4) With the increase in the number of Advanced Persistent Threats (APT) the entire network may be eventually be comprised to the attackers and it may become susceptible to launch other attacks.

1.1 Data Exfiltration Taxonomy

Stolfo et al. [3] and Fung [4] presents a detailed survey of insider attacks, detection techniques and future research direction towards detection and prevention of the same. Annarita [5] derived the data exfiltration taxonomy that was broadly classified into network, physical and cognitive. Here the network based data exfiltration includes covert channel and protocol exploits and the exploits using various tools and applications. Given the current day situation, there are exploits in the cloud

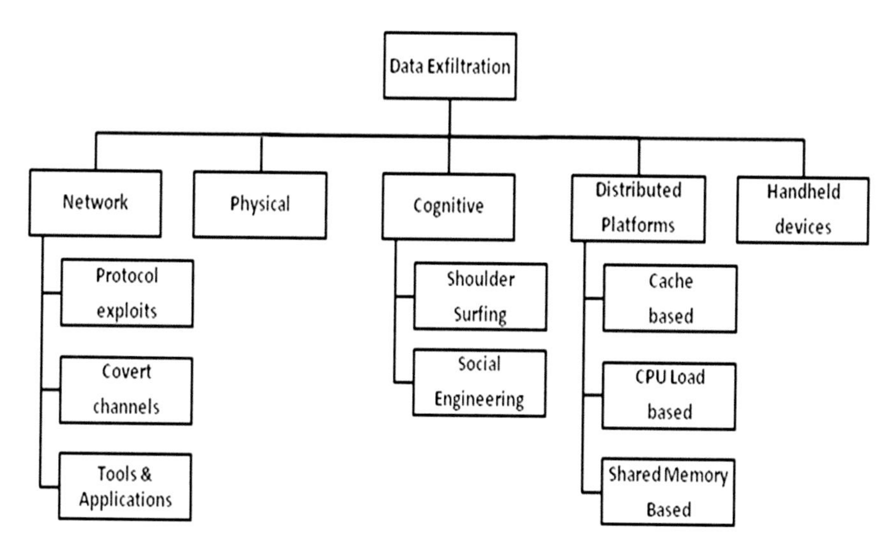

Fig. 1 Data exfiltration taxonomy

platforms, wireless and mobile communication networks, handheld devices. Figure 1 presents a data exfiltration taxonomy that includes distributed platforms and handheld devices to the taxonomy. Here the network exploits includes both wired and the wireless networks. Cognitive techniques includes shoulder surfing and other social engineering techniques.

Table 1 presents data exfiltration exploits under different category. Examples of common insider attacks are exploits, privilege escalation and DNS Poisoning etc., There are different protocol exploits through which covert channel exists. They are broadly classified as storage and timing based channels. Attacker apply precursor techniques such as rootkits, botnets, spyware, covert channels, phishing, pharming etc., to stage data exfiltration attacks later.

Table 1 Data exfiltration exploits

Type	Exploits	Attacks
Network (includes both wired and wireless networks)	Protocol	Different protocol based such as HTTP, TCP/IP. FTP, SMTP, SSH, ICMP, VoIP, RTP are exploited, DNS poisoning
	Covert channels	Timing and storage channels, side channels etc.
	Tools and applications	Malwares, spyware, rootkits, botnets, phishing and pharming
Distributed platforms	CPU load based, cache based and shared memory based exploits	
Physical	CD/DVD, USB and disks, digital media players	

Hence, we need a solution that detects data exfiltration in real time by monitoring the traffic, both network and the internal communications of the host. Though there are techniques that are behavior based they are not specific enough to detect data exfiltration.

This paper focuses on the data exfiltration happening through wired network as a medium. This paper is organized as follows. Section 2 presents the related work, challenges and gaps. Section 3 presents our proposed integrated data exfiltration framework. Section 4 presents experiments, analysis and results and Sect. 5 presents the conclusion and future work.

2 Related Work

Kweang [2] presents a complete coverage of the emerging cyber-attacks using malicious malwares, mobile malware signatures, smart devices, ATM and Point of sales, Phishing attacks. Areej [6] discusses the threat posed by botnets in data stealing and presented the statistics of the recent times. Garfinkel [7] discusses the information leakage over PDF. Bertino [8] explains four distinct dimensions that leads to data exfiltration in DBMS and this can be generalized. Cabuk et al. [9] presents the design and implementation of covert timing channels. Covert channels are between processes in the native network and between two virtual machines in a virtualized environment. Hypervisors are used to isolate the virtual machines running on shared hardware. Covert channels exploit the isolation to exfiltrate data as it is difficult to achieve perfect isolation. Ristenpart et al. [10] has demonstrated L2 cache channel in Amazon EC2. Percival [11, 12] demonstrated inter-process high bandwidth covert channels using L1 and L2 caches. Lee [13] proved that cross VM covert channels are due to the improper isolation in virtualization.

Over the years different frameworks, techniques have evolved that is capable of detecting data exfiltration. Each of these methods cater to a specific type of attack. Puneet [14] presents a detection framework that actively monitors the key parameters across the layers from hardware to the application. Shu [15] applies fuzzy fingerprint detection approach to detect accidental data leaks due to human error and application flaws. Won [16] presents a hybrid approach that combines signature matching and session behavior mapping for identifying application traffic. Wilson et al. [1] presents an access control approach for detecting the data exfiltration over physical media by monitoring files and processes and recording file access/transfer/modification by the host. Jung et al. [17] has architected a 3 level document control system for monitoring information leakage by insiders in the defense environment based on a document access control list (DACLs) mapping between the role and the user. But this model is incapable of identifying exfiltration happening by impersonating (masquerading) user access.

Liu et al. [18] presents SIDD framework that combines statistical and signal processing techniques over network flow data for detecting data exfiltration over the network. Areef [6] presents analysis and detection of data exfiltration over malware using a classifier algorithm base on entropy and Byte frequency distribution. Ramachandran [19] presents a behavior based model for detecting data exfiltration. Lavine [20] used honey pots to detect exploited systems in a network.

The main challenge in detecting data exfiltration attacks is the adversarial behavior attempts are made to mask patterns to make them appear normal to avoid detection. Network intrusion detection falls into two categories vise signature based and anomaly based approaches. Signature based approaches caters well to the needs of known attacks and they use pattern matching techniques to match from the pre-stored signatures. Researchers try out good number of approaches based on statistics, machine learning, and computational intelligence technique to detect intrusions. Figure 2 presents an overall view of the anomaly based intrusion detection approaches. These techniques have been applied on a variety of data such a network flow, data payload etc. to detect the network intrusions.

Thottan et al. [21] presents a detailed coverage and analysis of network anomaly based detection approaches. Hyunchul [22] classified the internet traffic based on the host-behavior based and the flow-feature based that avoided payload inspection. Elias [23] applies heuristics techniques for analyzing false positives alarms from IDS alerts, threat reports, host logging scans, vulnerability reports and search engine queries. Shahreza et al. [24] discusses an anomaly detection approach using Self organizing maps and particle swarm optimization. Hosts are vulnerable to threats application bugs, misconfigurations, missing patches, open ports, rootkits,

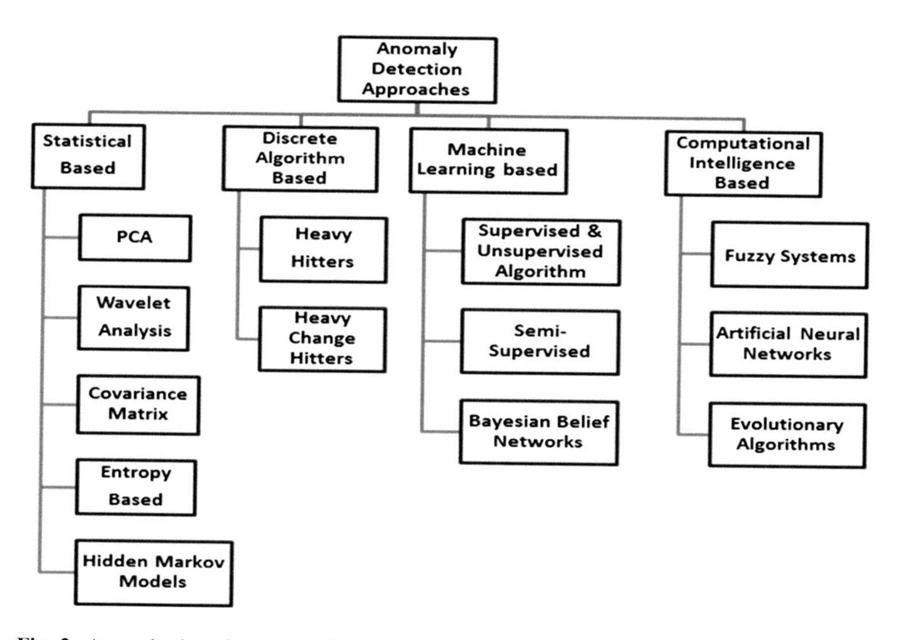

Fig. 2 Anomaly detection approaches

malware etc. Vulnerable hosts poses threat to the entire network. Subramanian [25] presents a threat aware network architecture that assigns a vulnerability threat score. [26–28] applied classification and correlation techniques for identifying IRC botnets. [29, 30] presents botnet detection using DNS traffic similarity. However, they are not designed to detect data exfiltration happening from the host.

2.1 Challenges and Gaps

The data exfiltration attacks exploit the loopholes in the protocols and exhibit the properties of the legitimate traffic that makes it difficult to distinguish between the legitimate and illegitimate traffic. Since the data exfiltration attacks emanate from hosts, it becomes necessary to monitor and profile the behavior of the host.

Present-day signature, policy and behavior based approaches are limited by their capabilities to detect specific insider attack such as data exfiltration, APTs and Zero day attacks. It is essential to devise a specific behavior based model for detecting data-exfiltration over the network. There is a need for a comprehensive solution that is capable of handling all types of attacks by understanding the network attacks as a whole and address the problem.

3 Integrated Data Exfiltration Detection Framework

The proposed integrated data exfiltration detection framework combines the strengths of both behavior based anomaly detection model and policy based approaches to detect data exfiltration from the host as shown in Fig. 3. It consists of three layers as (a) Data collection and preprocessing Layer, (b) Integrated data exfiltration detection engine and (c) the presentation layer. Each of these layers further consists of different components.

The data collection is the first step in data analysis. The data collection sensors are deployed at every host. These profilers, profile the host at different levels and collected data. Here the data is collected using the host, network and the vulnerability profilers. The captured data is then cleaned, preprocessed and then fed into the integrated framework detection engine. The detection engine consists of a set of models that can be applied on the data based on the requirement and preference. The proposed framework comprises of policy based analyzer, network behavior based model, vulnerability analyzer and the event correlator.

Different anomaly detection models are applied to the data. This framework is made flexible so as to accommodate new models and comparison be made against the existing models. Policy based analyzer checks and handles the violations against the thresholds that are set based on the user's behavior and the threat. System policies and thresholds can be changed by the user at any time. The engine

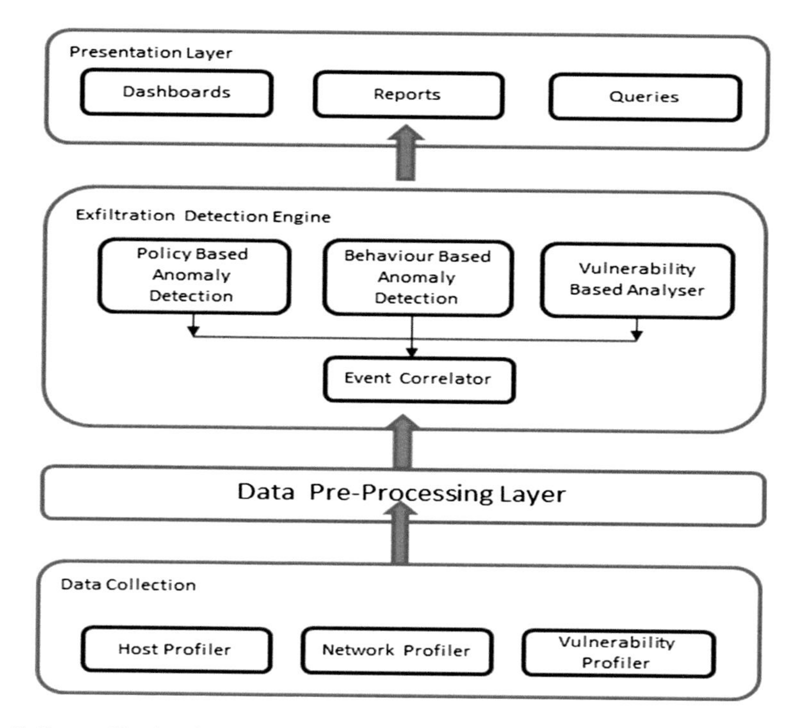

Fig. 3 Data exfiltration detection framework

then performs the collective analysis by event correlating the anomalies from individual components. The alerts are then generated. The same is stored for future references and reporting. The users and system administrators can then view the alerts and the threats using the dashboards. Each of these layers are discussed in detail in the following sections.

3.1 Data Profilers

The data profilers are intended to capture the behavior of the hosts in whole from multiple dimensions that represent the state of the system at given any moment of time. The profilers profile the host at equal interval of time depending on the requirement. The data profiler layer consists of the following components (a) Host Profiler, (b) Network traffic Profiler and (c) Vulnerability profiler.

Host profiler captures the system resources utilization and system logs that includes file system logs and the kernel logs. File system logs captures the unauthorized login and the file movements including external devices. Network profilers capture the network statistics (SIP, DIP, SP, DP and Protocol) along with other

Table 2 Data profilers and feature extraction

Profiler	Features profiled
Host profiler	System resources (CPU, memory and disk utilization)
	System logs (file systems, kernel logs)
	File and disk utilization statistics, device monitoring and tracing, user logs
Network traffic profiler	Incoming and outgoing packet bytes
	Control packets associated with teh host IP and MAC
Vulnerability profiler	Threats and vulnerabilities of the host, severity of the threats and the details of the open ports et as obtained using vulnerability scanners

parameters such as incoming and outgoing bytes and packets, control packets etc. Network traffic statistics and the system resource utilization are collected using SNMP MIBs. Vulnerability profilers collect vulnerable applications, processes and open ports running in the host. Table 2 presents the details of the data collected by each of these profilers.

The host is profiled for the interval specified (say every 1 s) and the values are aggregated by the preprocessor for further processing. The main intent is to capture every change in the host in terms of its behavior and the state of the system in terms of storage and memory. The preprocessed data is then fed into the detection engine for further processing.

3.2 Integrated Data Exfiltration Detection Engine

This engine is the heart of this system. It comprises of the following components (1) Network anomaly Analyzer (2) Policy based Analyzer, (3) Vulnerability analyzer, and (4) an Event correlator. This engine analyses the anomalies from individual component at a given time and decides whether any exfiltration has actually happened. It is important to note that all of these components here are based on the behavior of the host under study.

From the past research, it is clear that data leakage attacks emanate from hosts. It is necessary to understand the normal behavior of the host. Behavior based anomaly detection models compares the current behavior of the host with the normal behavior and alerts the deviations if any in behavior. Behavior based anomaly detection is divided into two phases viz., the training or the learning phase and the actual detection phase. Initially the data is collected and trained for a specific time period. It is assumed that the training data actually captured reflect user's behavior. During the detection phase, the detection engine compares the current behavior with the normal behavior and generated alerts. Figure 4 presents the representation of the same.

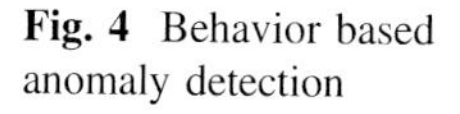

Fig. 4 Behavior based anomaly detection

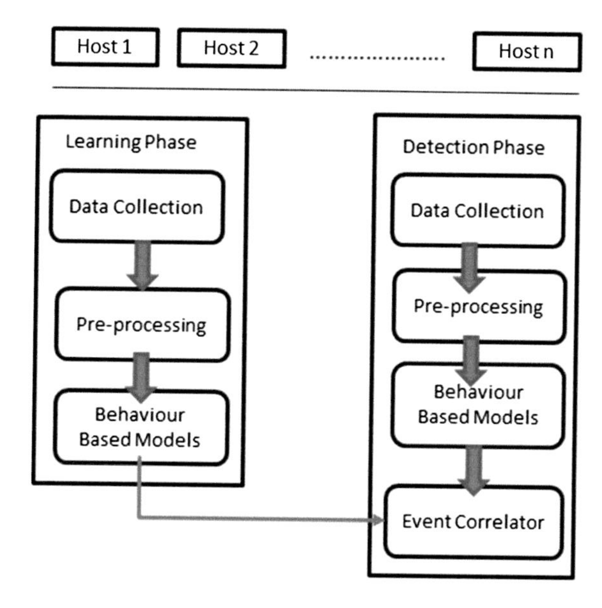

3.3 *Network Anomaly Analyzer*

Data Exfiltration takes place to and from hosts in the network. Considerable amount of system parameters are utilized when the data exfiltration is actually happening. Network anomaly combines the system and the network parameters to analyze if there is any anomaly. System and the Network parameters are captured using SNMP MIBs. The normal behavior of the host is obtained during the training phase. During the detection phase, the values obtained in training and the detection phases are compared for any deviation and to assert the anomalies. Different models can be applied at this point depending on the choice of the user. Here we tried two different algorithms namely the kernel density estimation technique and nearest neighborhood model.

The kernel density model was constructed as follows with the assumption that there are m hosts in a given network N. Let $H = \{h_1,h_2,h_3....h_m\}$ be the set of hosts with the learning period L and the profile period P. Let $X = \{x_1,x_2...x_n\}$ be the set of feature parameters that are profiled corresponding to the list (CPU utilization, memory utilization, source address, destination address, source port, destination port, incoming bytes, incoming packets, outgoing bytes and outgoing packets) of values.

Let $X_t = \{xt_1,xt_2,xt_3,...xt_n)$ be the set of values profiles at any time t for any host h_i $\forall$ i = 1 to m. For any given Xt Kernel density estimation for each of the parameter in Xt values are computed using the Eqs. (2) and (5). We compute the kernel density estimators at some time interval I (say every 2–3 min). The values are then stored along with the given time interval

$$f(x) = \frac{1}{n} \sum_{i=1}^{n} w(x - x_i, h) \tag{1}$$

where w(t, h) as defined by Gaussian Weighing function in Eq. (5)

$$w(t, h) = \frac{1}{\sqrt{2\pi}.h} e^{-t^2/2h^2} \tag{2}$$

In the detection phase, we again apply the same process as in the learning phase to compute the KDE values of all the hosts for the identified parameters for the interval I. Now the KDE values obtained in learning (A) and detection phase (B) compared using Karl Pearson's Correlation coefficient (r) for every parameter in X using the formula in Eq. 3.

$$r = \frac{\sum_{i=1}^{n} (X_i - \bar{X})(Y_i - \bar{Y})}{(n-1)\sigma_x \sigma_y} \tag{3}$$

In the similar fashion, alternate models can be used in place of kernel density estimation and evaluated. In case of a new model, there is a need to generate a new set of normal behavior model if so required. We also devised nearest neighborhood model that compares the computed averages of xi's in a moving window of time.

The outcomes from the model from both the training and the detection phase are compared using Karl Pearson's Correlation coefficient. As per the established theory, correlation coefficient shall lie between −1 and 1. Since all the values that are obtained are positive the range is between 0 and 1. If the values are closer to 1 then it implies that they establish a stronger relationship between the variables. If the value of r is close to 0 then it exhibits a weak relation or no correlation at all. This indicates that there is a disagreement between the learning and the current value, hence an anomaly. Here we define that if correlation value between KDEs of any parameter lie in the range 0.5 and 1, then it is an acceptable behavior. If the values fall in the range of 0–0.5 then this implies anomaly.

This way, we compute the correlation co-efficient for all the parameters. If we identify anomalies with respect to say one system based parameter like CPU utilization then we see if there exists anomalies identified based on other network parameters such as outgoing packets. Indeed if we find such cases then it signifies data-exfiltration. As mentioned earlier data-exfiltration anomalies are specific events on anomalous events which can be detected by considering both system and network parameters together.

If the result exhibits a weak correlation between the learnt and current phase for a given parameter, we term them as anomaly, otherwise the traffic is normal. Further, to confirm data exfiltration anomaly, we look into other parameters for similar deviations during the same interval. For example, if we detect anomalies in CPU utilization, then we see if the correlation of outgoing traffic KDE values is also anomalous during same time period. If that is true then we conclude that these anomalies reveal the data-exfiltration attack.

We categorize the anomalies based on the parameters taken into consideration for correlation. We group the anomalies as (i) infiltration, when the incoming traffic parameters are correlated with the system utilization (ii) exfiltration, when we correlate the outgoing traffic with the system utilization.

3.4 Policy Based Analyzer

Policy based analyzer computes the threshold values for file transfers, disk utilization, number of unsuccessful logins, using external devices etc. These values are set based on the threat scores of the host. Higher the risk of the host, lower threshold limits are set. The system and the kernel logs are analyzed for computing the unsuccessful logins, file transfer details along with the details of the files. This also captures the applications and the process that are run at a specific time. It also alerts the increase or decrease in disk utilization beyond the threshold values. Table 3 presents the sample policies on which the thresholds are set. These policies can be set either for individual hosts or for a group of hosts.

3.5 Vulnerability Analyzer

Vulnerability analyzer analyses the threats and the vulnerabilities present in the hosts. The threats vary between highly secured systems to the highly vulnerable systems. The threat scores can be assigned to the threats based on the nature of the

Table 3 Sample policies and descriptions

Policies	Description
DISK_SPACE_INCR_PERCENT	Alerts if the % disk space utilization increases beyond this value in proportion to the actual disk size
FAILED_LOGIN_ATTEMPT	Alerts if wrong password entered for this much times in a day
DISK_SPACE_RED_PERCENT	Alerts if the utilized disk space decreases more than this value of total disk size (in %)
SYS_REBOOT_COUNT	Alerts if the system rebooted more than this much times in a day
FAILURE_REMOTE_LOGIN	Alerts if the wrong password entered for this much times for accessing system remotely
KERNEL_VERSION	Alerts kernel version older than this version

Table 4 Event correlation classification chart

Policy based analyzer	Network analyzer	Vulnerability threat score	Class
Yes	Yes	High	Yes
Yes	No	High	Yes
Yes	No	Low	Yes
No	Yes	Low	No
No	No	No	No

threat. There are tools that perform vulnerability scans over the network and generate vulnerability reports for each scans. The values are then used for comparison. Comparison of each scan can be computed to evaluate the change in the threat state of the host.

3.6 *Event Correlator*

The event correlator correlates the anomalous events from one or more components. Depending on the threat score of the host, it correlates the anomalies and classify the same as anomalous or not. If either of the anomaly detection has an anomaly at the same time then it generates an alert classifies that as an anomaly. Table 4 shows sample classification for few combinations.

4 Experiments and Analysis

The data from a real time network consisting of 10 machines were collected and analyzed. Vulnerability scans were done on a daily basis that presents the vulnerability and threat score of the hosts. The hosts were trained for a period of a 7 days and further averaged to get the normal profile of the host. It is assumed these hosts were free from attacks during the training period. At the end of the training phase, respective behavior are stored. This profile include system and the network parameters. Based on the vulnerability and the threat score that were obtained, the policies were decided (Table 5).

For a highly secure host, change in the threshold values were set at 10 % which would allow change up to 10 % for file transfer and disk utilization. For a medium and highly vulnerable hosts, the threshold values were set at 7 and 4 % respectively. The data exfiltration attacks such as DNS attacks, DOS attacks, SYN flooding and Port scanning were performed. The results of the network analyzer for kernel density estimation and the nearest neighborhood techniques were compared (Table 6).

Table 5 Policy thresholds that are set for the users

Users	Disk space utilization (%)	Failed login attempt	System reboot count	Failure remote logins	External device access
User 1	8	6	7	2	20
User 2	4	6	7	8	9
User 3	10	8	15	9	6

Table 6 Comparison of KDE and nearest neighborhood model

Attacks performed	Total no of attacks	KDE model		NN model	
		Normal	Exfiltration	Normal	Exfiltration
DNS (without payload)	14	7	7	3	11
DNS (with Payloadof 50 bytes)	7	5	2	0	7
DNS (with Payloadof 200 bytes)	6	2	3	1	5
SYN	6	1	5	4	4

5 Conclusion

To conclude, our integrated framework for detecting data exfiltration based on host behavior, captures the behavior of the host with respect to the system and the network utilization. Also it captures every activity of the user with respective to file systems, captures the activities of the host with external devices. Based on our experiments, it was able to capture the file transfer happening from the host at relative medium and big files. Also we were able to capture the DDoS, port scanning attacks etc. accurately. However, we have to refine our model that is capable of detecting slow scan attacks and covert channels. Further we would like to extend this model to capture the covert channel attacks.

References

1. Wilson, D., Lavine, M.K.: A discretionary access control method for preventing data exfiltration (DE) via removable devices. In: ICDF2C, LNJCST, vol. 31, pp. 151–160 (2010)
2. Kwang, K., Choo, R.: The cyber threat landscape: challenges and future research directions. Comput. Secur. **30**, 719–731 (2011)
3. Stolfo, S.J., Bellovin, S.M., Hershkop, S., Keromytis, A.D., Sinclair, S., Smith, S.W.: Insider Attack and Cyber Security Beyond the Hacker. Advances in Information Security. Springer, New York (2008)

4. Fung, C.: Collaborative intrusion detection networks and insider attacks. J. Wirel. Mob. Netw. Ubiquit. Comput. Depend. Appl. 63–74 (2011)
5. Giani, A., Berk, V., Cybenko, G.V.: Data exfiltration and covert channels. In: SPIE 6201, Sensors, and Command, Control, Communications, and Intelligence (C3I) Technologies for Homeland Security and Homeland Defense V, 620103 (2006)
6. Al-Bataineh, A., Gregory White, G.: Analysis and detection of malicious data exfiltration in web traffic. In: IEEE Conference on Malicious and Unwanted Software (MALWARE) (2012)
7. Garfinkel, S.L.: Leaking Sensitive Information in complex document files—and how to prevent it. Secur. Priv. IEEE **12**(1), 20–27 (2014)
8. Bertino, E., Ghinita, G.: Towards mechanisms for detection and prevention of data exfiltration by insiders. In: ASIACCS '11 Proceedings of the 6th ACM Symposium on Information, Computer and Communications Security, pp 10–19 (2011)
9. Cabuk, S., Brodley, C.E., Shields, C.: IP Covert timing channels: design and detection. In: ACM Conference on Computer and Communications Security—CCS, pp. 178–187 (2004)
10. Ristenpart, T., Tromer, E., Shacham, H., Savage, S.: Hey, you, get off of my cloud: exploring information leakage in third-party compute clouds. In: Proceedings of the 16th ACM conference on Computer and communications security, pp. 199–212 (2009)
11. Xu, Y., Bailey, M., Jahanian, F., Joshi, K., Hiltunen, M., Schlichting, R.: An exploration of L2 cache covert channels in virtualized environments. In: ACM Cloud Computing Security Workshop, pp. 29–40 (2011)
12. Percival, C.: Cache missing for fun and profit. In: Proceedings of the BSDCan (2005)
13. Wang, Z., Lee, R.B.: Covert and side channels due to processor architecture. In: Proceedings of the 22nd Annual Computer Security Applications Conference, pp. 473–482 (2006)
14. Sharma, P., Joshi, A., Finin, T.: Detecting data exfiltration by integrating information across layers. In: IEEE information reuse and integration (2013)
15. Shu, X., Yao, D.: Data leak detection as a service. LNCS: Soc. Inf. Telecommun. Eng. **106**, 222–240 (2013)
16. Won, Y.J., Park, B.J., Ju, H., Kim, M., Hong, J.W.: A hybrid approach for accurate application traffic identification
17. Eom, J., Kim, N., Kim, S.H., Chung, T.M.: An architecture of document control system for blocking information leakage in military information system. Int. J. Secur. Appl. **6**(2) (2012)
18. Liu, Y., Corbett, C., Chiang, K., Archibald, R., Mukherjee, B., Ghosal, D.:SIDD: A framework for detecting sensitive data exfiltration by an insider attack. In: Proceedings of the 42nd Annual Hawaii International Conference on System Sciences, pp. 1–10 (2009)
19. Ramachandran, R., Neelakantan, S., Bidyarthy, A.S.: Behavior model for detecting data Exfiltration in network environment. In: IEEE IMSAA (2011)
20. Levine, J., LaBella, R., Owen.H., Contis, D., Culver, B.: The use of honeynets to detect exploited systems across large enterprise networks. In: Workshop on Information Assurance (2003)
21. Thottan, M., Liu, C., Ji, C.: In Anomaly Detection Approaches for Communication Networks. Computer Communications and Networks, pp 239–261 (2010)
22. Kim, H., Claffy, K.C., Fomenkov, M.: Internet traffic classification demystified: myths, caveats and the best practices. In: ACM Context (2008)
23. Raftopoulos, E., Dimitropoulos, X.: Detecting, validating and characterizing computer infections in the wild. In: ACM ICM (2011)
24. Shahreza, M.L., Moazzami, D., Moshiri, B., Delavar, M.R.: Anomaly detection using a self organizing map and particle swarm optimization. Sci. Itanica D **18**(6), 1460–1468 (2011)
25. Neelakantan, S., Rao, S.: A Threat-aware hybrid intrusion—detection architecture for dynamic network environments: CSI J. Comput. **1**(3) (2012)
26. Levine, J., LaBella, R., Owen, H., Contis, D., Culver, B.: The use of honeynets to detect exploited systems across large enterprise networks: In: IEEE IAW (2003)

27. Binkley, J.R., Singh, S. : An algorithm for anomaly-based botnet detection. In: USENIX SRUTI (2006)
28. Livadas, C., Walsh, R., Lapsley, D., Strayer, W.: Using machine learning techniques to identify botnet traffic. In: IEEE LCN (2006)
29. Strayer, W., Walsh, R., Livadas, C., Lapsley, D.: Detecting botnets with tight command and control. In: IEEE LCN (2006)
30. Choi, H., Lee, H., Lee, H., Kim, H.: Botnet detection by monitoring group activities in DNS traffic. In: IEEE CIT (2007)

Web Usages Mining in Automatic Detection of Learning Style in Personalized e-Learning System

Soni Sweta and Kanhaiya Lal

Abstract The e-learning system generates huge amount of data which contain hidden and valuable information and they are required to be explored for useful knowledge for decision making. Learner's activity related data and all behavioral vis-a-vis navigational data are stored in the log files. Extracting knowledgeable information from these data by using Web Usage Mining technique is a very challenging and difficult task. Basically, there are three steps of Web Usage Mining i.e. preprocessing, pattern discovery and pattern analysis. This paper proposes a Dynamic Dependency Adaptive Model (DDAM) based on Bayesian Network. This model mines learner's navigational accesses data and finds learner's behavioral patterns which individualize each learner and provide personalized learning path to them according to their learning styles in the learning process. Result shows that learners effectively and efficiently access relevant information according to their learning style which is useful in enhancing their learning process. This model is learner centric but it also discovers patterns for decision making process for academicians and people at top management.

Keywords Adaptive learning · Web usages mining · Bayesian network · Learning process · Personalized learning

1 Introduction

The term ''learning styles'' refers to the concept that individual differs in context of how they process information [1] and students' preferred ways to learn [2]. There are many learning style models described in literature [3–6], which show that every

S. Sweta (✉) · K. Lal
Department of Computer Science and Engineering, Birla Institute of Technology, Mesra, Ranchi (Patna), India
e-mail: soni.sweta16@gmail.com

K. Lal
e-mail: klal@bitmesra.ac.in

V. Ravi et al. (eds.), *Proceedings of the Fifth International Conference on Fuzzy and Neuro Computing (FANCCO - 2015)*, Advances in Intelligent Systems and Computing 415, DOI 10.1007/978-3-319-27212-2_27

individual prefers to learn in different learning style. According to above theories, Adaptive e-learning system can also strengthen above concept so that incorporating student learning style definitely enhances the learning process of learner. System that incorporate learning style can provide suggestion to students as well as instructor to optimize students' learning path. This automatic detection system also overcome the drawbacks of the traditional detection method in e-learning system which is mainly based on questionnaire. In this paper, authors describe how learning process influenced by learning styles and which parameters affect to evaluate personalized learning path according to individual learning style because preferred mode of input varies from individual to individual. Rest of the paper is divided into four sections. Second section describes background and all related work done so far in this domain. Third section explains the concept of proposed model. Fourth section is an experimental section which gives the results and related discussion. Last section concludes the whole work and gives the future aspects.

2 Background and Related Works

2.1 Background

Adaptive personalized e-learning is able to support different learning paths and contents according to learner's preferences so that it suits and fits into every individual learner's diverse needs and backgrounds [7, 8]. Many literature surveys showed that many work in domain of e-commerce have been done, but not much work done in e-learning domain. As advancement in technology and development of new tools, the concept of e-learning is highly demanded in view of how to know our students preferences and enhancing their learning processes. It is a challenging task in recent era. In traditional e-learning system, only few experts' opinions were responsible to provide learning paths and content in adaptive learning system so it was teacher centric. But here we have tried to provide learner centric adaptive system which enable to support learners with more self-control and efficient learning in the given e-learning environment. In context of adaptive e-learning system, Web Usages Mining is the application area of data mining techniques through which it discovers patterns from web data, targeted towards various applications of e-learning system. Figure 1 given below shows the different application of Web Usages Mining.

Web Usages Mining consists of mainly three components, preprocessing, and pattern discovery and pattern evaluation. This is show in flowchart given in Fig. 2.

There are many data mining techniques used in different processes of Web Usages Mining for example statistical analysis, association rule, clustering, classifications, sequential patterns and dependency model are used in pattern discovery.

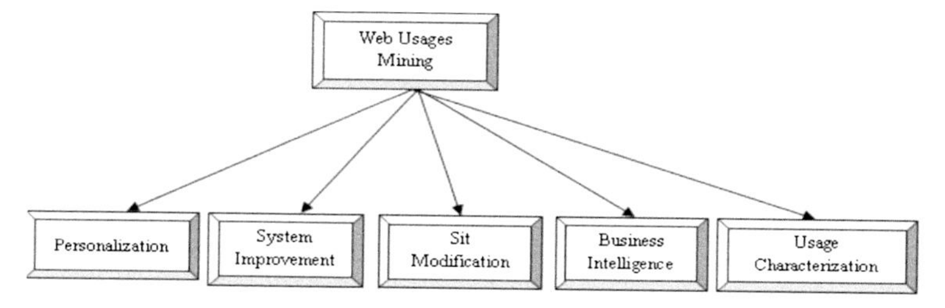

Fig. 1 Web usage mining application areas

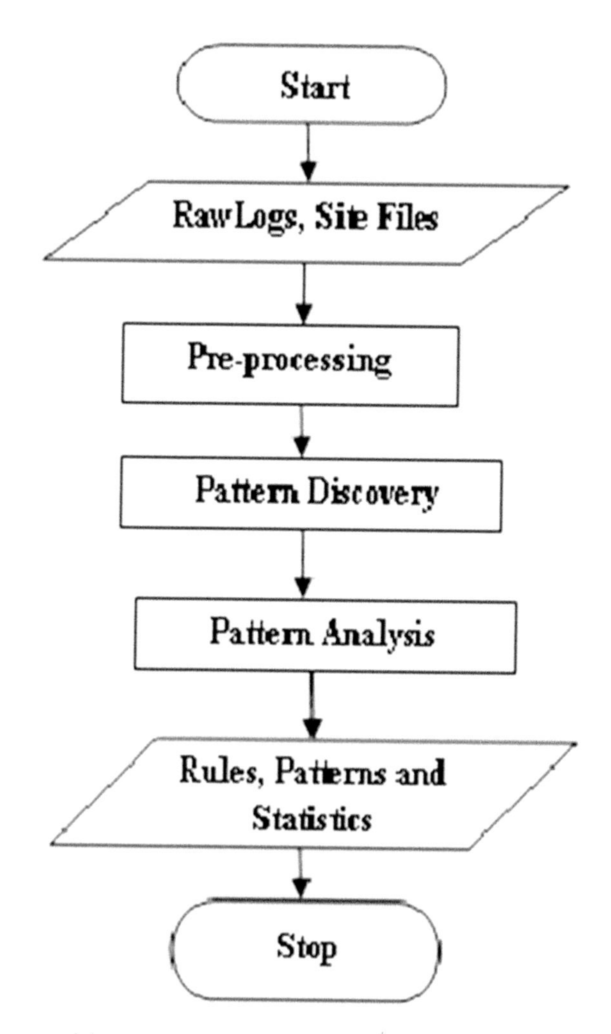

Fig. 2 Web usages mining process

2.2 Related Works

LS-AEHS is an adaptive e-learning system which incorporates learning style and show the effect of learning achievements of learners after adapting matching learning materials according to their learning style [9]. Increasing the effectiveness and efficiency of learning courses and learners' satisfaction by adapting to prior knowledge is an important approach [10, 11].

The Web Watcher [12], Site Helper [13], Krishnapuram [14], and clustering work by Mobasher et al. [15] and Yan et al. [16], all they have focused on providing Web Site personalization based on web usage mining. Yan et al. [16] used web server log data. They found and analyzed clusters of users having similar access patterns. Web Usage Mining [17, 18] is tool in the Internet community where data form online web is converted into meaningful knowledgeable utilities. There are many data mining techniques have been used to represent student models, such as rules [19], fuzzy logic [20], and case-based reasoning [21].

3 Proposed Approach for Detecting Learning Style

In the following subsections, about learning style, features of the patterns of navigational access behavioral data relevant to learning style and implementation details are presented.

3.1 Learning Style

Learning Style (LS) of a learner is a way how a learner collects, processes and organizes information [22]. This paper is based on Felder-Silverman learning style model. According to FSLSM, each learner has a preferred mode of learning style measured in four dimensions (active/reflective, sensing/intuitive, visual/verbal, sequential/global) [1, 5, 23]. The concept for providing adaptivity based on learning styles aims to enable LMSs to automatically generate personalized learning path according to the learner's learning style.

3.2 Relevant Features of Behavior Pattern

There are many parameters or variables values generated by user navigational pattern stored in log file on server. We only analyses those variables which directly or indirectly relevant and correlated with corresponding learning style of FSLS. These variable are described in Table 1.

Table 1 Relevant variables for each corresponding learning style dimension of FSLSM

Processing dimension (active/reflective)	Input dimension (verbal/visual)	Organization dimension (global/sequential)	Perception dimension (sensory/intuitive)
freq_Lo	time_Lo	t_overview	t_abstract
time_Lo	ques_graphic	freq_overview	t_illustration
type_Lo	ques_text	ques_detail	h_abstract
attamp_example	chat_part	ques_overview	h_concrete
t_example	t_chat	ques_interpret	t_lab
quiz_results	n_revision	ques_develop	h_lab
chat_part		navigation_skip	
t_chat		navigation_overview	
freq_assign		t_navigation_overview_	

The threshold values for all given parameters are set according to the literature [24, 25] and by using some statistical functions. Prefix used before the variables name i.e. t, and freq, are used for time and frequency.

3.3 Implementation

Implementation of this approach basically uses multiple resources because not only one resource has all features, which gives the best result. Moodle is one of the best open source software for providing to create powerful, flexible and engaging online courses and experiences in learning management system [26, 27]. The data gathered by Moodle LMS may require less amount of work in data pre-processing than data collected by other systems because it stored all the relevant and authenticate web usages data in database as well as in log file. Weka [28, 29] is open source software that provides a collection of machine learning and data mining algorithms. Now DDAM model is formed based on Bayesian Network with the help of WEKA tool. Weka supports ARFF file format which gives the additional benefit for using Moodle platform.

A Bayesian Network (BN) is based on Bays theorem which gives the formula of calculating conditional probability and is composed of two components: qualitative for defining structure and quantitative part for quantify the network [30]. This Bayesian network gives the conditional probabilities of all dependent variables which represent the strength of the dependencies among nodes represented by variables used to find the corresponding learning characteristics for a particular learning style. Moodle does not have any visualization tool. Therefore, in this paper Gismo tool [31] used for visualization. Weka, Moodle and Gismos all three have many common features like they support ARFF file format, may be implemented in Java and they can work on same dataset. So, these are the main reasons of using WEKA, Moodle and Gismo tools in combination.

3.4 Dynamic Dependency Adaptive Model (DDAM)

Generally learners were not very attentive and responsive about to fill ILS questionnaire due to many reasons. They may be influenced by others which can lead to wrong information about their learning style [32]. Our proposed approach based on [33, 34] is that the users' preferences can change due to many reasons. Factors that affects are the type and quality of the learning objects of the course.. Therefore, the dynamic user modeling based on students' behaviors in a course level or in a session level is strongly recommended.

Here, we only consider four dimensions of FSLS and eliminate organizational dimension because it is proved that induction is the natural human learning style. Experiments have also proved that most engineering students are inductive learners [1]. According to this, we collect all the relevant variables described above in Table 1 from LMS and set its threshold values in form of marginal probability distributions according to literature [24, 25]. By calculating the conditional probabilities of nodes which represent variables corresponding to the characteristics of particular learning style in reference to FSLS learning dimensions, Bayesian network is formed which shows the relationship among random variables. Arch provides the strength of relation among variables. Similarly based on this conditional probability distribution of parameters, we form all Bayesian Networks of all the four dimensions using relevant parameters which is shows in Fig. 3. For example for deciding input dimension, we set typ_Lo, t_t_Lo, no_visit_Lo and its threshold value as deciding factor. For example threshold values for chat, data access pattern and exam are given in Tables 2, 3, and 4.

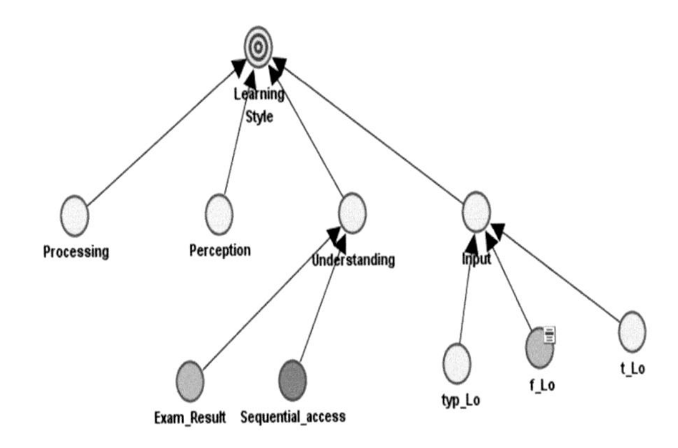

Fig. 3 Bayesian network model for learning style

Table 2 Chat probability threshold value

Chat	Probability
Participation	0.05
Listen	0.03
Not participation	0.00

Table 3 Data access probability threshold value

Data access pattern	Probability
Sequential	1.0
Global	0.03

Table 4 Exam result probability threshold value

Exam	Probability
High	0.07
Medium	0.03
Low	0.0

4 Experiment and Result

Using our GyanDarshan e-learning Tutorial made-up on Moodle platform, it is shown in Figs. 4 and 5, 40 participants logging data who are registered in a course C++, used for experiment purpose. Now with the help of WEKA software applied with DDAM classifier which analyses the patterns and automatically detects learning.styles. The relevant variables are based on many activities data associated with topics of a course and of a particular session shown in Table 1. This also gives the learner's navigational access and behavioral data. Gismo provides visualization

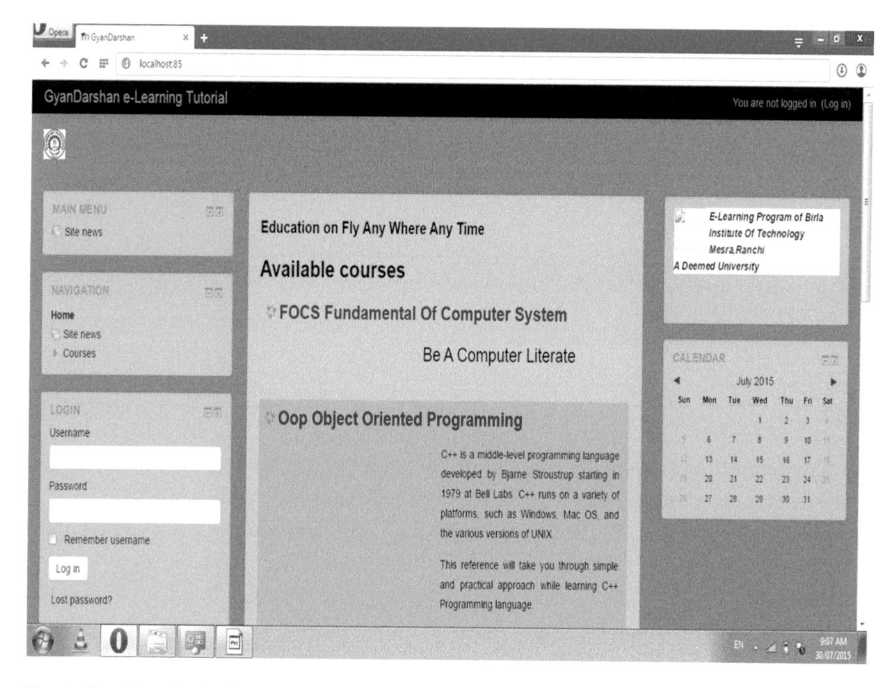

Fig. 4 Dashboard admin VIEW

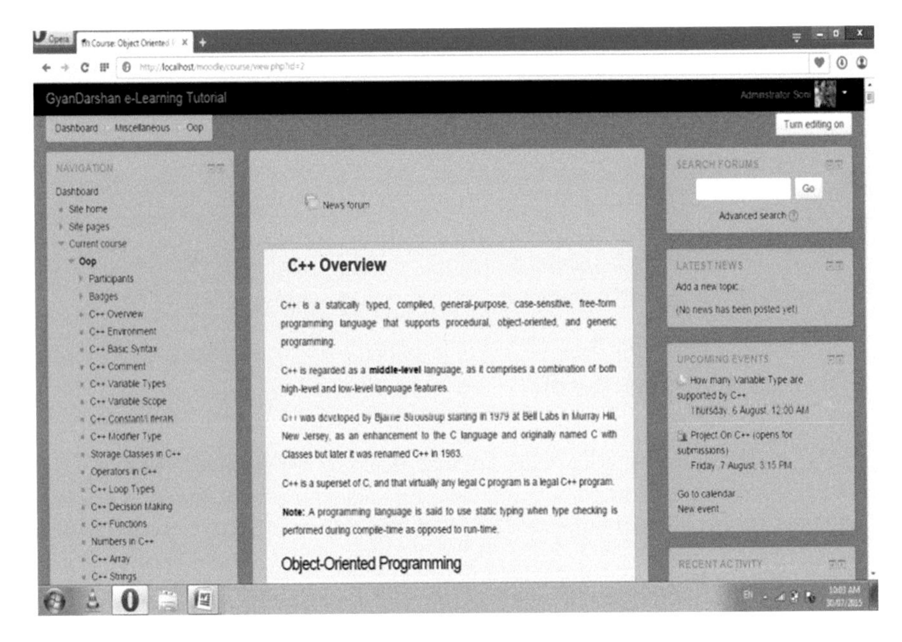

Fig. 5 Course content topic wise with various activities

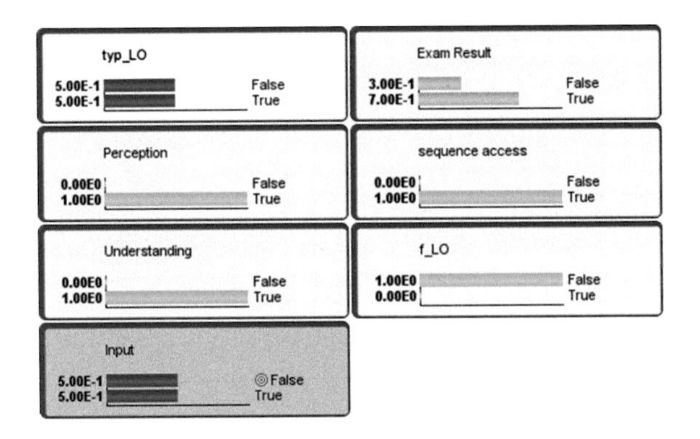

Fig. 6 CPT values of all learning dimension

tool for generating graphical representations that explores various learning aspects of students. It also shows the performance graph neatly.

Pre test and post test results show that if personalized learning path according to learners learning style is provided, they perform well in terms of their grade, time taken to acquire knowledge, understand learning process high precision value and get full satisfaction (Figs. 6, 7, and 8).

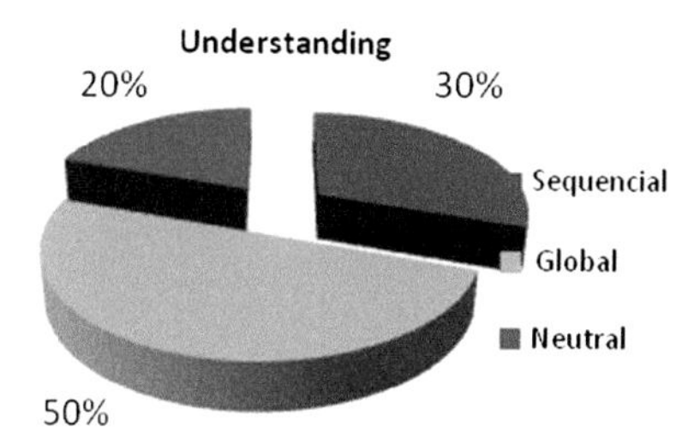

Fig. 7 Population of learners for understanding dimensions

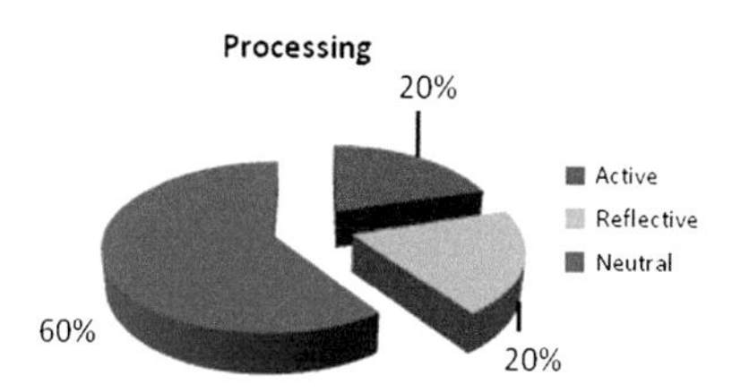

Fig. 8 Populations of learners for two dimensions

5 Conclusion and Future Works

This paper presents an approach to develop Dynamic Dependency Adaptive Model (DDAM) to integrate learning styles into Adaptive e-learning system to assess the effect of adapting educational system individualized to the student's learning style and learning path. It evaluates its effectiveness in the learning process. This system is a learner centric instead of teacher centric which could increase the students' autonomy. We are working on to obtain complete learner's information's and applying in descriptive data mining algorithms to obtain other hidden information of users so that students will get a personalized environment in near future. This model is based on the Felder Silverman learning theories but in future we will also explore other learning theories which may give other hidden and similar results which increases and strengthens of our research work and future prospects.

Acknowledgments We would like to show our gratitude towards the Prof. (Dr.) S.P. Lal Director, Birla Institute of Technology, Mesra, Ranchi (Patna) for sharing their pearls of wisdom with us during the course of this research.

References

1. Jeremić, Z., Devedžić, V.: Design pattern its: student model implementation. In: Null, pp. 864–865. IEEE (2004)

2. Srivastava, J., Cooley, R., et al.: Web usage mining: discovery and applications of usage patterns from web data. ACM SIGKDD Explor. Newsl. **1**(2), 12–23 (2000)
3. Kolb, A.Y., Kolb, D.A.: Learning styles and learning spaces: Enhancing experiential learning in higher education. Acad. Manage. Learn. Educ. **4**(2), 193–212 (2005). [Curry, L.: An organization of learning styles theory and constructs (1983)]
4. Felder, R.M., Spurlin, J.: Applications, reliability and validity of the index of learning styles. Int. J. Eng. Educ. **21**(1), 103–112 (2005)
5. Coffield, F., Moseley, D., et al.: Learning Styles and Pedagogy in Post 16 Learning: A Systematic and Critical Review. The Learning and Skills Research Centre, London (2004)
6. Pashler, H., McDaniel, M., et al.: Learning styles concepts and evidence. Psychol. Sci. Public Interest **9**(3), 105–119 (2008)
7. Huang, S.L., Shiu, J.H.: A user-centric adaptive learning system for e-learning 2.0. J. Educ. Technol. Soc. **15**(3), 214–225 (2012)
8. Joachims, T., Freitag, D., Mitchell, T.: Webwatcher: a tour guide for the World Wide Web. In: IJCAI, no. 1, pp. 770–777 (1997)
9. Boyle, C., Encarnacion, A.O.: Metadoc: an adaptive hypertext reading system. User Model. User-Adap. Interact. **4**(1), 1–19 (1994)
10. Weibelzahl, S., Weber, G.: Advantages, opportunities and limits of empirical evaluations: evaluating adaptive systems. KI **16**(3), 17–20 (2002)
11. Bra, P.D., Brusilovsky, P., Houben, G.J.: Adaptive hypermedia: from systems to framework. ACM Comput. Surv. **31**(4), 1–6 (1999)
12. Ngu, D.S.W., Wu, X.: Sitehelper: A localized agent that helps incremental exploration of the World Wide Web. Comput. Netw. ISDN Syst. **29**(8), 1249–1255 (1997)
13. Mobasher, B., Robert, C., Srivastav, J.: Automatic personalization based on web usage mining. Commun. ACM **43**(8), 142–151 (2000)
14. Lee, C.H., Fu, Y.H.: Web usage mining based on clustering of browsing features. In: Eighth International Conference on Intelligent Systems Design and Applications, vol. 1, pp. 281–286. ISDA'08. IEEE (2008)
15. Yang, W., et al.: Mining social networks for targeted advertising. In: Proceedings of the 39th Annual Hawaii International Conference on System Sciences, HICSS'06, vol. 6. IEEE (2006)
16. Joshi, K.P., et al.: Warehousing and mining web logs. In Proceedings of the 2nd International Workshop on Web Information and Data Management, pp. 63–68. ACM (1999)
17. Graf, S., et al.: Providing adaptive courses in learning management systems with respect to learning styles. In: Richards, G. (ed.) Proceedings of the World Conference on E-Learning in Corporate, Government, Healthcare, and Higher Education (e-Learn), pp. 2576–2583. AACE Press, Chesapeake, VA (2007)
18. Blöchl, M., Rumetshofer, H., Wob, W.: Individualized e-learning systems enabled by a semantically determined adaptation of learning fragments. In: Proceedings of 14th International Workshop on Database and Expert Systems Applications, pp. 640–645. IEEE (2003)
19. Xu, D., Wang, H., Su, K.: Intelligent student profiling with fuzzy models. In: Proceedings of the 35th Annual Hawaii International Conference on System Sciences, p. 8. HICSS. IEEE (2002)
20. Chrysafiadi, K., Virvou, M.: Fuzzy logic for adaptive instruction in an e-learning environment for computer programming. Fuzzy Systems, IEEE Trans. (99), 1 (2014)
21. Truong, H.M.: Integrating learning styles and adaptive e-learning system: current developments, problems and opportunities. Computers in Human Behavior (2015)
22. Garcia, P., et al.: Evaluating Bayesian networks' precision for detecting students' learning styles. Comput. Educ. **49**(2007), 794–808 (2007)
23. Carmona, C., Castillo, G., Millán, E.: Designing a dynamic bayesian network for modeling students' learning styles. In: Eighth IEEE International Conference on Advanced Learning Technologies, pp. 346–350. ICALT'08. IEEE (2008)
24. Popescu, E.: Diagnosing students' learning style in an educational hypermedia system. Cognitive and Emotional Processes in Web-based Education: Integrating Human Factors and

Personalization, Advances in Web-Based Learning Book Series, IGI Global, pp. 187–208 (2009)
25. Felder, R.M., Silverman, L.K.: Learning and teaching styles in engineering education. Eng. Educ. **78**(7), 674–681 (1988)
26. Witten, I.H., Frank, E.: Data Mining: Practical Machine Learning Tools and Techniques. Morgan Kaufmann, San Francisco (2005)
27. Romero, C., et al.: Data mining in course management systems: Moodle case study and tutorial. Comput. Educ. **51**(1), 368–384 (2008)
28. Mazza, R., Milani, C.: Exploring usage analysis in learning systems: Gaining insights from visualisations. In: Communication Dans: The Workshop on Usage Analysis in Learning Systems, the Twelfth International Conference on Artificial Intelligence in Education, pp. 65–72. Amsterdam, The Netherlands (2005)
29. Rice, W.H:. Moodle E-learning Course Development: A Complete Guide to Successful Learning using Moodle. Packt Publishing, Birmingham (2006)
30. Felder, R.M., Soloman, B.A.: Index of learning style questionnaire (ILSQ). http://www.engr.ncsu.edu/learningstyles/ilsweb.html. Accessed 01-2008
31. Graf, S., Lan, C.H., et al.: Investigations about the effects and effectiveness of adaptivity for students with different learning styles. In: Ninth IEEE International Conference on Advanced Learning Technologies, pp. 415–419. ICALT 2009. IEEE (2009)
32. Popescu, E.: Dynamic adaptive hypermedia systems for e-learning. Doctoral dissertation, Université de Technologie de Compiègne (2008)
33. Graf, S., Kinshuk, Liu, T.C.: Supporting teachers in identifying students' learning styles in learning management systems: an automatic student modelling approach. J. Educ. Technol. Soc. **12**(4), 3–14 (2009)
34. Samia, D., Abdelkrim, A.: An adaptive educational hypermedia system integrating learning styles: Model and experiment. In: 2012 International Conference on Education and e-Learning Innovations (ICEELI), pp. 1–6. IEEE (2012)

Author Index

V. Ravi et al. (eds.), *Proceedings of the Fifth International Conference on Fuzzy and Neuro Computing (FANCCO - 2015)*, Advances in Intelligent Systems and Computing 415, DOI 10.1007/978-3-319-27212-2

Printed in the United States
By Bookmasters